Nonlinear systems

Cambridge texts in applied mathematics

Maximum and minimum principles: a unified approach with applications
M. J. SEWELL

Introduction to numerical linear algebra and optimization
P. G. CIARLET

Solitons: an introduction
P. G. DRAZIN AND R. S. JOHNSON

The kinematics of mixing: stretching, chaos and transport
J. M. OTTINO

Integral equations: a practical treatment, from spectral theory to applications
D. PORTER AND D. S. G. STIRLING

Perturbation methods
E. J. HINCH

Nonlinear systems
P. G. DRAZIN

Nonlinear systems

P. G. DRAZIN
Professor of Applied Mathematics
University of Bristol

Published by the Press Syndicate of the University of Cambridge
The Pitt Building, Trumpington Street, Cambridge CB2 1RP
40 West 20th Street, New York NY 10011-4211, USA
10 Stamford Road, Oakleigh, Victoria 3166, Australia

First published 1992

Printed in the United Kingdom by Bell and Bain Ltd., Glasgow

British Library cataloguing in publication data available

Library of Congress cataloguing in publication data

Drazin, P. G.
Nonlinear systems / by P. G. Drazin.
p. cm. – (Cambridge texts in applied mathematics; 10)
Includes bibliographical references and index.
ISBN 0 521 40489 4 – ISBN 0 521 40668 4 (pbk)
1. Nonlinear theories. 2. Differential equations, Nonlinear.
3. Chaotic behavior in systems. I. Title. II. Series.
QA427.D7 1992
515′.355–dc20 91-27971 CIP

ISBN 0 521 40489 4 hardback
ISBN 0 521 40668 4 paperback

AO

To Susannah and Adam

University Book Store

4326 University Way N.E. P.O. Box C50009
Seattle, Washington 98105

NAME

ADDRESS 1 783 724 NO.

CLERK	CASH		CHG.

CITY JOY SABL STATE

93

9548 9 1/20 77

18 29.95

SUB TTL 29.95

8.2% SALES TAX 2.46

1783724

4/14/93 3 UBS CHARGE TTL 32.41

THE AMOUNT LISTED ABOVE IS TO BE CHARGED TO MY ACCOUNT

Nº 26638

Joy Sabl

AUTHORIZED SIGNATURE

Contents

Preface

This book is an introduction to the theories of bifurcation and chaos. It treats the solution of nonlinear equations, especially difference and ordinary differential equations, as a parameter varies. This is a fascinating subject of great power and depth, which reveals many surprises. It requires the use of diverse parts of mathematics – analytic, geometrical, numerical and probabilistic ideas – as well as computation. It covers fashionable topics such as symmetry breaking, singularity theory (which used to be commonly called catastrophe theory), pattern selection, chaos, predictability, fractals and Mandelbrot sets. But it is more than a fashionable subject, because it is a fundamental part of the theory of difference and differential equations and so destined to endure. Also the theory of nonlinear systems is applied to diverse and countless problems in all the natural and social sciences, and touches on some problems of philosophy.

The writing of the book evolved with lecture courses I have given to final-year undergraduates at the University of Bristol and to graduates at the University of Washington and Florida State University in the USA over the last decade. I hope that others will enjoy this book as our students have enjoyed the courses.

Most of the equations treated in traditional mathematics courses at university are linear. These linear algebraic, ordinary differential, partial differential and integral equations are solved by various powerful methods, which essentially depend upon the principle of superposition. However, this book is about nonlinear equations. Using little more than linear algebra and advanced calculus, we shall introduce the theories of bifurcation, imperfections, free oscillations, forced oscillations, and chaos. (A knowledge of the elements of the phase plane is a desirable pre-requisite of Chapter 1, but the phase plane is treated from the beginning in Chapter 6.) The treatment is suitable for final-year undergraduates or first-year postgraduates, whether they major in mathematics, physics, chemistry, engineering, meteorology, oceanography or economics, provided that they have already mastered linear algebra and advanced calculus and they are eager to learn.

My primary aim is to introduce simply the mathematical properties of nonlinear systems as an integrated theory, rather than to present isolated fashionable topics. A secondary aim is to give an impression of the diverse applications of the theory, without detracting from the primary aim. The approach is to discuss topics in as concrete a way as possible, using worked examples and problems to motivate and illustrate general principles. Few general results are proved, and many results are merely made plausible. I have tried to tell the truth and nothing but the truth, not to tell the whole truth, by quoting results and simplifying where it seems desirable; I hope that I have not oversimplifed the material. For a single volume or lecture course to cover so much requires a superficial treatment of many points. I am conscious particularly that the analytic rigour is deficient and that the treatment of nonlinear ordinary differential equations is less thorough and systematic than in several good books. This is a price to pay for including so much other material.

Chapter 1 is an elementary introduction to, and summary of much of, Chapters 2, 3, 5 and 6. So an instructor might base a short course on Chapter 1 with occasional excursions into other chapters, might go straight through the book from beginning to end, or might start with Chapter 2 and then go straight on to the end. Chapter 4 contains miscellaneous topics which, although of relevance and interest, might be omitted from a course. There are, no doubt, other strategies on which a successful course may be based.

More-advanced parts of the text, which might well be omitted at a first reading, are denoted by asterisks; these parts are paragraphs, examples, problems, subsections, sections, Chapter 4 and the Appendix. The end of a worked example is denoted by a small square. The system of equation numbers is such that (1.2.3) denotes the third equation of §2 of Chapter 1, and so forth; the first number, denoting the chapter, is omitted when the equation cited is in the same chapter as the citation, and the second number, denoting the section, is omitted when the equation is in the same section as the citation. Thus equation (1.2.3) would be described as (3) in §1.2, as (2.3) in §1.3, and as (1.2.3) in Chapter 2. The second problem of Chapter 3 is denoted by Q3.2, and its answer by A3.2, and so forth.

The problems are of very variable difficulty, but are ordered according to the subject matter of the main text rather than according to their difficulty. Asterisks, hints, references and brief answers are provided to help readers recognize and overcome the difficulties. Many applications of nonlinear systems are introduced in the problems without any attempt to study the modelling carefully. The aim is to emphasize the mathematical aspects, but to motivate the mathematics by use of models and to lead readers towards the modelling.

I thank Dr C. J. Budd (for some suggestions), Mr W. M. Challacombe (for help in preparing Figs. 8.2, 8.3), Dr K. J. Falconer (for analytical advice and criticism of a draft of the book), Miss Alice James and Mr M. Woodgate (for help in preparing Fig. 3.11), Mr B. Joseph (for help in preparing Fig. 8.8), Mr. T. Milac (for help in preparing Fig. 3.6), Ms C. R. Pharoah (for drafting and re-drafting so patiently almost all the figures), Dr P. L. Read (for help in preparing Figs. 8.18, 8.19), Dr D. S. Riley (for help in proof-reading), Dr Susan C. Ryrie (for much criticism of drafts of this book and for help in preparing Figs. 8.14–17), Mr P. Shiarly (for help in word processing and preparing Figs. 3.12, 3.13), Mr. K. Slater (for help in preparing Figs. 4.5, 6.7), Dr M. Slater (for copious and careful criticism of a draft of the book), and to the students of the University of Bristol, the University of Washington and Florida State University with whom I have learnt so much. Please do not blame them for the defects of the book because, of course, the responsibility is mine alone.

Bristol P. G. Drazin

1 Introduction

Begin: to have commenced is half the deed. Half yet remains: begin again on this and you will finish all.

Ausonius (*Epigrams* no. xv)

1 Nonlinear systems, bifurcations and symmetry breaking

A *nonlinear system* is a set of nonlinear equations, which may be algebraic, functional, ordinary differential, partial differential, integral or a combination of these. The system may depend on given parameters. *Dynamical system* is now used as a synonym of nonlinear system when the nonlinear equations represent evolution of a solution with time or some variable like time; the name dynamical system arose, by extension, after the name of the equations governing the motion of a system of particles, even though the nonlinear system may have no application to mechanics. We may also regard a nonlinear system as representing a feedback loop in which the output of an element is not proportional to its input. Nonlinear systems are used to describe a great variety of phenomena, in the social and life sciences as well as the physical sciences, earth sciences and engineering. The theory of nonlinear systems has applications to problems of economics, population growth, the propagation of genes, the physiology of nerves, the regulation of heart-beats, chemical reactions, phase transitions, elastic buckling, the onset of turbulence, celestial mechanics, electronic circuits and many other phenomena. This introduction to nonlinear systems, then, is an introduction to a great variety of mathematics and to diverse and numerous applications. We shall emphasize the mathematical aspects of nonlinearity which arise so often in applications, many of which give rise to surprising results. We shall see, in particular, that as a parameter changes slowly a solution may change either slowly and continuously or abruptly and discontinuously. A metaphor for such an abrupt change is the proverb 'It is the last straw that breaks the camel's back.'

In applications of the theory of nonlinear systems we are usually interested in enduring rather than transient phenomena, and so in steady states. Thus *steady solutions* of the governing equations are of special importance.

Of these steady solutions only the *stable* ones, i.e. those which, when disturbed slightly at some instant, are little changed for ever afterwards, correspond to states which persist in practice, and so are usually the only ones observable. It follows that a state may change abruptly not only if it ceases to exist but also if it becomes unstable as a parameter changes slowly.

We shall also see that as a result of instability a small cause may have a large effect in the sense that a small disturbance at a given instant may grow and become significant such that after a long time the behaviour of the system depends substantially on the nature of the disturbance, however small the disturbance was. For example, a spherical pendulum with the bob finely balanced directly above its point of suspension may be destabilized by a jog, even by a gentle breath on it; further, the direction and timing of the ensuing motion of the pendulum depend strongly on the very small disturbance of the unstable position of equilibrium. Lorenz described this in a metaphor in which the unstable prairie atmosphere might be triggered by the flutter of the wings of a butterfly in a distant jungle, and thereby a devastating tornado might arise; this is called the *butterfly effect*.

A *bifurcation* occurs where the solutions of a nonlinear system change their qualitative character as a parameter changes. In particular, bifurcation theory is about how the number of steady solutions of a system depends on parameters. The theory of bifurcation, therefore, concerns all nonlinear systems and thence has a great variety of applications. A bifurcation, contradicting Linnaeus's assertion that 'Nature does not proceed by jumps', may confound intuition, so many applications of the theory are important. We shall see how a bifurcation of a nonlinear system and the onset of instability of a solution usually occur at the same critical value of a parameter governing the system. Bifurcation may also occur at other values of the parameter.

Bifurcation is often associated with what is called *symmetry breaking*. For this, the system has some sort of symmetry, i.e. the nonlinear problem is invariant under some group of transformations. The symmetry implies that the set of all the solutions is invariant under the transformations, but not that each solution is invariant. Symmetry is broken at bifurcation if all solutions are symmetric when a parameter is greater (or less) than a critical value but some are asymmetric when the parameter is less (or greater) than that value. For example, when a circular pond dries out, the mud at the bottom shrinks, and the pattern of the cracks of the mud which develops does not have circular symmetry. The causes of pattern selection are closely related to the bifurcations and the stability of the solutions of the governing equations.

Also oscillations occur in many applications, so periodic solutions and their stability are important too. Further, aperiodic unsteady solutions may occur as seemingly random solutions with stationary *statistical* properties; these are *chaotic* solutions or, simply, *chaos.* A chaotic solution also may be stable in the sense that it persists even when the solution is perturbed slightly at some time; this is a dynamic equilibrium. Much of the book concerns chaos, although no formal definition of chaos is given for all systems. It seems that, for the present at least, chaos is best treated like a beard in the following sense. For most chins, there is no doubt whether a chin is clean-shaven or sports a beard; but an occasional chin provokes controversy as to whether it is bearded or its owner has temporarily omitted to shave. In short, there is general, but not complete, agreement on what is and what is not a beard, and the word 'beard' is found useful in describing chins. In a similar way, we shall learn how to recognize chaos when we see it, but may not agree precisely what to call it.

The theory of stability and bifurcations is introduced in this chapter by giving first some informal definitions and a little historical background and then a few illustrative examples. Next, nonlinear oscillations and difference equations are introduced. Thus this chapter provides an overview of the whole book by detailed consideration of some, albeit elementary, examples. These examples may at first be studied as separate problems, but their importance lies in their being canonical cases typical of wide classes of bifurcations. Most of the ideas and methods used later will be raised in this chapter, although their later treatment will be more systematic, more general and more thorough. However, chaos will not be introduced until Chapter 3; it will be seen not as an isolated phenomenon but as a common property of an important subclass of solutions of nonlinear systems. Indeed, ideas arising in all the chapters will be seen to be linked in many different ways. They will be built up bit by bit and drawn together finally in Chapter 8. An overall theme is how the stable solutions of a nonlinear system vary as a parameter is increased.

2 The origin of bifurcation theory

The ideas of bifurcation theory arose slowly and imperceptibly at first, being almost as old as algebra itself. At the simplest, we may view the quadratic equation,

$$x^2 - a = 0,$$

as an example. The roots $\pm\sqrt{a}$ are real for $a > 0$ and are a complex conju-

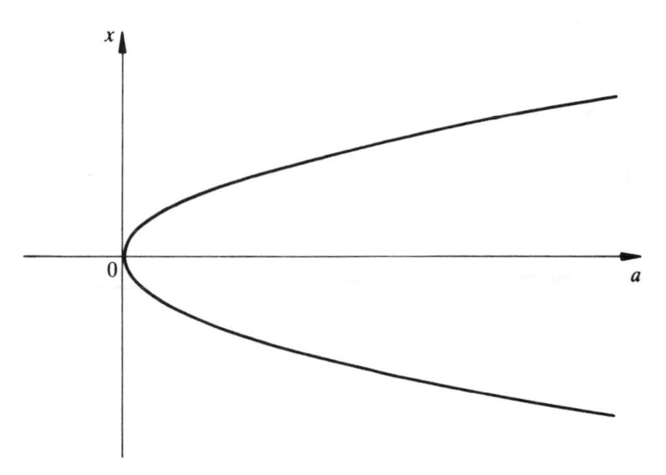

Fig. 1.1 A sketch of the bifurcation diagram for the equation $x^2 - a = 0$.

gate pair for $a < 0$. We may say that there is a change in the character of the solutions at $a = 0$, where there is a repeated root $x = 0$. There is also symmetry breaking in the sense that the quadratic equation is invariant under complex conjugation, and that each solution is invariant under complex conjugation (i.e. is real) for $a > 0$ but no solution is invariant under complex conjugation for $a < 0$. If we confine our attention to real solutions, then there are two for $a > 0$, one for $a = 0$ and none for $a < 0$. We have sketched the parabola in the (a, x)-plane to illustrate the real solutions in Fig. 1.1; this is an example of what is called a *bifurcation diagram.* The horizontal axis gives the value of the parameter and the vertical axis gives the variable to measure the solutions. We shall seen in §3 that this diagram may be useful in describing the solutions of a closely related ordinary differential equation.

Less-trivial examples arose naturally in the solution of problems of particle dynamics. Thus bifurcations came to be considered in the seventeenth century, although our point of view is rather different now. For one example (Andronow & Chaikin 1949, p. 75), the number and character of the positions of equilibrium of a bead, constrained to move on a smooth circular wire in a uniformly rotating vertical plane under the action of gravity, depend on the rate of rotation, as is shown in Fig. 1.2 and in Q1.8.

It may, however, be said that the significance of bifurcations came to be recognized first in the eighteenth century. Euler's (1744) work on the equilibrium and buckling of an elastic column under a load and D'Alembert's work on the figures of equilibrium of a rotating mass of self-gravitating fluid are the foundations of bifurcation theory.

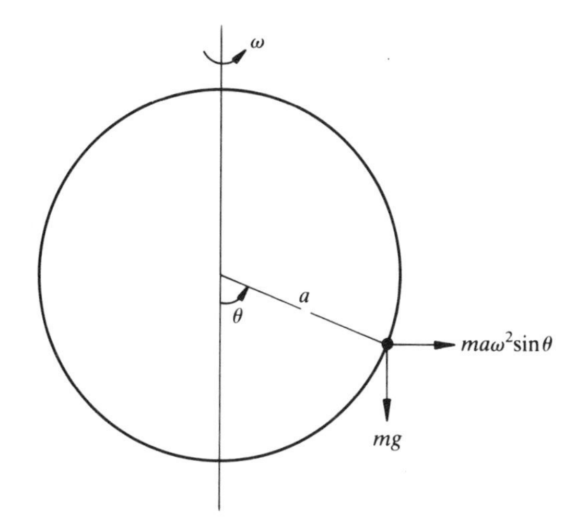

Fig. 1.2 The bead on the smooth wire has one position (at $\theta = 0$) of stable equilibrium if $\omega^2 < g/a$ and two (with equal and opposite angles $\theta = \pm\arccos(g/a\omega^2)$) if $\omega^2 > g/a$.

The problem which D'Alembert worked on has been studied from the seventeenth to the twentieth century. Consider a mass of a uniform incompressible fluid which has constant uniform rotation and is subject to its own gravitational attraction. If the fluid is in dynamic equilibrium as a balance between pressure, centrifugal force and self-gravitation, then what is the figure of the body formed by the fluid? Answers to this question have been used to model a star and a planetary system in formation. Newton, Maclaurin, Jacobi, Liouville, Dirichlet, Dedekind, Riemann, Poincaré, Liapounov, Elie Cartan and many other great mathematicians and scientists have worked on the problem. It has stimulated a lot of enduring mathematics, although the physical applications of the problem have proved disappointing because important physical processes (such as the relation between the pressure and density of the star, and the generation of energy in the interior) were neglected in the model. When the rotation is weak, the figure is an oblate spheroid, first calculated by Maclaurin. As the rate of rotation increases beyond a critical value (which, by dimensional analysis, may be seen to be proportional to the square-root of the product of the gravitational constant and the density of the fluid) ellipsoidal figures of equilibrium, first calculated by Jacobi, occur with three different semi-axes. Symmetry about the axis of rotation is thereby broken. Poincaré (1885, p. 270) first used the French word *bifurcation* in its present sense. He found that a sequence of pear-shaped figures of equilibrium branches off

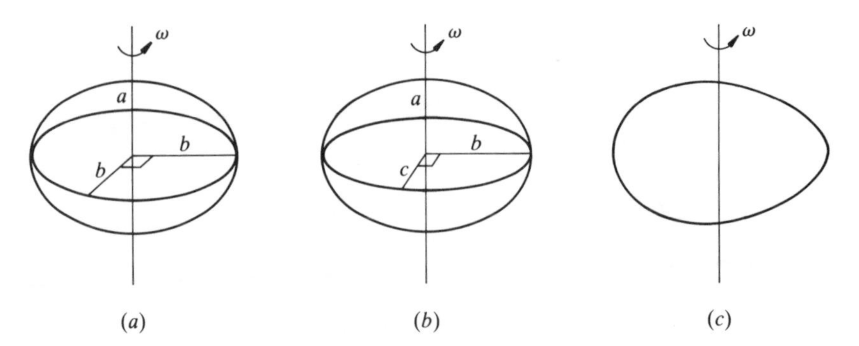

Fig. 1.3 Sketches of some figures of equilibrium of a rotating self-gravitating mass of fluid. (*a*) Maclaurin spheroid. (*b*) Jacobi ellipsoid. (*c*) Poincaré pear-shaped body.

the sequence of Jacobi ellipsoids, just as the Jacobi ellipsoids branch off the Maclaurin spheroids (see Fig. 1.3). Poincare also did a lot of work on the general mathematical theory of bifurcation. Later Liapounov (1906) worked on the problem, and was thereby led to develop a general method of perturbation in bifurcation theory, now known as the Liapounov–Schmidt method.

3 A turning point

In this section and the next four sections we treat the most important types of bifurcation by describing the steady solutions of some simple ordinary differential equations, their dependence on a parameter, and their stability. Stability is defined formally in Chapter 5; for the present an intuitive notion of stability will suffice. The notion and importance of stability were recognized by Lagrange in the eighteenth century, but it is difficult to summarize them more clearly than Clerk Maxwell did in a lecture in 1873 (cf. Campbell & Garnett 1882, p. 440):

> When ... an infinitely small variation of the present state will alter only by a small quantity the state at some future time, the condition of the system, whether at rest or in motion, is said to be stable; but when an infinitely small variation in the present state may bring about a finite difference in the state of the system in a finite time, the condition of the system is said to be unstable.
>
> It is manifest that the existence of unstable conditions renders impossible the prediction of future events, if our knowledge of the present state is only approximate, and not accurate.

For our first example to consider in detail, we take the ordinary differential equation,

$$\frac{dx}{dt} = a - x^2. \tag{1}$$

We shall treat this only as a simple mathematical problem here, although the importance and many applications of essentially similar problems may be appreciated later. We shall often regard t as the time, although it is more properly regarded as the independent variable of the equation.

An *equilibrium point*, or a *critical point*, of equation (1) is a steady solution, i.e. a value, X say, such that there exists a solution x for which $x(t) = X$ for all t. Therefore $dx/dt = 0$ and it follows at once from equation (1) that

$$X = \pm\sqrt{a}. \tag{2}$$

Note that we seek only real solutions, so that there are two steady solutions for $a > 0$, one for $a = 0$, and none for $a < 0$. To examine the stability of the solution $x = X$ for $a > 0$, we rewrite equation (1) without approximation as

$$\frac{dx'}{dt} = -(2X + x')x', \tag{3}$$

where we define the *perturbation* of the equilibrium point as $x' = x - X$. It is plausible that we may find the stability of the solution $x = X$ for $a > 0$ by supposing that x' is small, linearizing the equation, and solving it. (In fact, this method can be rigorously justified for quite general systems, except at the *margin* of stability, i.e. except for those values of a separating linearly stable and unstable solutions. Here we can see that a small nonlinear term cannot change the sign of the right-hand side of equation (3), and so cannot change the stability, unless $X = 0$.) Therefore we consider the linearized system,

$$\frac{dx'}{dt} = -2Xx'. \tag{4}$$

This equation has the solution $x'(t) = x'_0 e^{st}$, where x'_0 is the given initial value $x'(0)$ of x' and the exponent is seen at once to be

$$s = -2X. \tag{5}$$

It follows that if $X = \sqrt{a}$ then $x'(t) \to 0$ as $t \to \infty$ for all x'_0. Therefore all small initial perturbations of the equilibrium point $X = \sqrt{a}$ remain small for all time and the point is stable. We can similarly see that the point

$X = -\sqrt{a}$ is unstable because a small initial disturbance grows until it becomes no longer small.

The solution $X = 0$ for $a = 0$ is at the margin of stability of the linearized system, in the sense that there are arbitrarily close values of a for which the solution is stable, and there are arbitrarily close values of a for which it is unstable. However, the linearized system of a given nonlinear system is not in general sufficient to determine the stability of the null solution at the margin of stability itself, because the terms nonlinear in x' might make unstable the solution which is just linearly stable. For this simple equation (1), indeed, we may contradict the stability given by the linear theory, because we may solve the nonlinear problem explicitly at the margin $a = 0$ of stability. In that case equation (1) becomes

$$\frac{\mathrm{d}x}{\mathrm{d}t} = -x^2,$$

and we deduce readily that if $x(0) = x_0$ then

$$x(t) = \frac{x_0}{1 + x_0 t} \qquad \text{when } x_0 t + 1 > 0 \tag{6}$$

$$\to \left\{ \begin{array}{ll} 0 & \text{as } t \to \infty \text{ for all } x_0 \geqslant 0 \\ -\infty & \text{as } t \to 1/(-x_0) \text{ for all } x_0 < 0 \end{array} \right\}.$$

Therefore the equilibrium solution $x = 0$ for $a = 0$ is unstable, because a small perturbation *may* grow so much that it eventually ceases to be small. Indeed, this is an example of *blow-up*, for which a solution attains an infinite limit in a finite time.

It is convenient to illustrate the steady solutions in a bifurcation diagram, i.e. to plot the equilibrium points versus the value of the parameter in the (a, x)-plane. The diagram for equation (1) is shown in Fig. 1.4. It is conventional to draw stable steady solutions as continuous curves and unstable ones as dashed curves. We have also added broad arrows to indicate the evolution of a solution $x(t)$ with time for a fixed value of a.

The point $(0, 0)$ is an example of what is called a *simple turning point*, a *fold*, or a *saddle-node bifurcation* (cf. Q6.3). The number of steady solutions changes as a increases through zero, and the stability of the solution changes as X increases through zero in this example. The function $a(X)$, defined by $0 = a - X^2$, is well-behaved for all X, even though $X(a)$ is a 2–1 function for $a > 0$ and does not exist for $a < 0$. We shall see that turning points with the same qualitative character arise frequently in many kinds of nonlinear systems.

This example is so simple that even if $a \neq 0$ we may again confirm our results by explicit integration of equation (1). By separation of variables,

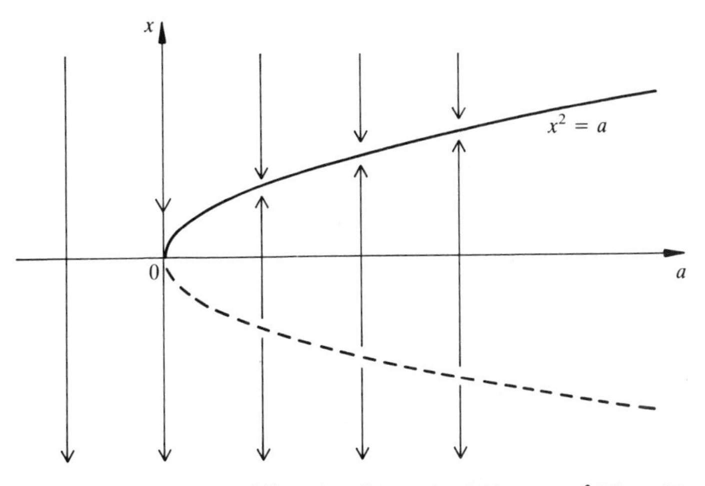

Fig. 1.4 A turning point: the bifurcation diagram for $\mathrm{d}x/\mathrm{d}t = a - x^2$. The stable solutions are denoted by a continuous curve and the unstable by a dashed curve. The arrows indicate the directions of change of solutions with time.

we see that

$$t = \int_{x_0}^{x} \frac{\mathrm{d}u}{a - u^2}.$$

It follows that

$$x(t) = \left\{ \begin{array}{ll} a^{1/2}\left(\dfrac{x_0 + a^{1/2}\tanh(a^{1/2}t)}{a^{1/2} + x_0\tanh(a^{1/2}t)}\right) & \text{if } a > 0 \\ (-a)^{1/2}\left(\dfrac{x_0 - (-a)^{1/2}\tan\{(-a)^{1/2}t\}}{(-a)^{1/2} + x_0\tan\{(-a)^{1/2}t\}}\right) & \text{if } a < 0 \end{array} \right\}. \tag{7}$$

This gives the solution $x(t)$ for all $t \geqslant 0$, or, if the solution becomes singular, until the solution becomes singular, in terms of the given initial value x_0 of x. It follows that

$$x(t) \to \left\{ \begin{array}{ll} a^{1/2} \quad \text{as } t \to \infty & \text{if } a > 0, x_0 > -a^{1/2} \\ -a^{1/2} \quad \text{as } t \to \infty & \text{if } a > 0, x_0 = -a^{1/2} \\ -\infty \quad \text{as } t \to (a)^{-1/2}\operatorname{arctanh}(-a^{1/2}/x_0) & \text{if } a > 0, x_0 < -a^{1/2} \\ -\infty \quad \text{as } t \to (-a)^{-1/2}\arctan\{-(-a)^{1/2}/x_0\} & \text{if } a < 0 \end{array} \right\}. \tag{8}$$

Note that the limits are independent of the initial value x_0 of x, although which limit is attained does depend on x_0. Note also that the solution may become infinite after a finite time t, the value of t at the singularity depending on x_0 as well as the parameter a which occurs in the differential equation; such a movable singularity is characteristic of nonlinear equations, in contrast to the fixed singularities of linear equations. Fig. 1.4 shows the limits more easily, but the formulae provide extra information about the time that is required for $x(t)$ to progress to its infinite limit.

4 A transcritical bifurcation

The *logistic equation*,

$$\frac{dn}{dt} = an - bn^2, \tag{1}$$

arose as a simple model in the theory of population growth (Verhulst 1838, Pearl & Reed 1920). It may represent the growth of a population of a given species, the number of individuals of the species being approximated by the real variable n. When n is sufficiently small the population grows or dies out exponentially according to whether a is positive or negative respectively. If the population grows then, after a while, it will become so numerous that its food supply becomes inadequate or its predators thrive and therefore its growth rate will be reduced. It follows that the exponential growth rate will be moderated in a way represented, at least approximately, by the nonlinear term for some positive value of b. However, for a mathematical exposition, we shall here not restrict the signs of n and b.

The equilibrium points are

$$n = 0 \qquad \text{for all } b, \qquad \text{and} \qquad n = a/b \qquad \text{for all } b \neq 0, \text{for all } a. \tag{2}$$

To examine the stability of the null solution, we linearize equation (1) and find

$$\frac{dn}{dt} = an. \tag{3}$$

Therefore we take $n(t) \propto e^{st}$ and deduce that

$$s = a. \tag{4}$$

This at once gives the general solution

$$n(t) = n_0 e^{at}, \tag{5}$$

where n_0 is the value of n at $t = 0$. This solution decays exponentially to

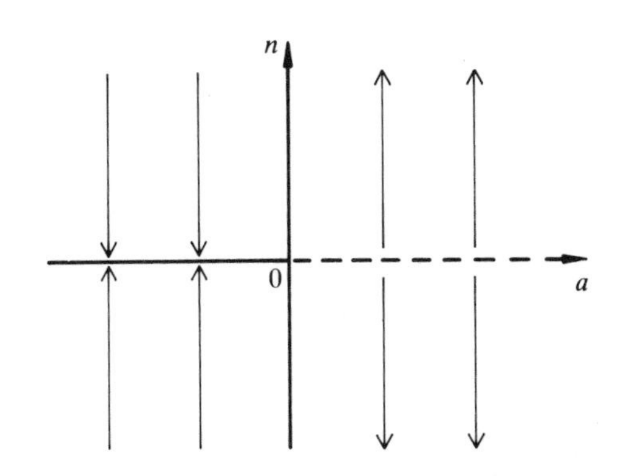

Fig. 1.5 The bifurcation diagram of the linearized system $\mathrm{d}n/\mathrm{d}t = an$ in the (a, n)-plane. Note that the n-axis represents all the stable solutions $n = n_0$ for $a = 0$. Arrows indicate the directions of change of some solutions with time.

zero if $a < 0$, grows exponentially to infinity if $a > 0$, and remains constant if $a = 0$. Therefore the null solution of the linearized system (3) is stable if $a \leqslant 0$ and unstable if $a > 0$, as is illustrated in the bifurcation diagram of Fig. 1.5.

We may similarly show that the solution $n = a/b$ is unstable if $a < 0$ and stable if $a > 0$.

In fact we may solve the nonlinear equation (1) explicitly in simple terms. It may be solved by separation of variables. However, it is a Bernoulli equation which may also be solved by rewriting it without approximation as

$$-\frac{1}{n^2}\frac{\mathrm{d}n}{\mathrm{d}t} + \frac{a}{n} = b,$$

i.e.

$$\frac{\mathrm{d}(1/n)}{\mathrm{d}t} + \frac{a}{n} = b$$

(It can easily be verified that our conclusion (6) is true even if $n = 0$ and $1/n$ does not exist.) This is a linear equation in $1/n$ with elementary solution,

$$\frac{1}{n(t)} = \left\{ \begin{array}{ll} \dfrac{b}{a} + \left(n_0^{-1} - \dfrac{b}{a}\right)\mathrm{e}^{-at} & \text{if } a \neq 0 \\ n_0^{-1} + bt & \text{if } a = 0 \end{array} \right\},$$

where $n_0 = n(0) \neq 0$. Therefore

$$n(t) = \begin{cases} \dfrac{an_0}{bn_0 + (a - bn_0)\mathrm{e}^{-at}} & \text{if } a \neq 0 \\[2ex] \dfrac{n_0}{1 + bn_0 t} & \text{if } a = 0 \end{cases} \tag{6}$$

$$\to \begin{cases} a/b & \text{if } a > 0 \text{ and } bn_0 > 0 \\ 0 & \text{if } a \leqslant 0 \text{ and } bn_0 > a \end{cases} \quad \text{as } t \to \infty. \tag{7}$$

Also

$$n(t) \to \mp\infty \qquad \text{as } t \to \begin{cases} a^{-1}\ln\{1 + a/(-bn_0)\} & \text{if } a \neq 0, bn_0 < \min(0, a) \\ 1/(-bn_0) & \text{if } a = 0, bn_0 < 0 \end{cases}, \tag{8}$$

the limit being $-\infty$ if $b > 0$ and $+\infty$ if $b < 0$. Note that, as in §3, the limits are independent of the initial value n_0 of n, although which limit is attained does depend on n_0, and that the solution may become infinite after a finite time, the value of t at the singularity depending on n_0 as well as the coefficients a and b which occur in the differential equation.

In summary, we draw the bifurcation diagrams for the two cases $b > 0$ and $b < 0$ in Fig. 1.6. Again, the bifurcation diagrams give most of the essential results easily and clearly. This type of bifurcation, characterized by the intersection of two bifurcation curves, is called a *transcritical bifurcation.* Examples of transcritical bifurcation occur for many types of nonlinear system, not only for ordinary differential equations. Of course, the case $b = 0$ is given in Fig. 1.5. Also, if the model were of the growth of a population of n individuals, then we would impose the condition that $n \geqslant 0$, and so only the upper halves of the bifurcation diagrams would be relevant.

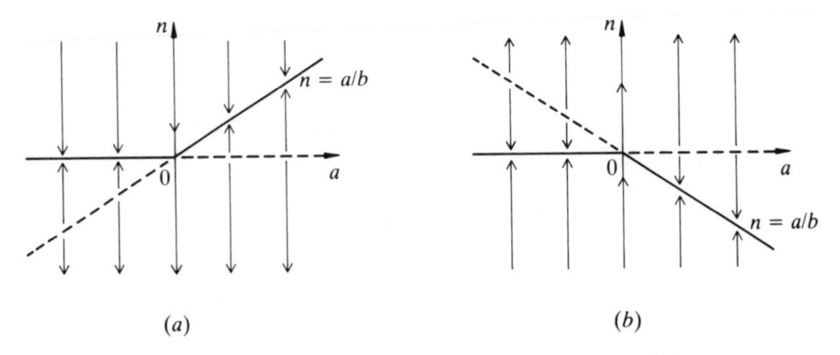

Fig. 1.6 Transcritical bifurcations: the bifurcation diagrams of the system $\mathrm{d}n/\mathrm{d}t = an - bn^2$ in the (a, n)-plane for (a) $b > 0$ and (b) $b < 0$.

You may have noticed that equation (1) has the same form as equation (3.3), which is an exact transformation of equation (3.1). The parameters a and b in equation (1) are the analogues of $-2\sqrt{a}$ and 1 respectively in equation (3.3). However, we use a as the abscissa in each bifurcation diagram, and insist on the reality of the abscissa. Thus our taking the positive square-root $\sqrt{a}$ in §3 gives a turning point, whereas there is a transcritical bifurcation in this section.

5 A pitchfork bifurcation

Landau (1944), investigating nonlinear stability of steady flow of a Newtonian fluid, postulated a model equation which is essentially

$$\frac{\mathrm{d}x}{\mathrm{d}t} = ax - bx^3, \tag{1}$$

where a, b are real parameters. Landau's model was substantiated in the 1960s, but we shall not dwell on the modelling here. We shall merely take equation (1) as an instructive mathematical example and solve it. Note that it differs from equation (4.1) in having a cubic rather than a quadratic nonlinearity.

The equilibrium points are easily seen to be

$$x = 0 \qquad \text{for all } a, \tag{2}$$

and

$$x = \pm A \qquad \text{for all } a, b \text{ such that } a/b > 0, \tag{3}$$

where we define $A = (a/b)^{1/2}$.

To examine the stability of the null solution, take the linearized system,

$$\frac{\mathrm{d}x}{\mathrm{d}t} = ax, \tag{4}$$

and deduce the general solution,

$$x(t) = x_0 \mathrm{e}^{at}, \tag{5}$$

as before, where x_0 is the initial value of x. Therefore the null solution of the linearized system is stable if and only if $a \leqslant 0$. This result is illustrated in Fig. 1.5. However, we shall see very soon that when $a = 0$ the null solution of the nonlinear equation (1) is stable if $b > 0$ and unstable if $b < 0$.

In fact we can again solve the nonlinear equation explicitly, replacing the dependent variable x of equation (1) by $1/x^2$, finding a linear equation in $1/x^2$, and deducing the elementary solution. Thus we rewrite equation (1) as

$$\frac{1}{x^3}\frac{dx}{dt} = \frac{a}{x^2} - b,$$

i.e.

$$\frac{dx^{-2}}{dt} + 2ax^{-2} = 2b,$$

and deduce that

$$x^{-2}(t) = \begin{cases} \dfrac{b}{a} + \left(x_0^{-2} - \dfrac{b}{a}\right)e^{-2at} & \text{if } a \neq 0 \\ x_0^{-2} + 2bt & \text{if } a = 0 \end{cases},$$

where x_0 is the value of x at $t = 0$. Therefore

$$x^2(t) = \begin{cases} \dfrac{ax_0^2}{bx_0^2 + (a - bx_0^2)e^{-2at}} & \text{if } a \neq 0 \\ \dfrac{x_0^2}{1 + 2bx_0^2 t} & \text{if } a = 0 \end{cases}. \quad (6)$$

This gives $x(t) = \{x^2(t)\}^{1/2} \operatorname{sgn} x_0$, where the sign function is defined by $\operatorname{sgn} x = 1$ if $x > 0$, $\operatorname{sgn} x = 0$ if $x = 0$, and $\operatorname{sgn} x = -1$ if $x < 0$. Now various cases arise for separate treatment.

(i) $b > 0$ *and* $a > 0$. In this case it follows that

$$x(t) \to A \operatorname{sgn} x_0 \qquad \text{as } t \to \infty.$$

The limit of x has the same sign as x_0 but is otherwise independent of the initial value x_0 of x. It can be seen that the nonlinear term $-bx^3$ of equation (1) reduces the exponential growth of small unstable disturbances of the null solution eventually, and leads to attainment by $x(t)$ of the constant value A or $-A$. This is called *equilibration.* The temporal development of x is illustrated in Fig. 1.7. This case is an example of what is called *bistability*, i.e. of the existence of two stable equilibrium points of the same system.

(ii) $b > 0$ *and* $a < 0$. In this case the nonlinear term in equation (1) reinforces the exponential damping of the linearized system, albeit very weakly, and

$$x(t) \to 0 \qquad \text{as } t \to \infty \qquad \text{for all } x_0.$$

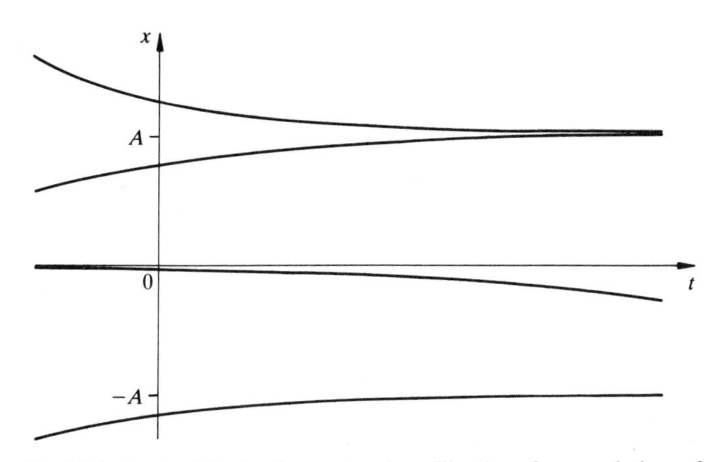

Fig. 1.7 A sketch of the development and equilibration of some solutions of equation (1) when $a, b > 0$.

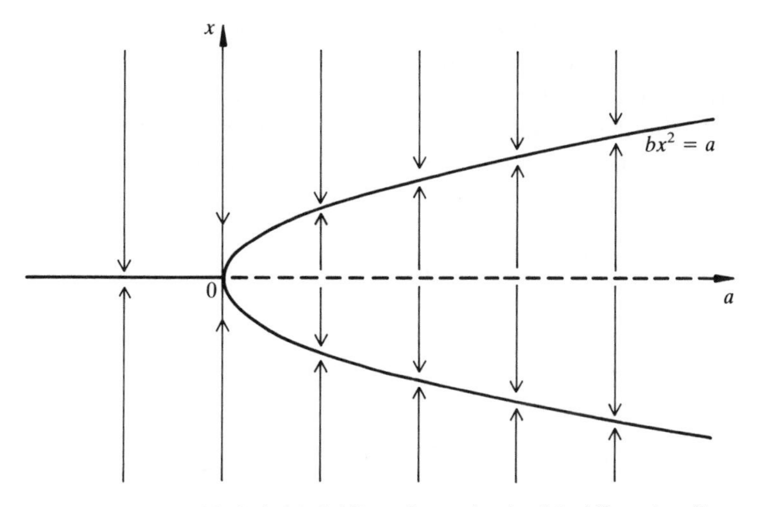

Fig. 1.8 A supercritical pitchfork bifurcation: a sketch of the bifurcation diagram in the (a, x)-plane for $b > 0$. Arrows indicate the directions of change of some solutions with time.

In summary of cases (i) and (ii) for $b > 0$, we draw the bifurcation diagram of Fig. 1.8. There is a bifurcation at the origin, with a unique real point $x = 0$ of equilibrium for $a \leqslant 0$ but three points of equilibrium for $a > 0$. The null solution is stable if and only if $a \leqslant 0$, and the solutions $x = \pm A$ are stable whenever they exist, i.e. for all $a > 0$. There is said to be *supercritical bifurcation* in this case, because the bifurcated solutions arise as a *in*creases above its critical value, zero, at which the null solution is marginally stable.

(iii) $b < 0$ *and* $a > 0$. In this case the nonlinear term on the right-hand side of equation (1) has the same sign as the linear one, and reinforces the exponential growth of small disturbances of the null point. It can be seen from equation (6) that the solution 'blows up' after a finite time, i.e.

$$x(t) \to \infty \qquad \text{as } t \to (2a)^{-1} \ln(1 - a/bx_0^2).$$

(iv) $b < 0$ *and* $a < 0$. In this case the two terms on the right-hand side of equation (1) have different signs and so may have zero sum. Therefore the equilibrium points $x = \pm A$ exist, and they act as 'thresholds' so that if $-A < x_0 < A$ then $x(t) \to 0$ as $t \to \infty$ but if $\pm x_0 > A$ then $x(t) \to \pm\infty$ respectively as $t \to (-2a)^{-1} \ln\{x_0^2/(x_0^2 - A^2)\}$. This is illustrated in Fig. 1.9.

In summary of cases (iii) and (iv) with $b < 0$, we draw the bifurcation diagram Fig. 1.10. There is a unique steady solution $x = 0$ if $a \geqslant 0$ but three steady solutions $x = 0, \pm A$ if $a < 0$. The null solution is stable if and only if $a < 0$, and the two solutions $x = \pm A$ are unstable whenever they exist, i.e. for all $a < 0$. There is said to be *subcritical bifurcation*. There is also said

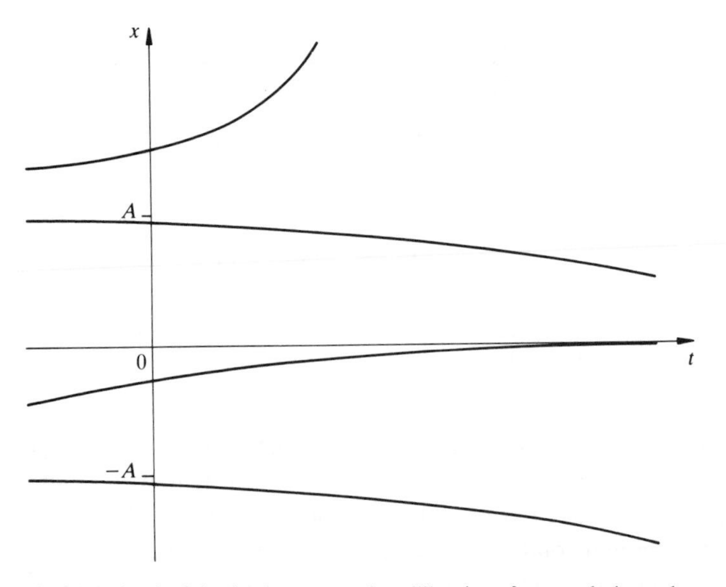

Fig. 1.9 A sketch of the development and equilibration of some solutions when $a, b < 0$.

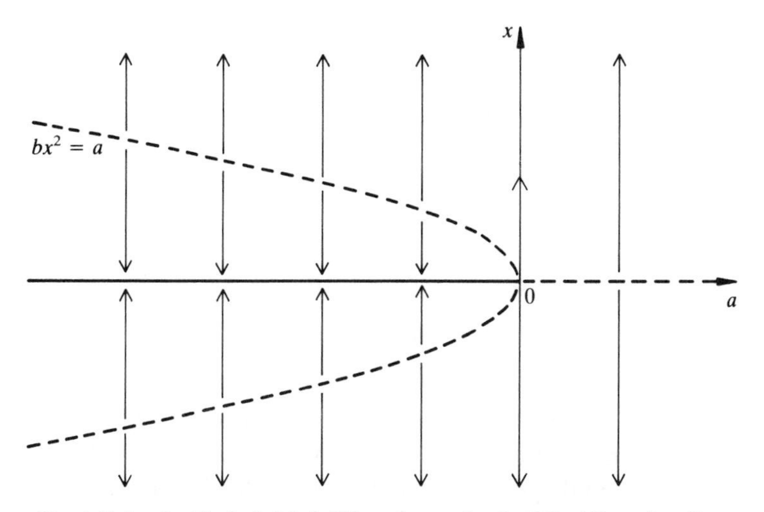

Fig. 1.10 A subcritical pitchfork bifurcation: a sketch of the bifurcation diagram in the (a, x)-plane for $b < 0$. Arrows indicate the directions of change of some solutions with time.

to be *subcritical instability* of the null solution to disturbances of finite amplitude, because a solution grows if $x_0^2 > A^2$, although the null solution is stable to disturbances for which $x_0^2 < A^2$, when the parameter a is *less* than its critical value, zero, for instability.

These are examples of *pitchfork bifurcations*, named after the appearance of the bifurcation diagrams in Figs. 1.8 and 1.10. Pitchfork bifurcations often occur for systems with some symmetry, as a manifestation of symmetry breaking. Equation (1) is of invariant form if x is replaced by $-x$, so is the pair of solutions $x = A, -A$, but a single solution, either $x = A$ or $x = -A$, is not invariant. We have in fact met an example of a pitchfork bifurcation already in §2: the bead on the circular wire in a vertical plane which rotates has one or three positions θ of equilibrium according to whether the dimensionless parameter $a\omega^2/g$ is respectively less or greater than one; the bifurcation is described locally by a pitchfork although the differential equation is of second order (Q1.8).

Equation (1) can readily be seen to be equivalent to the equation,

$$\frac{\mathrm{d}x^2}{\mathrm{d}t} = 2ax^2 - 2bx^4, \tag{7}$$

and therefore to

$$\frac{\mathrm{d}y}{\mathrm{d}t} = 2ay - 2by^2,$$

where $y = x^2 \geqslant 0$. Thus there is the upper half of a transcritical bifurcation in the (a, y)-plane for fixed b, as is shown in §4.

In summary of §§3–5 we note that each example is a treatment of a first-order differential equation of the form

$$\frac{dx}{dt} = F(a, x), \tag{8}$$

which admits a steady solution such that $x(t) = X$ for all t if

$$F(a, X) = 0. \tag{9}$$

This determines the equilibrium points X for each value of a; there may be no zero X of F, one zero or many zeros according to both F and the value of a. It is informative to plot all the curves $x = X(a)$ in the (a, x)-plane as the bifurcation diagram. In considering the stability of an equilibrium point X, the value of a is fixed, and so it is convenient to suppress a in the analysis below. Suppose then that

$$\frac{dx}{dt} = F(x) \tag{10}$$

for a well-behaved function F, and $F(X) = 0$. To find the evolution of all small perturbations $x' = x - X$ of X, expand as a Taylor series

$$\begin{aligned} F(x) &= F(X + x') \\ &= F(X) + x'F'(X) + O(x'^2) \qquad \text{as } x' \to 0 \\ &= x'F'(X) + O(x'^2), \end{aligned}$$

and neglect the very small terms of order x'^2. Thus the linearized form of equation (10) is

$$\frac{dx'}{dt} = x'F'(X). \tag{11}$$

This has general solution

$$x(t) = x'_0 e^{st}, \tag{12}$$

where $s = F'(X)$. It follows that if $F'(X) < 0$ then all small perturbations of the equilibrium point decay monotonically as $t \to \infty$, and X is stable, and also that if $F'(X) > 0$ then all small perturbations grow monotonically and so X is unstable. The evolution of all solutions of equation (10) can be traced globally by considering the signs of $F(x)$.

The illustrations of a turning point, a transcritical bifurcation and a pitchfork bifurcation in §§3–5 by use of first-order equations give the essence of these bifurcations of steady solutions for higher-order equations. However, the time dependence of the solutions we have found is rather special; for example, we shall see in §7 that the equation

$$\frac{d^2x}{dt^2} = ax - bx^3$$

has unsteady solutions which are rather different from those of equation (1). But first we shall see in the next section that some bifurcations which may occur for second- and higher-order systems do not occur for first-order systems.

6 A Hopf bifurcation

For the next example we shall consider the system of ordinary differential equations,

$$\frac{dx}{dt} = -y + (a - x^2 - y^2)x, \qquad \frac{dy}{dt} = x + (a - x^2 - y^2)y, \tag{1}$$

for all real a. The solution will introduce an important aspect of bifurcation: the branching of a time-periodic solution from a steady one.

First seek the equilibrium points and test their stability, as before. On putting $dx/dt = dy/dt = 0$, it follows readily that the only steady solution is the null solution,

$$x = 0, \qquad y = 0. \tag{2}$$

Investigating the stability of this solution, we linearize equations (1) for small x and y to find

$$\frac{dx}{dt} = -y + ax, \qquad \frac{dy}{dt} = x + ay. \tag{3}$$

The general solution of this system is a linear combination of the normal modes with $x, y \propto e^{st}$, say $x(t) = e^{st}u$, $y(t) = e^{st}v$. These satisfy the equation

$$\mathbf{Ju} = s\mathbf{u} \tag{4}$$

to determine the eigenvalues s and eigenvectors $\mathbf{u} = [u, v]^{\mathrm{T}}$ of the 2×2 matrix

$$\mathbf{J} = \begin{bmatrix} a & -1 \\ 1 & a \end{bmatrix}.$$

Therefore

$$\begin{aligned} 0 &= \det(\mathbf{J} - s\mathbf{I}) \\ &= \begin{vmatrix} a - s & -1 \\ 1 & a - s \end{vmatrix} \\ &= (a - s)^2 + 1. \end{aligned}$$

Therefore

$$s = a \pm \mathrm{i}. \tag{5}$$

Therefore there is linear stability if $\mathrm{Re}(s) < 0$ for both eigenvalues, i.e. if $a < 0$, and instability if $a > 0$.

This example has been chosen so that again it may be solved explicitly in simple terms. If we use polar coordinates so that $x = r\cos\theta$, $y = r\sin\theta$ for $r \geqslant 0$, then $x + \mathrm{i}y = r\mathrm{e}^{\mathrm{i}\theta}$ and so

$$\begin{aligned} \frac{\mathrm{d}(r\mathrm{e}^{\mathrm{i}\theta})}{\mathrm{d}t} &= \frac{\mathrm{d}x}{\mathrm{d}t} + \mathrm{i}\frac{\mathrm{d}y}{\mathrm{d}t} \\ &= -y + \mathrm{i}x + (a - x^2 - y^2)(x + \mathrm{i}y), \end{aligned} \tag{6}$$

i.e.

$$\left(\frac{\mathrm{d}r}{\mathrm{d}t} + \mathrm{i}r\frac{\mathrm{d}\theta}{\mathrm{d}t}\right)\mathrm{e}^{\mathrm{i}\theta} = \mathrm{i}r\mathrm{e}^{\mathrm{i}\theta} + (a - r^2)r\mathrm{e}^{\mathrm{i}\theta}.$$

Therefore, on dividing by $\mathrm{e}^{\mathrm{i}\theta}$ and equating real and imaginary parts,

$$\frac{\mathrm{d}r}{\mathrm{d}t} = r(a - r^2), \qquad \frac{\mathrm{d}\theta}{\mathrm{d}t} = 1. \tag{7}$$

It follows from the method of §5 that

$$r^2(t) = \left\{\begin{array}{ll} \dfrac{ar_0^2}{r_0^2 + (a - r_0^2)\mathrm{e}^{-2at}} & \text{if } a \neq 0 \\ \dfrac{r_0^2}{1 + 2r_0^2 t} & \text{if } a = 0 \end{array}\right\}, \tag{8}$$

$$\theta = t + \theta_0, \tag{9}$$

where $r = r_0$, $\theta = \theta_0$ at $t = 0$. Note the similarity of equation (5.1) and equation (6), which can be written as

$$\frac{\mathrm{d}z}{\mathrm{d}t} = \mathrm{i}z + (a - |z|^2)z, \tag{10}$$

where $z = x + \mathrm{i}y$ (see also Q1.11).

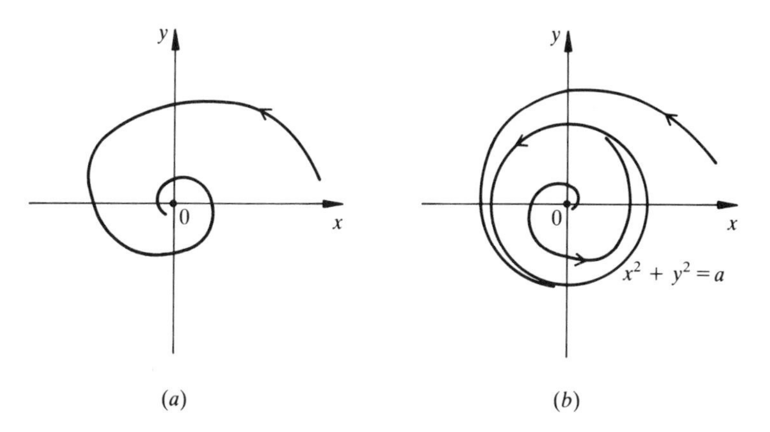

Fig. 1.11 A Hopf bifurcation: sketches of phase portraits of the system (1) in the (x, y)-plane. Arrows indicate the directions of some orbits as time increases. (*a*) $a \leqslant 0$. (*b*) $a > 0$.

A solution can be represented as a point $\mathbf{x} = (x, y)$ by use of Cartesian coordinates in a plane. Then, for $a \leqslant 0$, *all* solutions $\mathbf{x}(t) \to \mathbf{0}$ as $t \to \infty$. Each *trajectory* or *orbit*, i.e. each locus of the point $\mathbf{x}(t)$, spirals into the origin of the (x, y)-plane counter-clockwise as $t \to \infty$. See Fig. 1.11 for a *phase portrait* of the orbits in the (x, y)-plane. For $a > 0$ the origin becomes an unstable *focus*. A new stable periodic solution, namely

$$x = \sqrt{a}\cos(t + \theta_0), \qquad y = \sqrt{a}\sin(t + \theta_0), \tag{11}$$

arises as a increases through zero. Such a solution is called a *limit cycle*, because it is a periodic solution approached by other solutions in the limit as $t \to \infty$. It is represented by a closed curve, in this case the circle $x^2 + y^2 = a$, in the *phase plane* of the orbits. It can be seen in Fig. 1.12 that there is *super*critical bifurcation in this example, the periodic solutions bifurcating from the null solution as a *in*creases through zero. (For details of the theory of the phase plane see Chapter 6, where a more thorough account of two-dimensional differential systems is given *ab initio*.)

The essence of a transcritical bifurcation and a pitchfork bifurcation is that the real eigenvalue s of the unique least stable mode increases through zero as a parameter a increases (or decreases) through a critical value, say a_c, and one or two new steady solutions arise. In contrast, for a *Hopf bifurcation*, exemplified in this section, the real part of a pair of eigenvalues of the least stable complex conjugate modes increases through zero as a increases or decreases through a_c, and a time-periodic solution arises; these points are illustrated schematically in Fig. 1.13. Hopf (1942) showed that

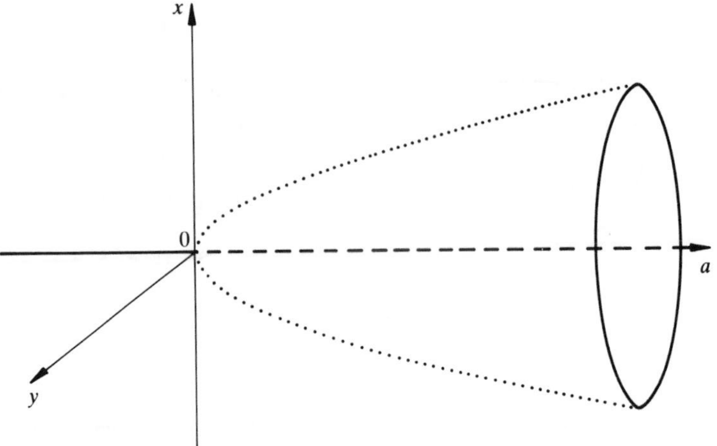

Fig. 1.12 Sketch of the bifurcation diagram in the (a, x)-plane for the system (1). The dotted curves $x = \pm\sqrt{a}$ give the maximum and minimum of $x(t)$ over a period of the stable oscillation.

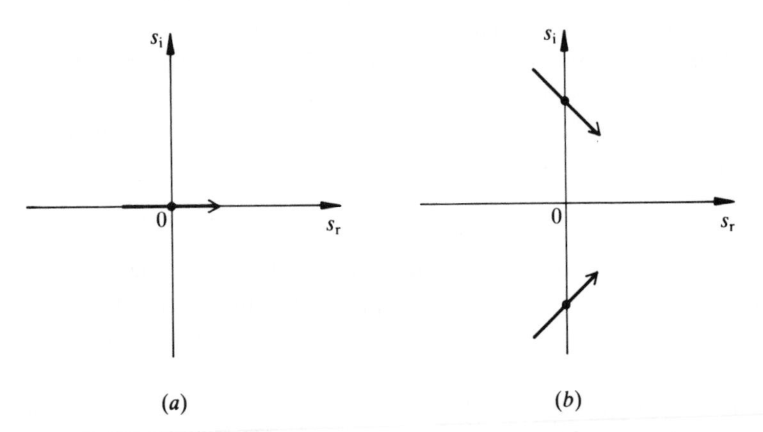

Fig. 1.13 Behaviour of eigenvalues s of the linearized systems at bifurcation as the parameter a moves through its critical value a_c. Dots denote the location of s when $a = a_c$, and arrows the directions of change of s as a increases (or decreases). (*a*) Saddle-node, transcritical or pitchfork bifurcation. (*b*) Hopf bifurcation.

this type of bifurcation occurs quite generally for systems of nonlinear differential equations.

7 Nonlinear oscillations of a conservative system

We have noted that first-order differential equations may contain the essence of saddle-node, transcritical and pitchfork bifurcations but are too simple to exhibit other bifurcations, or to describe oscillations which often occur with higher-order systems. Indeed, we have already seen that a Hopf bifurcation occurs only for a system of at least two first-order equations. (It is shown in Chapter 5 that a second-order equation is equivalent to a system of two first-order equations.) To introduce nonlinear oscillations now, we shall take another example, namely equations of the form

$$\frac{d^2x}{dt^2} = f(x). \tag{1}$$

We shall assume that f is as smooth as the arguments below require. This equation may be regarded as the equation of motion of a particle moving in a line under the force field f per unit mass with potential V, where

$$V(x) = -\int_{x_0}^{x} f(u)\,du, \tag{2}$$

on taking some lower bound x_0 for definiteness.

An equilibrium point X is a zero of f, i.e. a local extremum of V. To consider the stability of the point X, define $x' = x - X$, and linearize equation (1) for small perturbations x' to get

$$\frac{d^2x'}{dt^2} = f'(X)x'. \tag{3}$$

This gives simple harmonic motion, and therefore stability, if $f'(X) < 0$, but exponentially growing solutions in general, and therefore instability, if $f'(X) > 0$. These results may be compared with those for the first-order equations of the form $dx/dt = f(x)$ in §§3–5. In each case the equilibrium solutions X are the zeros of f. However, the solution $x(t)$ approaches or leaves X monotonically for a first-order equation according as $f'(X)$ is negative or positive respectively, whereas $x(t)$ oscillates about or (in general) leaves X according as $f'(X)$ is negative or positive respectively. For the second-order system (1) the point $(X, 0)$ is a centre in the phase plane of $(x, dx/dt)$ if $f'(X) < 0$, and a saddle point if $f'(X) > 0$.

Equation (1) has the elementary integral

$$\frac{1}{2}\left(\frac{dx}{dt}\right)^2 + V(x) = E, \tag{4}$$

where E is a constant of integration such that $E = \frac{1}{2}\dot{x}_0^2$ if $x = x_0$ and $dx/dt = \dot{x}_0$ at $t = t_0$. This is the energy equation of a particle in the force field f. It shows that $V(x) \leqslant E$. Also equation (4) can be written as

$$\frac{1}{2}\left(\frac{dx}{dt}\right)^2 = F(x), \tag{5}$$

where $F(x) = E - V(x)$. This specifies the magnitude of dx/dt in terms of x for each solution but not the sign of dx/dt. Therefore x increases or decreases monotonically until it approaches a zero of F; this shows that the behaviour of $x(t)$ near a zero of F is important.

Suppose then that x approaches a simple zero x_1 of F, so $F(x_1) = 0$ and $f(x_1) = F'(x_1) \neq 0$. Therefore

$$\frac{dx}{dt} = \pm\{2f(x_1)(x - x_1)\}^{1/2} + O\{(x - x_1)^{3/2}\} \qquad \text{as } x \to x_1,$$

so $x - x_1$ has the same sign as $f(x_1)$. We can integrate this iteratively to deduce that

$$x(t) = x_1 + \tfrac{1}{2}f(x_1)(t - t_1)^2 + O\{(t - t_1)^4\} \qquad \text{as } t \to t_1, \tag{6}$$

where t_1 is the time when $x = x_1$, i.e. $x(t_1) = x_1$. Therefore x has a simple minimum or maximum x_1 at t_1 according as $f(x_1)$ is positive or negative respectively.

Next suppose that x approaches a double zero x_1 of F, so $F(x_1) = F'(x_1) = 0$ and $f'(x_1) = F''(x_1) \neq 0$. Now equation (5) gives $f'(x_1) > 0$ and

$$\frac{dx}{dt} = \pm\{f'(x_1)\}^{1/2}(x - x_1) + O(|x - x_1|^2) \qquad \text{as } x \to x_1.$$

On integration this gives

$$x(t) - x_1 \sim \text{constant} \times \exp[\pm\{f'(x_1)\}^{1/2}t] \qquad \text{as } t \to \mp\infty. \tag{7}$$

respectively. Thus a solution may approach the limit x_1, the double zero of F, in an infinite time. Note that in this case x_1 is an unstable point of equilibrium which may be approached, by a *special* solution, as time tends to infinity (see in Example 1.2 how a simple pendulum *may* come to rest at its unstable point of equilibrium directly above its point of suspension if its initial velocity has a special relationship to its initial angle).

These ideas give the qualitative character of many solutions. Suppose, for example, that $F(x) > 0$ for $x_1 < x < x_2$ and F has simple zeros x_1 and x_2. Taking $x_1 < x_0 < x_2$ and $\dot{x}_0 > 0$, without loss of generality, we see that x increases monotonically until it reaches x_2. There it changes direction in accord with equation (6), and then decreases monotonically until it reaches x_1. Thereafter it changes direction at x_1, increases monotonically, and passes through x_0. When it passes through x_0 it will have the same value $\dot{x}_0$ of dx/dt as it did initially, so that the same motion will repeat itself, giving a periodic solution. The period is

$$T = \int_0^T \mathrm{d}t$$

$$= \oint \frac{\mathrm{d}x}{\mathrm{d}x/\mathrm{d}t},$$

where the circle on the integral sign denotes that the integration is to be taken 'around' the orbit for one period,

$$= \oint \frac{\mathrm{d}x}{\pm\{2F(x)\}^{1/2}}$$

$$= 2\int_{x_1}^{x_2} \frac{\mathrm{d}x}{\{2F(x)\}^{1/2}}, \tag{8}$$

because $\mathrm{d}x/\mathrm{d}t$ is positive (at the same point x) over one half of the period and negative but with the same magnitude over the other half.

We can picture the solutions of the system (1) by regarding it as the equation of motion of a particle of unit mass on a smooth plane wire with equation $z = V(x)/g$, where z is the height, x is the distance along the wire, and g is the acceleration due to gravity. This gives the energy equation (4), shows that the equilibrium points where $\mathrm{d}V/\mathrm{d}x = 0$ are stable if V has a minimum and unstable otherwise, and makes apparent the occurrence of nonlinear oscillations about stable points. This is illustrated in Fig. 1.14 of the graph of $V(x) - E = -F(x)$; note that the motion is confined to values of x where $F \geqslant 0$.

Note that the equations $\mathrm{d}x/\mathrm{d}t = f(x)$ and $\mathrm{d}^2x/\mathrm{d}t^2 = f(x)$ have the same

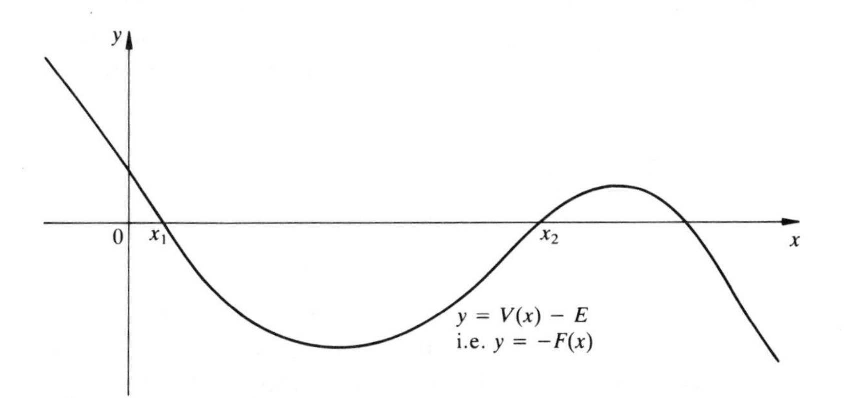

Fig. 1.14 Sketch of a graph of a 'typical' function $V(x) - E$ for a given motion. Motion may occur only where $V(x) \leqslant E$. Other motions for the same field may occur for different values of E.

equilibrium points, namely the zeros of f. Moreover a zero X is stable for each equation if $f'(X) < 0$ and unstable for each equation if $f'(X) > 0$. It is for these reasons that the equations $\mathrm{d}x/\mathrm{d}t = F(a, x)$ and $\mathrm{d}^2x/\mathrm{d}t^2 = F(a, x)$ have the same bifurcation diagram. So we meet turning points, transcritical points and pitchforks for second- as well as for first-order equations. However, the temporal development of solutions of second-order equations is different from that of solutions of first-order equations. In particular, stability of an equilibrium point is monotonic for a first-order equation but simple harmonic motion for the corresponding second-order equation.

Example 1.1: a turning point. Suppose that

$$\frac{\mathrm{d}^2x}{\mathrm{d}t^2} = a - x^2.$$

Then there are equilibrium points at $x = \pm\sqrt{a}$ for all $a \geqslant 0$, as in §3, but the solutions do not all vary monotonically with time. The 'energy' integral,

$$\tfrac{1}{2}v^2 + \tfrac{1}{3}x^3 - ax = E,$$

gives the orbits in the phase plane of x and $v = \mathrm{d}x/\mathrm{d}t$, different orbits being in general given by different values of E. If $a > 0$ then $E = \mp\tfrac{2}{3}a\sqrt{a}$ at the equilibrium points $(\pm\sqrt{a}, 0)$ respectively, and $(\sqrt{a}, 0)$ is a *centre* and $(-\sqrt{a}, 0)$ is a *saddle point*. The closed orbit through $(-\sqrt{a}, 0)$ is said to be a *separatrix*, because it separates the closed orbits, which represent periodic solutions, from unbounded orbits. Note also that the saddle-point is connected with itself by the separatrix; for this reason the separatrix is an example of what is called a *homoclinic orbit*. The separatrix has equation

$$\tfrac{1}{2}v^2 = \tfrac{1}{3}(x + \sqrt{a})^2(2\sqrt{a} - x).$$

A periodic solution has period

$$T = 2\int_{x_1}^{x_2} \frac{\mathrm{d}x}{\{2(E + ax - \tfrac{1}{3}x^3)\}^{1/2}},$$

where x_1 and x_2 are the zeros of $E + ax - \tfrac{1}{3}x^3$ for $-\tfrac{2}{3}a\sqrt{a} < E < \tfrac{2}{3}a\sqrt{a}$. Note that $T \to \infty$ as $E \to \pm\tfrac{2}{3}a\sqrt{a}$, the separatrix being a limit of periodic orbits as the period tends to infinity. If $a \leqslant 0$ then all the orbits are unbounded and $x(t) \to -\infty$ as $t \to \infty$, except for the unstable point of equilibrium at the origin when $a = 0$. The phase portraits for $a > 0$, $a = 0$ and $a < 0$ are sketched in Fig. 1.15. □

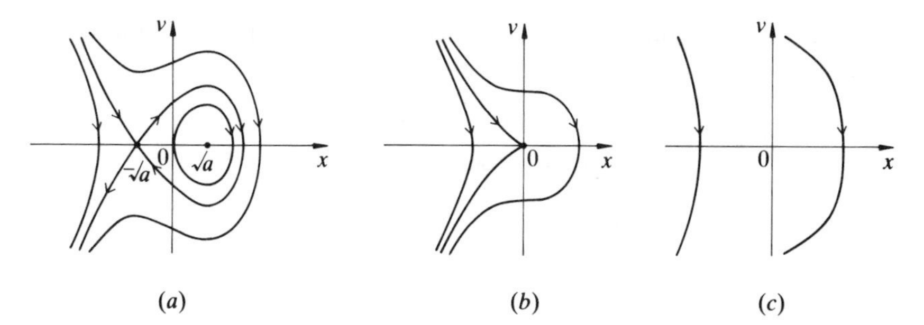

Fig. 1.15 Sketch of the phase portrait of $d^2x/dt^2 = a - x^2$ for (a) $a > 0$, (b) $a = 0$, and (c) $a < 0$.

Example 1.2: the simple pendulum. The equation of motion of a simple pendulum of length l is

$$l\frac{d^2\theta}{dt^2} = -g\sin\theta, \tag{9}$$

where θ is the angle the pendulum makes with the downward vertical, and has energy equation

$$\frac{1}{2}\left(\frac{d\theta}{dt}\right)^2 = g(\cos\theta - \cos a)/l, \tag{10}$$

where the amplitude a of the solution is defined as the positive value of θ where $d\theta/dt = 0$. See the phase portrait in Fig. 1.16. Note the saddle points at $((2n+1)\pi, 0)$ for $n = 0, \pm 1, \pm 2, \ldots$, and the closed orbits in the region between each pair of orbits, called *heteroclinic orbits*, connecting neigh-

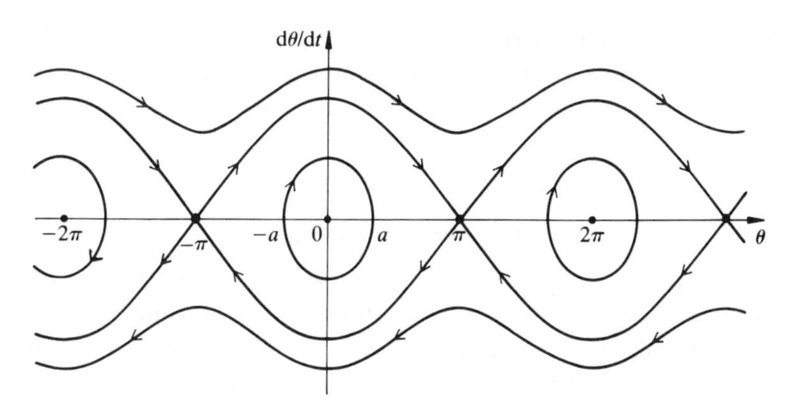

Fig. 1.16 Sketch of the phase portrait of $ld^2\theta/dt^2 = -g\sin\theta$ in the $(\theta, d\theta/dt)$-plane.

bouring saddle points. Now each closed orbit represents a periodic solution. This gives a nonlinear oscillation of period

$$T = 2\int_{-a}^{a} \frac{\mathrm{d}\theta}{\{2g(\cos\theta - \cos a)/l\}^{1/2}}$$

$$= 4(l/g)^{1/2}\int_{0}^{a} \frac{\mathrm{d}\theta}{\{4(\sin^2\frac{1}{2}a - \sin^2\frac{1}{2}\theta)\}^{1/2}}$$

$$= 4(l/g)^{1/2}\int_{0}^{\pi/2} \frac{2\sin\frac{1}{2}a\cos\phi\,\mathrm{d}\phi}{\cos\frac{1}{2}\theta\{4(\sin^2\frac{1}{2}a - \sin^2\frac{1}{2}a\sin^2\phi)\}^{1/2}},$$

on substituting $\sin\frac{1}{2}\theta = \sin\frac{1}{2}a\sin\phi$,

$$= 4(l/g)^{1/2}\int_{0}^{\pi/2} \frac{\mathrm{d}\phi}{\cos\frac{1}{2}\theta}$$

$$= 4(l/g)^{1/2}K(\sin^2\tfrac{1}{2}a), \tag{11}$$

where the *complete elliptic integral of the first kind* is defined as

$$K(m) = \int_{0}^{\pi/2} \frac{\mathrm{d}\phi}{(1 - m\sin^2\phi)^{1/2}} \qquad \text{for } 0 \leqslant m < 1. \tag{12}$$

Therefore

$$T = 4(l/g)^{1/2}\int_{0}^{\pi/2} \{1 + \tfrac{1}{2}\sin^2\tfrac{1}{2}a\sin^2\phi + O(\sin^4\tfrac{1}{2}a)\}\,\mathrm{d}\phi \qquad \text{as } a \to 0$$

$$= 4(l/g)^{1/2}\{\tfrac{1}{2}\pi + \tfrac{1}{32}a^2\pi + O(a^4)\}$$

$$= 2\pi(l/g)^{1/2}\{1 + \tfrac{1}{16}a^2 + O(a^4)\}. \tag{13}$$

This gives the leading term in the correction to the period $2\pi(l/g)^{1/2}$ of oscillations of small amplitude. Note that the period increases with amplitude, as we might have anticipated, because the pendulum moves more slowly as it rises. □

8 Difference equations

To emphasize that there are nonlinear systems other than ordinary differential equations, we shall next introduce nonlinear difference equations. We shall see that they have bifurcations which are very similar to those of ordinary differential equations, and that many concepts and methods are useful in understanding both kinds and, indeed, other kinds of nonlinear systems. Difference equations are mathematically interesting and have many important applications. They also are, in many ways, more elementary and fundamental than differential equations, involving a discrete independent variable rather than a continuous one; therefore they perhaps deserve more attention than they usually get from instructors and students, although the applications of differential equations are even more important.

A *difference equation, recurrence equation* or *iterated map* is in general of the form

$$\mathbf{x}_{n+1} = \mathbf{F}(\mathbf{x}_n, n) \qquad \text{for } n = 0, 1, 2, \ldots, \tag{1}$$

where $\mathbf{x}_n \in \mathbb{R}^m$ and $\mathbf{F}\colon \mathbb{R}^m \times \mathbb{Z} \to \mathbb{R}^m$. Thus difference equations are functional, sometimes algebraic, systems which correspond to differential equations. We regard the integral variable n as corresponding to the independent real variable t of a differential equation, take $\mathbf{x}_0$ as a given initial value of the dependent variable $\mathbf{x}_n$, and consider the behaviour of $\mathbf{x}_n$, especially as $n \to \infty$. We shall further study functions $\mathbf{F}$ which also depend on parameters, and how the solutions $\{\mathbf{x}_n\}$, i.e. the sequences $\{\mathbf{x}_0, \mathbf{x}_1, \ldots\}$, change qualitatively with those parameters.

In fact a difference equation may arise from taking a finite-difference approximation to a differential equation, or a differential equation may arise as a continuum approximation to a discrete process. For example, if $x(t)$ satisfies the logistic differential equation,

$$\frac{\mathrm{d}x}{\mathrm{d}t} = ax - bx^2, \tag{2}$$

then we may choose h as a suitable small positive number, define $x_n = x(nh)$ for $n = 0, 1, \ldots$, and approximate the derivative $\mathrm{d}x/\mathrm{d}t$ at $t = nh$ by the Euler forward difference $(x_{n+1} - x_n)/h$. It follows that

$$\frac{x_{n+1} - x_n}{h} = ax_n - bx_n^2$$

approximately, i.e. that

$$x_{n+1} = F(a, b, h, x_n), \tag{3}$$

where $F(a, b, h, x) = x + h(ax - bx^2)$. This little example illustrates the important point that most computers are digital rather than analogue, dealing with discrete rather than continuous variables, and so it is the rule for a differential equation to be 'solved' by first taking a difference equation which approximates it and then solving the difference equation. In this way we can compute the *stable* equilibrium points of a differential equation of the form

$$\frac{dx}{dt} = f(x), \tag{4}$$

where $x \in \mathbb{R}$ and $f: \mathbb{R} \to \mathbb{R}$, by iterating the solutions of the difference equation

$$x_{n+1} = x_n + hf(x_n). \tag{5}$$

Numerical analysts have studied the efficiency of this method and its convergence as the 'time' step $h \to 0$. The method may be easily generalized for systems of ordinary differential equations with $\mathbf{x} \in \mathbb{R}^m$ and $\mathbf{f}: \mathbb{R}^m \to \mathbb{R}^m$, and for functions $\mathbf{f}$ which depend on t as well as $\mathbf{x}$. Also other finite-difference approximations to the differential equation may be used successfully.

Yet if we seek only equilibrium solutions of equation (4) then we need seek only the zeros of f, which may be computed by use of any one of many other difference equations. A well-known equation follows by the *Newton-Raphson method.* For this we assume that f is continuously differentiable, estimate x_n as a zero X of f, and construct an improved estimate x_{n+1} by approximating the curve with equation $y = f(x)$ by its tangent $y = f(x_n) + (x - x_n)f'(x_n)$ at the point $(x_n, f(x_n))$. This gives the next approximation to X as

$$x_{n+1} = x_n - f(x_n)/f'(x_n) \tag{6}$$

if $f'(x_n) \neq 0$. We guess x_0 from what knowledge of f we have at the start, and compute successive approximations $x_1, x_2, \ldots,$ stopping when we have reason to believe that x_n approximates the exact zero X with the desired accuracy. The choice of x_0 and the convergence of the process as $n \to \infty$ are considered in many textbooks of numerical analysis.

Differential equations also determine difference equations in a very different way. Suppose that we consider, for example, the solutions $\mathbf{x}(t) \in \mathbb{R}^3$ of a given differential system,

$$\frac{d\mathbf{x}}{dt} = \mathbf{G}(\mathbf{x}), \tag{7}$$

where $\mathbf{G}: \mathbb{R}^3 \to \mathbb{R}^3$. Note that this system is *autonomous*, i.e. $\mathbf{G}$ does not depend on the independent variable t explicitly. Then we may take any point $\mathbf{x}_n = (x_n, y_n, 0)$ in the plane with equation $z = 0$, and define another point $\mathbf{x}_{n+1}$ in the plane as follows. Set $\mathbf{x}(0) = \mathbf{x}_n$, integrate the differential system to find the *orbit* of $\mathbf{x}$ in the *phase space*, i.e. find the locus of $\mathbf{x}(t)$ in $\mathbb{R}^3$ for $t > 0$. Let $\mathbf{x}_{n+1}$ be the point where the orbit next crosses the plane

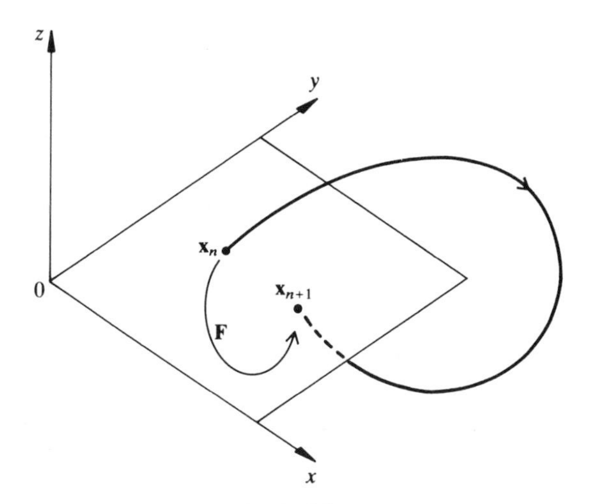

Fig. 1.17 A sketch of the return map.

$z = 0$ in the same sense. Of course, the orbit may never cross the plane again, so $\mathbf{x}_{n+1}$ may not exist. In this way we define a map $\mathbf{F}: \mathbb{R}^2 \to \mathbb{R}^2$ such that $\mathbf{x}_{n+1} = \mathbf{F}(\mathbf{x}_n)$. It is called the *Poincaré map* or the *first return map* of the plane $z = 0$ for the differential system (7), as illustrated in Fig. 1.17. We can similarly define a return map for any non-planar smooth surface in $\mathbb{R}^3$. We can also define return maps for a differential system of order higher than three, for which the phase space has dimension higher than three. Return maps are very useful as distillations of the complicated orbits in a higher-dimensional space, even though they are found by discarding much information about those orbits.

Example 1.3: a 'stroboscopic' return map. A Poincaré section may also be taken in a line. Thus consider the system

$$\frac{\mathrm{d}x}{\mathrm{d}t} = -y + (a - x^2 - y^2)x, \qquad \frac{\mathrm{d}y}{\mathrm{d}t} = x + (a - x^2 - y^2)y,$$

and define the Poincaré map $P: \mathbb{R} \to \mathbb{R}$ as the map of successive intersections of the orbit $\mathbf{x}(t)$ and the x-axis in the same sense. We solved the system in §6, and so can see that if $\mathbf{x}(0) = (x_n, 0)$ then $\theta = t$ if $x_n > 0$ and $\theta = t + \pi$ if $x_n < 0$. Therefore the orbit cuts the x-axis in the same sense, i.e. from below if $x_n > 0$, next when $t = 2\pi$. This gives

$$\begin{aligned} x_{n+1} &= r(2\pi)\mathrm{sgn}(x_n) \\ &= x_n[a/\{x_n^2 + (a - x_n^2)\mathrm{e}^{-4\pi a}\}]^{1/2} \end{aligned}$$

if $a \neq 0$, i.e. $x_{n+1} = P(x_n)$ where

$$P(x) = x[a/\{x^2 + (a - x^2)e^{-4\pi a}\}]^{1/2}. \square$$

Again, difference equations arise directly from modelling natural phenomena. A famous example is the *logistic difference equation*, or *quadratic difference equation*,

$$N_{n+1} = aN_n - bN_n^2. \tag{8}$$

This equation is often used with $a > 1$, $b > 0$ to model the growth of a population whose nth generation has N_n individuals. The ideas of the modelling are similar to those for the logistic differential equation treated in §4. It follows that if $b = 0$ then

$$N_n = aN_{n-1} = a^2 N_{n-2} = \cdots = a^n N_0, \tag{9}$$

i.e. then the population grows exponentially with n. The nonlinear term, $-bN_n^2$, however, represents the moderation of growth due, perhaps, to the limitation of the food supply or the concomitant increase in the number of predators. We can simplify the nonlinear equation (8) a little by transforming the dependent variable to $x_n = bN_n/a$, so that

$$\begin{aligned} x_{n+1} &= bN_{n+1}/a \\ &= b(aN_n - bN_n^2)/a \\ &= ax_n(1 - x_n). \end{aligned} \tag{10}$$

The theory of linear difference equations is quite general and systematic. It is analogous to the theory of linear ordinary differential equations in many ways. On the other hand the theory of nonlinear difference equations is largely a collection of special methods. We shall show, however, that bifurcation theory provides a framework with which some aspects of nonlinear difference equations may be examined systematically.

The analogue of a steady solution of a differential equation is a *fixed point* of a difference equation. We say that $\mathbf{X}$ is a fixed point of the difference equation $\mathbf{x}_{n+1} = \mathbf{F}(\mathbf{x}_n)$ or of the map $\mathbf{F}: \mathbb{R}^m \to \mathbb{R}^m$ if

$$\mathbf{F}(\mathbf{X}) = \mathbf{X}. \tag{11}$$

It follows that if $\mathbf{x}_0 = \mathbf{X}$ then

$$\mathbf{x}_n = \mathbf{X} \qquad \text{for } n = 1, 2, \ldots. \tag{12}$$

By extension, we allow points at infinity to be fixed points; for example, $-\infty$ is a fixed point if $m = 1$ and $F(x) = -x^2$.

Example 1.4: the fixed points of the logistic map. Consider the logistic equation (10). It is associated with the *logistic map* $F: \mathbb{R} \times \mathbb{R} \to \mathbb{R}$ such that $F(a, x) = ax(1 - x)$. If $x \in [0, 1]$ and $a \in [0, 4]$ then $0 \leqslant F(a, x) \leqslant 1$, so we usually shall take $0 \leqslant x_0 \leqslant 1$ and $0 \leqslant a \leqslant 4$, and confine our attention to $0 \leqslant x_n \leqslant 1$. Now equation (10) has fixed points X such that

$$X = aX(1 - X)$$

and therefore

$$X = 0, \qquad \text{or} \quad (a-1)/a \qquad \text{if } a \neq 0. \square \tag{13}$$

Next let us examine, by analogy with the stability of an equilibrium point of a differential equation, the stability of a fixed point X of a one-dimensional difference equation of the form

$$x_{n+1} = F(x_n) \qquad \text{for } n = 0, 1, \ldots, \tag{14}$$

where $F: \mathbb{R} \to \mathbb{R}$ is continuously twice-differentiable. We have $F(X) = X$ by definition of the fixed point. Define the perturbation of the fixed point as

$$x_n' = x_n - X.$$

Therefore equation (14) becomes

$$\begin{aligned} X + x_{n+1}' &= F(X + x_n') \\ &= F(X) + x_n' F'(X) + O(x_n'^2) \qquad \text{as } x_n' \to 0. \end{aligned}$$

Therefore the linearized system is

$$x_{n+1}' = F'(X) x_n'. \tag{15}$$

Its solution is

$$x_n' = F'(X) x_{n-1}' = \ldots = \{F'(X)\}^n x_0'.$$

Therefore, as $n \to \infty$, we find $x_n' \to 0$ if $|F'(X)| < 1$, x_n' is bounded if $|F'(X)| = 1$, and $x_n' \to \infty$ if $|F'(X)| > 1$. Therefore the null solution of the linearized system is stable if and only if $|F'(X)| \leqslant 1$. We deduce that the fixed point of the nonlinear system (14) is stable if $|F'(X)| < 1$ and unstable if $|F'(X)| > 1$. The linearized system gives $x_n' = x_0'$ if $F'(X) = 1$ and $x_n' =$

$(-1)^n x_0'$ if $F'(X) = -1$ and hence stability if $|F'(X)| = 1$; but in fact the linearized system is not sufficient to determine the stability of the fixed point of the nonlinear system in this case, because nonlinear terms may cause growth or decay of solutions when $|F'(X)| = 1$ (as they do for solutions of an ordinary differential equation in §§3–5 when $F'(X) = 0$).

Example 1.5: stability of the fixed points of the logistic map. For the logistic map $F(x) = ax(1-x)$, we find $F'(X) = a(1-2X)$. For the fixed point $X = 0$ we find $F'(X) = a$ and deduce that it is stable for $-1 < a < 1$; in fact (see Q1.32) it is also stable for $a = 1$. For $X = (a-1)/a$, we find $F'(X) = a\{1 - 2(a-1)/a\} = 2 - a$ and deduce that it is stable for $1 < a < 3$; in fact it is also stable for $a = 3$ (see §3.4). □

We have seen in §§6, 7 that differential equations may have periodic solutions as well as equilibrium points. Similarly, difference equations may have periodic solutions as well as fixed points. To see that, suppose that there exist a function $F: \mathbb{R} \to \mathbb{R}$ and distinct points X_1, X_2 such that $X_2 = F(X_1)$, $X_1 = F(X_2)$. Then, choosing $x_0 = X_2$, we find that $x_1 = F(x_0) = F(X_2) = X_1$, $x_2 = F(x_1) = F(X_1) = X_2, \ldots, x_{2r-1} = X_1, x_{2r} = X_2, \ldots$ for $r = 2, 3, \ldots$. We say that the set $\{X_1, X_2\}$ of two points is a *solution of period two* or a *two-cycle* of F. There may similarly occur solutions of periods three, four etc.

Example 1.6: a two-cycle of the logistic difference equation. If there exists a solution of the logistic equation (10) with period two then

$$X_1 = aX_2(1 - X_2)$$

and

$$\begin{aligned} X_2 &= aX_1(1 - X_1) \\ &= a\{aX_2(1 - X_2)\}[1 - \{aX_2(1 - X_2)\}], \end{aligned}$$

i.e.

$$aX_2(X_2 - 1 + 1/a)\{a^2X_2^2 - a(a+1)X_2 + a + 1\} = 0.$$

This gives the two fixed points (with $X_2 = X_1$), 0 and $(a-1)/a$, which we have found in Example 1.4, and also two other real roots,

$$X_1, X_2 = [a + 1 \pm \{(a+1)(a-3)\}^{1/2}]/2a, \tag{16}$$

for $a < -1$ or $a > 3$. This solution of period two bifurcates from the fixed point $X = \frac{2}{3}$ at $a = 3$; this is an example of what is called a *flip bifur-*

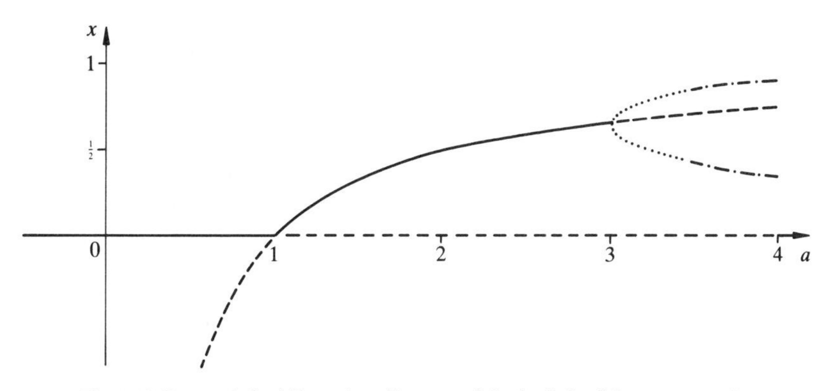

Fig. 1.18 Some of the bifurcation diagram of the logistic difference equation $x_{n+1} = ax_n(1 - x_n)$. Stable fixed points are denoted by a continuous curve, unstable fixed points by a dashed curve, stable two-cycles by a dotted curve, and unstable two-cycles by a dot-dashed curve.

cation. (There is also a flip bifurcation at $X = 0$, $a = -1$.) The periodic solution is in fact stable for $3 < a \leqslant 1 + \sqrt{6} = 3.4495$, and bifurcates further when it becomes unstable at $a = 3.4495$. Therein lies a long story which we shall take up in Chapter 3. The story so far is summarized in the bifurcation diagram in the (a, x)-plane shown in Fig. 1.18. Note the transcritical bifurcation at $(1, 0)$ and the flip bifurcation at $(3, \frac{2}{3})$. □

9 An experiment on statics

We shall next describe a simple experiment, which you may easily perform yourself, and interpret it qualitatively in terms of bifurcations in order to illustrate a general way of thinking about nonlinear systems. The experiment concerns elasticity, although we shall use only physical intuition for our interpretation. Consider the experiment which is shown in Fig. 1.19. It was proposed for the present purpose by Benjamin (cf. Iooss & Joseph 1990, §II.11). As the length l of the arch of wire increases, the strength of gravity relative to the elastic forces will increase. Therefore, when l is small, the upright symmetric position $\theta = 0$ of the arch of wire is stable. However, as l exceeds some critical value, l_c say, the arch will become unstable and lean on one side or the other. Looking at the experiment, you may see that the arch falls quite quickly to its new asymmetric position of equilibrium, and that the new position cannot be made very close to the upright by making $l - l_c$ sufficiently small. If the board and the wire have a configuration symmetric in $\pm\theta$ then the side on which the arch leans will be determined by the initial conditions. As l increases further, the arch will

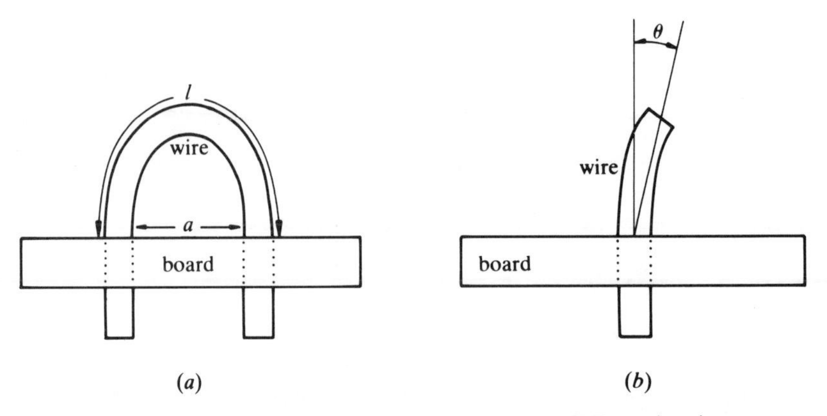

Fig. 1.19 A sketch of the experiment. (*a*) Side elevation. (*b*) Front elevation.

lean more steeply. For these lengths l the only stable positions of equilibrium are the two in which the arch leans on one side or the other, with equal but opposite angles $\pm\theta$. Further experiments will show that there is *hysteresis*, in the sense that if l is decreased slowly from a value greater than l_c then the arch will remain in an asymmetric position when l is decreased below l_c. Finally, when l is decreased below another critical value, l_0 say, the arch will spring back to its original upright position $\theta = 0$. This verbal description corresponds to the bifurcation diagram of Fig. 1.20(*a*). We may plausibly regard it as a manifestation of an equation of motion of the form,

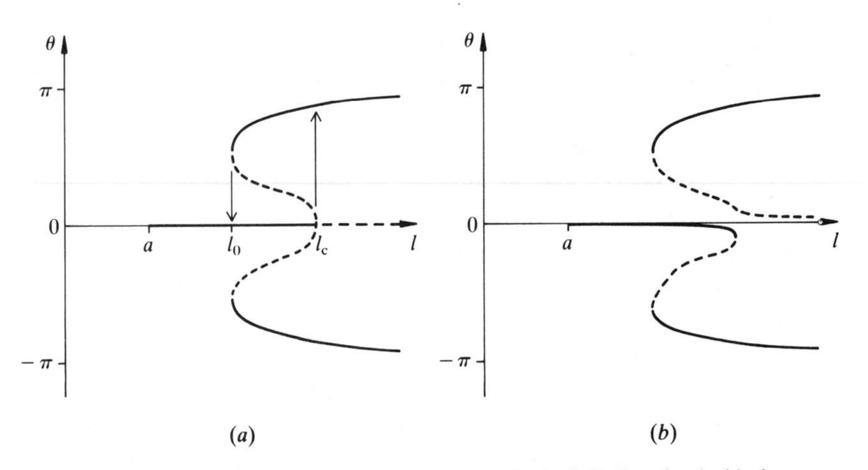

Fig. 1.20 (*a*) A sketch of the bifurcation diagram in the (l, θ)-plane for the idealized symmetric experiment. The arrows denote a hysteresis loop. (*b*) A sketch of the bifurcation diagram for a slightly asymmetric experiment.

$$\frac{d\theta}{dt} = F(l, \theta),$$

with steady solutions $l = L(\theta)$ such that $F(L(\theta), \theta) = 0$ for $-\pi < \theta < \pi$, say (although this differential equation is a crude model).

The assumed symmetry of the experiment implies that $F(l, -\theta) = F(l, \theta)$ for all θ, and therefore that the bifurcation curves $l = L(\theta)$ in Fig. 1.20(*a*) are symmetric about the l-axis. The symmetry breaking which occurs where $l = l_c$ is associated with the subcritical pitchfork bifurcation there. The two turning points at $l = l_0$ are symmetrically placed. You may observe, however, that in your experiment the wire is a little bent so that the arch always leans on the same side, and so deduce that the configuration of your apparatus is not exactly symmetric. We say that a system which is only slightly asymmetric is *imperfect*, whereas the idealized symmetric system is *perfect*. In practice, the experiment will be imperfect however carefully the apparatus is built. So it is instructive and useful to consider bifurcation diagrams for imperfect systems. A bifurcation diagram for the imperfect system is suggested in Fig. 1.20(*b*); test it with your own experiment. You could improvise an experiment with a shoe-lace instead of the wire and your hand instead of the board.

The simple examples of this chapter have been used to introduce several important concepts and methods of bifurcation theory. In the following chapters we shall develop the theory more systematically and in greater detail, and also introduce some advanced concepts and methods. However, the types of bifurcation we have already met will be seen to occur often in a wide variety of mathematical problems which may represent a wider variety of applications. They even occur for nonlinear partial differential systems, in which the dependent variables belong to a function space of infinite dimension. We shall see that the essential properties of many bifurcations may be represented in a phase space of dimension as low as one or two. It is for this reason that the simple bifurcations we have already exemplified are so common and so useful. There are, however, important bifurcations which must be represented in phase spaces of dimensions higher than two, and these bifurcations require deeper understanding and describe more complicated phenomena.

Further reading

§1.1 This book is an introduction to the theory of nonlinear systems from the point of view of an applied mathematician. Pippard (1985) introduces the subject from the point of view of a physicist, and Devaney (1989) from the point of view of a pure mathematician. In addition to these clear and illustrative general texts we shall recommend papers and books as they are needed, point by point and section by section. Devaney's (V1989) video is a vivid introduction to the mathematical theory of chaos.

The 'butterfly effect' was described by Lorenz in a lecture at Washington DC in 1970 (but note also the remarkable story by Bradbury (1953)).

§1.2 For an extensive account of the figures of equilibrium of a rotating mass of gravitating fluid and their stability, read Chandrasekhar (1969).

§1.3 Stability is formally defined in §5.1. We shall show more carefully in Chapter 6 why the linearized system gives the stability of the solution of the nonlinear second-order differential system except when the linearized system is at the margin of stability. However, a rigorous proof for ordinary differential systems of arbitrary order is given by, e.g., Coddington & Levinson (1955, Chap. 13).

§1.6 We assume that the reader is familiar with the elements of the theory of differential equations of second order, of the phase plane and of the types of singularity at a point of equilibrium; these will be discussed further in Chapter 6.

§1.8 For an introduction to the theory of linear difference equations, the book by Bender & Orszag (1978) is recommended.

Problems

Q1.1 *A transcritical bifurcation.* Sketch the bifurcation diagram of the equation,

$$\frac{\mathrm{d}x}{\mathrm{d}t} = x(a - c - abx),$$

in the (a, x)-plane for given positive constants b and c, indicating which steady solutions are stable.

Q1.2 *A logistic equation with a cubic term.* Show that if

$$\frac{\mathrm{d}x}{\mathrm{d}t} = ax - bx^2 + cx^3$$

for $x(0) > 0$, a, $b > 0$, $b^2 > 4ac$, then $x \uparrow X$ as $t \to \infty$ provided that $x(0)$ is sufficiently small, where $X = a/b + O(a^2)$ as $a \downarrow 0$ for fixed b, c.

[Drazin & Reid (1981, p. 458).]

Q1.3 *Another transcritical bifurcation.* Sketch the bifurcation diagram of the equation,

$$\frac{\mathrm{d}x}{\mathrm{d}t} = -x(x^2 - 2bx - a),$$

in the (a, x)-plane for a given positive constant b, indicating which steady solutions are stable.

Discuss the states of the system which could be observed in practice when a oscillates very slowly between $-2b^2$ and $2b^2$ if solutions with $x < 0$ are prohibited. [Assume that the system has 'noise', i.e. that small perturbations occur frequently.]

Q1.4 *Slow passage through a transcritical bifurcation.* Show that if

$$\frac{dx}{dt} = ax - x^2,$$

$a = \varepsilon t$, $\varepsilon > 0$, and $x = x_0$ at $t = t_0$ then

$$x(t) = x_0 \exp(\tfrac{1}{2}\varepsilon t^2) \Big/ \left\{ \exp(\tfrac{1}{2}\varepsilon t_0^2) + x_0 \int_{t_0}^{t} \exp(\tfrac{1}{2}\varepsilon s^2)\, ds \right\}.$$

Deduce that if $x_0 > 0$ and $t_0 = T_0/\sqrt{\varepsilon} < 0$ then $x(t) \to 0$ as $\varepsilon \downarrow 0$ for fixed t, $T_0 < 0$, and $x(t) \to a$ as $\varepsilon \downarrow 0$ for fixed $a > 0$.

Discuss the behaviour of such a solution for which ε is small, t_0 is large and negative, and $x_0 > 0$, interpreting it as giving the stable quasi-static solution $x(t) = 0$ for $t < 0$ and $x(t) = a$ for $t > 0$ with a time of order $\varepsilon^{-1/2}$ for transition from the former to the latter.

[Note that when $t < 0$ the solution $x(t)$ approaches so close to the quasi-statically stable null solution that when $t > 0$ it takes a long time to leave the quasi-statically unstable null solution and approach the quasi-statically stable solution $X = a$. This would not occur in practice if there were noise in the system. Lebovitz & Schaar (1975), Haberman (1979).]

Q1.5 *The lemniscate.* Sketch the bifurcation diagram of the equation,

$$\frac{dx}{dt} = 2(a^2 - x^2) - (a^2 + x^2)^2,$$

in the (a, x)-plane, indicating which steady solutions are stable, and identifying two turning points and a transcritical bifurcation.

Q1.6 *The folium of Descartes.* Sketch the bifurcation diagram of the equation,

$$\frac{dx}{dt} = x^3 + a^3 - 3ax,$$

in the (a, x)-plane, indicating which steady solutions are stable, and identifying a turning point and a pitchfork bifurcation.

Q1.7 *The Landau equation: a cautionary example.* Show that if

$$\frac{dx}{dt} = a(a/6b)^{1/2} \sin\{(6b/a)^{1/2} x\},$$

$a/b > 0$ and $x(0) > 0$ is sufficiently small, then $x(t) \to \pi(a/6b)^{1/2}$ as $t \to \infty$. Sketch the bifurcation diagram in the (a, x)-plane for fixed $b > 0$.

[Note that $dx/dt = ax - bx^3 + 3b^2x^5/40a + \ldots$ but that $x(t) \not\to A = (a/b)^{1/2}$ as $t \to \infty$ and $a \downarrow 0$, because then the coefficients of the terms in the fifth and higher powers of x become large. Drazin & Reid (1981, p. 458).]

Q1.8 *A particle on a rotating circular wire.* A particle moves around a smooth circular wire of radius a which is fixed relative to a vertical plane. Gravity acts on the particle, and the plane rotates with constant angular velocity ω about a vertical diameter of the circle. It is given that

$$a\frac{d^2\theta}{dt^2} = -g\sin\theta + a\omega^2\cos\theta\sin\theta,$$

where the radius through the particle makes an angle θ with the downward vertical.

Find the number of the positions of equilibrium (i.e. steady solutions) and their stability as the dimensionless parameter $a\omega^2/g$ varies. Sketch the bifurcation diagram in the plane of $a\omega^2/g$ and θ.

[Andronow & Chaikin (1949, p. 75).]

Q1.9 *Action of a spring on a particle on a circular wire.* A particle of mass m is constrained to move around a smooth circular wire of radius a fixed in a vertical plane. The particle is acted on by gravity and is attached to one end of a spring of natural length l $(< 2a)$ and stiffness constant k, the other being fixed to the highest point of the circle. The equation of motion is given to be

$$am\frac{d^2\theta}{dt^2} = k(2a\cos\tfrac{1}{2}\theta - l)\sin\tfrac{1}{2}\theta - mg\sin\theta,$$

where θ is the angle between the downward vertical and the radius to the particle.

(a) Find the number of the equilibrium points and their stability as k varies. Sketch the bifurcation diagram in the (k, θ)-plane

(b) Re-examine the solution to the problem, regarding $l/2a$ and ak/mg as independent dimensionless parameters. Sketch the bifurcation diagram in the $(l/2a, \theta)$-plane (for fixed $ka/mg > 0$). Describe the bifurcation surface in the three-dimensional space of ak/mg, $l/2a$ and θ.

Q1.10 *The rotation of a weighted pulley.* A light wheel of radius a has a uniform semicircular rim of mass M, and may rotate freely in a vertical plane about a horizontal axis through its centre. A light string passes around the wheel and suspends a mass m. It is given that this system is governed by the equation,

$$(M - m)a^2\frac{d^2\theta}{dt^2} = ag(m - 2\pi^{-1}M\sin\theta),$$

where θ is the angle between the downward vertical and the diameter through the centre of mass of the heavy rim.

Find the equilibrium points, and conditions for their existence and stability. Sketch the bifurcation diagram in the (k, θ)-plane, where $k = m/M$. What happens to the wheel when k is large?

Q1.11 *The complex Landau equation.* Show that if

$$\frac{dz}{dt} = az - b|z|^2 z$$

for complex z, a, b, then

$$\frac{d|z|^2}{dt} = 2a_r|z|^2 - 2b_r|z|^4, \qquad \frac{d(\text{ph}\, z)}{dt} = a_i - b_i|z|^2,$$

where $z = |z|e^{i\,\text{ph}\, z}$, $a = a_r + ia_i$, $b = b_r + ib_i$ in terms of modulus, phase, and real and imaginary parts.

Q1.12 *A stable limit cycle.* Show that if

$$\frac{dx}{dt} = -y + x(1 - x^2 - y^2)/(x^2 + y^2)^{1/2},$$

$$\frac{dy}{dt} = x + y(1 - x^2 - y^2)/(x^2 + y^2)^{1/2}$$

then $x(t) = \cos(t + \theta_0)$, $y(t) = \sin(t + \theta_0)$ represents a stable limit cycle.
[Nemytskii & Stepanov (1960, p. 24).]

Q1.13 *A semistable limit cycle.* Show that if

$$\frac{dx}{dt} = -y + x(1 - x^2 - y^2)^2 \qquad \frac{dy}{dt} = x + y(1 - x^2 - y^2)^2$$

then $x(t) = \cos(t + \theta_0)$, $y(t) = \sin(t + \theta_0)$ represents a semistable limit cycle, i.e. it is a periodic solution which is stable on one side and unstable on the other (and so is unstable).
[Nemytskii & Stepanov (1960, p. 25).]

Q1.14 *An infinity of limit cycles.* Show that if

$$\frac{dx}{dt} = -y + xf((x^2 + y^2)^{1/2}), \qquad \frac{dy}{dt} = x + yf((x^2 + y^2)^{1/2}),$$

where f is defined by $f(r) = \sin\{1/(r^2 - 1)\}$ for $r \neq \pm 1$ and $f(\pm 1) = 0$, then limit cycles are given by $x(t) = r_j \cos(t + \theta_0)$, $y(t) = r_j \sin(t + \theta_0)$ for $j = 0, 1, \ldots$, where $r_0 = 1$ and $r_j = (1 + 1/j\pi)^{1/2}$ for $j = 1, 2, \ldots$. Which are stable?
[Nemytskii & Stepanov (1960, p. 26).]

Q1.15 *A Hopf bifurcation.* Given that $x = r\cos\theta$, $y = r\sin\theta$ for $r \geqslant 0$, and

$$\frac{dx}{dt} = -y + x\{1 - f(x, y)/a\}, \qquad \frac{dy}{dt} = x + y\{1 - f(x, y)/a\}$$

for $a > 0$ and a continuous function f, show that

$$\frac{dr}{dt} = \{1 - f(r\cos\theta, r\sin\theta)/a\}r, \qquad \frac{d\theta}{dt} = 1.$$

Deduce that the null solution is stable if $a < f(0,0)$ and unstable if $0 < f(0,0) < a$. Show that if $0 < a < f(x,y)$ for all x, y then the solution $x(t)$, $y(t) \to 0$ as $t \to \infty$ for all initial conditions.

Verify that if $f(x,y) = g(r)$ and $a = g(r_0)$ for some positive differentiable function g and some positive constant r_0 then there exist solutions of the form

$$x(t) = r_0 \cos(t + \theta_0), \qquad y(t) = r_0 \sin(t + \theta_0).$$

Show that the orbit of these periodic solutions is stable if $g'(r_0) = 0$ and unstable if $g'(r_0) < 0$.

Further, given that $g(r) = 1 + 1/\{1 + (r-1)^2\}$, sketch the bifurcation diagram in the first quadrant of the (a, r_0)-plane, and sketch the phase portraits in the (x, y)-plane for each of the qualitatively different cases which arise.

Q1.16 *A conservative system with periodic solutions.* Show that if

$$\frac{d^2x}{dt^2} + \operatorname{sgn} x = 0$$

then $\frac{1}{2}(dx/dt)^2 + |x| = E$ for some constant $E \geqslant 0$. Sketch the phase portrait in the $(x, dx/dt)$-plane. Show that each orbit has period $4(2E)^{1/2}$ for $E > 0$.

[Note that, by definition, $\operatorname{sgn} x = 1$ if $x > 0$, $= 0$ if $x = 0$, and $= -1$ if $x < 0$.]

Q1.17 *The oscillation of a nonlinear spring.* Show that if

$$\frac{d^2x}{dt^2} + x - \varepsilon x^3 = 0$$

then there are oscillations of amplitude a with period

$$T = \frac{4K\{\epsilon a^2/(2 - \epsilon a^2)\}}{(1 - \frac{1}{2}\epsilon a^2)^{1/2}},$$

where K is the complete elliptic integral of the first kind defined by

$$K(m) = \int_0^{\pi/2} \frac{d\phi}{(1 - m\sin^2\phi)^{1/2}} \qquad \text{for } m < 1.$$

Deduce that

$$T = 2\pi\{1 + \tfrac{3}{8}\epsilon a^2 + O(\epsilon^2 a^4)\} \qquad \text{as } \epsilon a^2 \to 0.$$

[Cf. Example 6.6.]

Q1.18 *The relativistic harmonic oscillator.* It is given that the relativistic one-dimensional motion of a particle of rest mass m_0 and velocity $v = dx/dt$ is governed by the equation

$$\frac{d}{dt}\left\{\frac{m_0 v}{(1 - v^2/c^2)^{1/2}}\right\} + kx = 0,$$

where c is the speed of light and k a positive constant. Supposing that a is the amplitude of an oscillation, so that $x = \pm a$ where $v = 0$, deduce the mass-energy integral in the form,

$$\frac{m_0 c^2}{(1 - v^2/c^2)^{1/2}} + \tfrac{1}{2}kx^2 = m_0 c^2 + \tfrac{1}{2}ka^2.$$

Show further that there is an oscillation of period

$$T = \frac{4}{c}\int_0^a \frac{f\,\mathrm{d}x}{(f^2 - 1)^{1/2}},$$

where $f = 1 + \epsilon(a^2 - x^2)$ and $\epsilon = k/2m_0c^2$. Hence show that

$$T = 2\pi(m_0/k)^{1/2}\{1 + \tfrac{3}{8}\epsilon a^2 + O(\epsilon^2 a^4)\} \qquad \text{as } \epsilon a^2 \to 0,$$

i.e. in the Newtonian limit as $c \to \infty$.

Q1.19 *Solitary waves.* Seek a wave solution of permanent form for the *Korteweg–de Vries equation,*

$$\frac{\partial u}{\partial t} - 6u\frac{\partial u}{\partial x} + \frac{\partial^3 u}{\partial x^3} = 0,$$

by assuming that $u(x, t) = f(x - ct)$ for some constant c and function f. Deduce that

$$\tfrac{1}{2}f'^2 = f^3 + \tfrac{1}{2}cf^2 + Af + B$$

for some constants A, B of integration.

Seeking a solitary wave, such that $f(X)$, $f'(X)$, $f''(X) \to 0$ as $X \to \infty$, show that

$$f(X) = -\tfrac{1}{2}c\,\mathrm{sech}^2\{\tfrac{1}{2}\sqrt{c}(X - X_0)\} \qquad \text{for } -\infty < X < \infty,$$

where X_0 is some real constant and $c \geqslant 0$.

[Korteweg & de Vries (1895); cf. Drazin & Johnson (1989, Chap. 2).]

Q1.20 *Smooth step-like waves of permanent form.* It is given that the substitution of $u(x, t) = f(x - ct)$ into some nonlinear partial differential equation for u gives rise, on integration, to an ordinary differential equation of the form $\tfrac{1}{2}f'^2 = F(f)$; where c is taken to be a constant, F'' is continous, F has double zeros at f_1, f_2, $F(f) > 0$ for $f_1 < f < f_2$, and f_1 and f_2 are some constants. Show that there exist two monotonic solutions f such that $f(X) \to f_1$ as $X \to \mp\infty$ and $f(X) \to f_2$ as $X \to \pm\infty$ respectively.

[Such solutions are called *topological solitons.*]

Q1.21 *The elastica.* (a) It is given that the angle ψ of deflection of a static uniform elastic beam at a distance s from one end is governed by the equation,

$$\frac{\mathrm{d}^2\psi}{\mathrm{d}s^2} + \omega^2 \sin\psi = 0,$$

where ω is a positive parameter. Show that

$$\tfrac{1}{2}\psi'^2 - \omega^2 \cos\psi = \text{constant}$$

for each solution, where $\psi' = d\psi/ds$. Sketch the phase portrait of the solutions in the (ψ, ψ')-plane.

(b) Suppose, moreover, that the beam has length l, is clamped at one end, so that $\psi = 0$ at $s = 0$, and free at the other end, so that $\psi' = 0$ at $s = l$. Solving the linearized equation for small ψ together with the boundary conditions, show that $\omega = (n + \tfrac{1}{2})\pi/l$ is an eigenvalue for $n = 0, 1, \ldots$.

Show that if ψ increases monotonically from 0 to α (where α is a given positive angle, not necessarily small) as s increases from 0 to l, then

$$\omega l = \int_0^{\pi/2} \frac{d\phi}{(1 - k^2 \sin^2 \phi)^{1/2}},$$

where $k = \sin\tfrac{1}{2}a$. [Hint: substitute $\sin\tfrac{1}{2}\psi = k\sin\phi$.] Deduce that

$$\omega l = \tfrac{1}{2}\pi\{1 + \tfrac{1}{4}\sin^2\tfrac{1}{2}\alpha + O(\alpha^4)\} \qquad \text{as } \alpha \to 0.$$

Considering next solutions which are not monotonic, show that

$$\omega l = (2n+1)\int_0^{\pi/2} \frac{d\phi}{(1 - k^2 \sin^2 \phi)^{1/2}} \qquad \text{for } n = 1, 2, \ldots$$

for orbits which go around the origin of the phase plane and for which $\psi = \alpha$ at $s = l$. Sketch the bifurcation diagram in the (ω^2, α)-plane, where $\psi = \alpha$ at $s = l$ but ψ need not be a monotonic function of s.

Q1.22 *An equation transformable to a conservative system.* Show that if

$$\frac{d^2x}{dt^2} + k(x)\left(\frac{dx}{dt}\right)^2 = f(x)$$

then

$$\frac{1}{2}m(x)\left(\frac{dx}{dt}\right)^2 + V(x) = E,$$

where $V(x) = -\int_0^x m(u)f(u)\,du$, $m(x) = \exp\{2\int_0^x k(u)\,du\}$. Hence show that

$$\frac{d^2y}{dt^2} = g(y),$$

where $y = \int_0^x (m(u))^{1/2}\,du$, $g(y) = (m(x))^{1/2}f(x)$.

Q1.23 *An elementary proof of a fixed-point theorem.* Prove that if $F\colon \mathbb{R} \to \mathbb{R}$ is continuous and there exists a closed interval I such that $\mathrm{I} \subset F(\mathrm{I})$ then there exists $X \in \mathrm{I}$ such that $F(X) = X$.

[Hint: either use the intermediate-value theorem or sketch the graphs of $y = F(x)$ and $y = x$ in the (x, y)-plane for $x \in \mathrm{I}$.]

Q1.24 *Iterated approximation to a square-root.* Using the Newton–Raphson method to approximate a zero of $f(a, x) = x^2 - a$ for $a > 0$, show that successive approximations to $\sqrt{a}$ may be given by

$$x_{n+1} = \tfrac{1}{2}(x_n + a/x_n) \qquad \text{for } n = 0, 1, \ldots.$$

Hence evaluate the square-root of 55 to three significant figures.

Q1.25 *Convergence of the Newton–Raphson method.* Given that f is a twice continuously differentiable function on $[a, b]$ such that $f(a)f(b) < 0$ and $f'(x) \neq 0$ for all $x \in (a, b)$, show that f has a unique zero X in (a, b).

Show that if $x_{n+1} = F(x_n)$ for $n = 0, 1, \ldots$, where $F(x) = x - f(x)/f'(x)$, and $x_n, x_{n+1} \in (a, b)$, then

$$|x_{n+1} - X| \leqslant \frac{M}{2m}|x_n - X|^2,$$

where $m = \min_{a \leqslant x \leqslant b}|f'(x)|$ and $M = \max_{a \leqslant x \leqslant b}|f''(x)|$.

Q1.26 *Fibonacci numbers, the golden mean and a continued fraction.* Defining the *Fibonacci numbers* x_n by

$$x_{n+2} = x_{n+1} + x_n \qquad \text{for } n = 0, 1, \ldots,$$

$x_0 = 0$ and $x_1 = 1$, and defining $y_n = x_n/x_{n+1}$, deduce that

$$y_{n+1} = 1/(1 + y_n).$$

Defining the *golden mean* as $\alpha = \frac{1}{2}(\sqrt{5} - 1), = 0.618$, prove that

$$y_n \to \alpha \qquad \text{as } n \to \infty.$$

Deduce that the continued fraction

$$\cfrac{1}{1 + \cfrac{1}{1 + \cfrac{1}{1 + \cdots}}} = \alpha.$$

Q1.27 *Another continued fraction.* Given $a > 0$, seek to compute $X = \sqrt{a}$ as follows. Estimate $r > 0$ as a first approximation to $\sqrt{a}$ and define $\delta = a - r^2$ so that $X = r + \delta/(r + X)$. Then iterate $x_{n+1} = r + \delta/(r + x_n)$ with $x_0 = r$.

Find the fixed points of the difference equation and whether they are stable.

Hence express $\sqrt{a}$ as a continued fraction.

Use the difference equation also to evaluate the square-root of 101 to six significant figures.

Q1.28 *Enhancement of stability of a fixed point.* Define $G: \mathbb{R} \times \mathbb{R} \to \mathbb{R}$ by

$$G(a, x) = aF(x) + (1 - a)x,$$

given the map $F: \mathbb{R} \to \mathbb{R}$. Show that, for $a \neq 0$, X is a fixed point of F if and only if it is a fixed point of G. Find those values of a for which X is a stable fixed point of G when F is continuously differentiable.

In particular, take $F(x) = 50 - \cosh x$. Show graphically that there is a unique positive fixed point X of F, and numerically that $4 < X < 5$. Deduce that X is an unstable fixed point of F, but a stable fixed point of G if $a = 0.02$. [You are given that $\cosh 4 = 27.31$, $\sinh 4 = 27.29$, $\cosh 5 = 74.21$ and $\sinh 5 = 74.20$.]

Define H over an interval where $F' \neq 1$ by

$$H(x) = x - \{x - F(x)\}/\{1 - F'(x)\}.$$

Deduce that, in this interval, X is a fixed point of H if and only if it is a fixed point of F. Show that X is a stable fixed point of H for all twice differentiable F.

Q1.29 *Stability of the null solution of a finite-difference approximation to a differential equation.* Suppose that

$$x_{n+1} = x_n + hf(x_n) \qquad \text{for } n = 0, 1, \ldots,$$

where $f: \mathbb{R} \to \mathbb{R}$ is continuously differentiable, $f(0) = 0$ and $h > 0$. Show that the fixed point $X = 0$ is stable if $f'(0) < 0$ and unstable if $f'(0) > 0$ when h is sufficiently small.

[Note that this agrees with the condition for stability of the null solution of the differential equation $\mathrm{d}x/\mathrm{d}t = f(x)$.]

Q1.30 *The logistic map.* Define $F: \mathbb{R} \times \mathbb{R} \to \mathbb{R}$ by $F(a, x) = ax(1 - x)$. Show that if $0 \leqslant a \leqslant 4$ then F maps the interval $[0, 1]$ into itself and if $a = 4$ then F maps the interval onto itself. Show that if $a > 1$ and either $x_0 < 0$ or $x_0 > 1$ then $x_n \to -\infty$ as $n \to \infty$, where $x_{n+1} = F(a, x_n)$.

Q1.31 *Elementary transformation of the logistic equation.* Show that if $y_{n+1} = 1 - by_n^2$ and $x_n = (\frac{1}{4}a - \frac{1}{2})y_n + \frac{1}{2}$ for $n = 0, 1, \ldots$, where $b = \frac{1}{4}a^2 - \frac{1}{2}a$, then $x_{n+1} = ax_n(1 - x_n)$. Show that if $z_n = by_n$ then $z_{n+1} = b - z_n^2$.

Q1.32 *The asymptotic solution of the logistic equation in a special case.* Show that if

$$x_{n+1} = x_n(1 - x_n) \qquad \text{for } n = 0, 1, \ldots,$$

and $0 < x_0 < 1$ then $x_n \sim 1/n$ as $n \to \infty$.

[Putnam (1967).]

Q1.33 *The explicit solution of the logistic equation in a special case.* Show that if

$$x_{n+1} = 2x_n(1 - x_n) \qquad \text{for } n = 0, 1, \ldots,$$

and $y_n = 1 - 2x_n$, then $y_{n+1} = y_n^2$. Hence or otherwise show that

$$x_n = \tfrac{1}{2}\{1 - (1 - 2x_0)^{2^n}\}.$$

Deduce that $\lim_{n\to\infty} x_n = \frac{1}{2}$ if and only if $0 < x_0 < 1$.

[Bender & Orszag (1978), p. 53).]

Q1.34 *The limit of the solution of a difference equation.* Show that if $x_n \neq 2$ and

$$x_{n+1} = \frac{1}{2 - x_n} \qquad \text{for } n = 0, 1, \ldots,$$

then

$$x_n = \frac{n - (n-1)x_0}{n + 1 - nx_0}$$

and therefore $\lim_{n\to\infty} x_n = 1$.

[Putnam (1947).]

Q1.35 *A linear difference equation.* Suppose that

$$x_{n+2} + 2bx_{n+1} + cx_n = 0 \qquad \text{for } n = 0, 1, \ldots,$$

where real $b, c \neq b^2, x_0$ and x_1 are given. To solve this problem, show that

$$\mathbf{x}_{n+1} = \mathbf{A}\mathbf{x}_n,$$

where the column vector $\mathbf{x}_n = \begin{bmatrix} x_n \\ x_{n+1} \end{bmatrix}$ and the 2×2 matrix $\mathbf{A} = \begin{bmatrix} 0 & 1 \\ -c & 2b \end{bmatrix}$.

Deduce that $\mathbf{x}_n = \mathbf{A}^n\mathbf{x}_0$.

Show that there exist ξ_1, ξ_2 such that $\mathbf{x}_0 = \xi_1\mathbf{u}_1 + \xi_2\mathbf{u}_2$ for all $\mathbf{x}_0$, where s_j is the eigenvalue of $\mathbf{A}$ belonging to the eigenvector $\mathbf{u}_j$ for $j = 1, 2$, and thence that $\mathbf{x}_n = \xi_1 s_2^n\mathbf{u}_1 + \xi_2 s_2^n\mathbf{u}_2$. Hence or otherwise prove that

$$x_n = \xi_1\{-b + (b^2 - c)^{1/2}\}^n + \xi_2\{-b - (b^2 - c)^{1/2}\}^n,$$

where $\xi_1, \xi_2 = \frac{1}{2}\{x_0 \pm (bx_0 + x_1)/(b^2 - c)^{1/2}\}$ respectively.

Deduce that if $x_{n+2} = x_n + x_{n+1}$, $x_0 = 0$ and $x_1 = 1$ then the Fibonacci numbers $x_n = [\{\frac{1}{2}(\sqrt{5} + 1)\}^n - \{-\frac{1}{2}(\sqrt{5} - 1)\}^n]/\sqrt{5} \sim \{\frac{1}{2}(\sqrt{5} + 1)\}^n/\sqrt{5}$ as $n \to \infty$.

Q1.36 *Equality of slopes of a second-generation map at its fixed points.* Define the second-generation map G of F by $G(x) = F(F(x))$, where F is a continuously differentiable function, i.e. $F \in \mathbf{C}^1(-\infty, \infty)$, and suppose that F has a two-cycle $\{X, Y\}$ such that $X = F(Y)$ and $Y = F(X)$. Deduce that X and Y are fixed points of G, and that $G'(X) = F'(X)F'(Y) = G'(Y)$.

Q1.37 *Some cycles of a map.* Defining F by $F(x) = (1 - x)/(1 + ax)$ for $a > 0, x \neq -1/a$, show that F maps the interval $[0, 1]$ onto itself. Find all the fixed points of F and ascertain whether each is stable, according to the values of a. Find all the two-cycles of F. What other cycles of F are there?

2

Classification of bifurcations of equilibrium points

Not chaos-like together crush'd and bruis'd,
But, as the world, harmoniously confused:
Where order in variety we see,
And where, tho' all things differ, all agree.

Alexander Pope (*Windsor Forest*)

1 Introduction

We have met *autonomous* differential equations, i.e. equations which do not involve the independent variable explicitly, of the form

$$\frac{d\mathbf{x}}{dt} = \mathbf{F}(\mathbf{x}), \tag{1}$$

where $\mathbf{F}: \mathbb{R}^m \to \mathbb{R}^m$. Then the equilibrium points, sometimes called the *critical points*, are the zeros $\mathbf{X}$ of $\mathbf{F}$, i.e. the points $\mathbf{X}$ such that $\mathbf{F}(\mathbf{X}) = \mathbf{0}$. Likewise we have met difference equations of the form

$$\mathbf{x}_{n+1} = \mathbf{G}(\mathbf{x}_n) \qquad \text{for } n = 0, 1, \ldots, \tag{2}$$

where $\mathbf{G}: \mathbb{R}^m \to \mathbb{R}^m$, and sought fixed points $\mathbf{X}$ such that $\mathbf{X} = \mathbf{G}(\mathbf{X})$. Therefore finding the fixed points of a difference equation is equivalent to finding the equilibrium points of an ordinary differential system, on identifying $\mathbf{x} - \mathbf{G}(\mathbf{x}) = \mathbf{F}(\mathbf{x})$ for all $\mathbf{x} \in \mathbb{R}^m$. Further, the bifurcations of the fixed points of a difference equation, as a parameter changes, are equivalent to the bifurcations of an ordinary differential system. Indeed, we have met a transcritical bifurcation for a difference equation, and will soon meet turning points and pitchfork bifurcations in examples of difference equations with a parameter.

So we shall in this chapter seek the solutions $\mathbf{x}$ of nonlinear functional equations of the form

$$\mathbf{F}(\mathbf{a}, \mathbf{x}) = \mathbf{0} \tag{3}$$

as l parameters $\mathbf{a}$ vary, where $\mathbf{F}: \mathbb{R}^l \times \mathbb{R}^m \to \mathbb{R}^m$. We define a *bifurcation point*, or a *branch point*, as a solution $(\mathbf{a}_0, \mathbf{X}_0)$ of equation (3) such that the number of solutions $\mathbf{x}$ of (3) in a small neighbourhood of $\mathbf{X}_0$ changes when $\mathbf{a}$ varies within a small neighbourhood of $\mathbf{a}_0$. You may verify that this definition gives turning points, transcritical points and pitchforks as bifurcation points. (The definition needs extension for a nonlinear system with solutions which evolve with time, to allow for the change of the qualitative character of the unsteady solutions as a parameter changes.)

The zeros of $\mathbf{F}$ cannot, of course, in general be given explicitly as elementary functions of $\mathbf{a}$ by solving equation (3). We may be able to find the zeros analytically in a few special cases, but usually must resort to numerical methods of solution. However, we can examine the *local* properties of the solutions systematically and thereby classify the types of bifurcation. The classification becomes complicated as the number of types grows with l and m, but nonetheless there are important particular and general results known.

2 Classification of bifurcations in one dimension

First treat the case $m = 1$, beginning with the subcase $l = 1$. Thus we seek solutions $x = X(a)$ of the equation

$$F(a, x) = 0,$$

for the map $F: \mathbb{R} \times \mathbb{R} \to \mathbb{R}$. These solutions may be represented by curves in the bifurcation diagram of the (a, x)-plane.

To apply local analysis, we suppose that one point (a_0, X_0) of the solution is known somehow, i.e. that

$$F(a_0, X_0) = 0. \tag{1}$$

Then we may expand F as a Taylor series about this point in order to find neighbouring solutions, provided that F is infinitely differentiable with respect to a and x at (a_0, X_0):

$$\begin{aligned} 0 &= F(a, X) && (2)\\ &= F(a_0, X_0) + (X - X_0)F_x(a_0, X_0) + (a - a_0)F_a(a_0, X_0) \\ &\quad + \tfrac{1}{2}(X - X_0)^2 F_{xx}(a_0, X_0) + \cdots \\ &= (X - X_0)F_{x0} + (a - a_0)F_{a0} + \tfrac{1}{2}(X - X_0)^2 F_{xx0} + \cdots, && (3) \end{aligned}$$

when we use equation (1) and denote evaluation of the partial derivatives of F at (a_0, X_0) by the subscript zero. Therefore

$$X(a) = X_0 - (a - a_0)F_{a0}/F_{x0} + o(a - a_0) \qquad \text{as } a \to a_0, \tag{4}$$

provided that $F_{x0} \neq 0$. This heuristic argument gives the equilibrium solutions near (a_0, X_0) in the (a, x)-plane, in fact the tangent to the curve at (a_0, X_0) in the bifurcation diagram. It can be *proved*, by use of the implicit function theorem, that the curve $x = X(a)$ exists in a neighbourhood of (a_0, X_0) if both F is continuously differentiable with respect to a and x in a neighbourhood of (a_0, X_0) and $F_{x0} \neq 0$.

If $F_{x0} = 0$ then equation (3) gives

$$X(a) = X_0 \pm \{-2(a - a_0)F_{a0}/F_{xx0}\}^{1/2} + o((a - a_0)^{1/2}) \tag{5}$$

as $(a - a_0)\,\mathrm{sgn}(F_{a0}/F_{xx0}) \uparrow 0$, provided that $F_{xx0} \neq 0$. Note that we require $a - a_0$ to have the appropriate sign to ensure that the two solutions (5) (representing a turning point at (a_0, X_0)) are real.

An infinity of further special cases can be seen to arise according to which of the leading partial derivatives of F vanish at (a_0, X_0).

Example 2.1: an exponential map. Consider the difference equation $x_{n+1} = G(a, x_n)$, where $G(a, x) = a\mathrm{e}^x$; then the fixed points X of G are the zeros of $F(a, x) = a\mathrm{e}^x - x$. We see, by inspection, that one zero is given by $a = 0$, $X = 0$. So, expanding about this point, we find

$$\begin{aligned} 0 &= F(a, X) \\ &= a(1 + X + \tfrac{1}{2}X^2 + \tfrac{1}{6}X^3 + \cdots) - X. \end{aligned}$$

Therefore

$$\begin{aligned} X &= a + aX + \tfrac{1}{2}aX^2 + \cdots \\ &= a + a^2 + O(a^3) \qquad \text{as } a \to 0. \end{aligned}$$

Note also that $F(a, x) = 0$ is equivalent to $x = \ln x - \ln a$, and that $x/\ln x \to \infty$ as $x \to \infty$. So there is another fixed point for which

$$X(a) \sim -\ln a \qquad \text{as } a \downarrow 0.$$

We also see by inspection that a zero of F is given by $X = 1$ when $a = 1/\mathrm{e}$. To find other solutions nearby, we expand

$$\begin{aligned} 0 &= F(a, X) \\ &= F(1/\mathrm{e}, 1) + (X - 1)F_x(1/\mathrm{e}, 1) + (a - 1/\mathrm{e})F_a(1/\mathrm{e}, 1) \\ &\quad + \tfrac{1}{2}(X - 1)^2 F_{xx}(1/\mathrm{e}, 1) + (X - 1)(a - 1/\mathrm{e})F_{ax}(1/\mathrm{e}, 1) + \cdots \\ &= 0 + 0 + (a - 1/\mathrm{e})\mathrm{e} + \tfrac{1}{2}(X - 1)^2 + \mathrm{e}(X - 1)(a - 1/\mathrm{e}) \\ &\quad + \tfrac{1}{6}(X - 1)^3 + \cdots. \end{aligned}$$

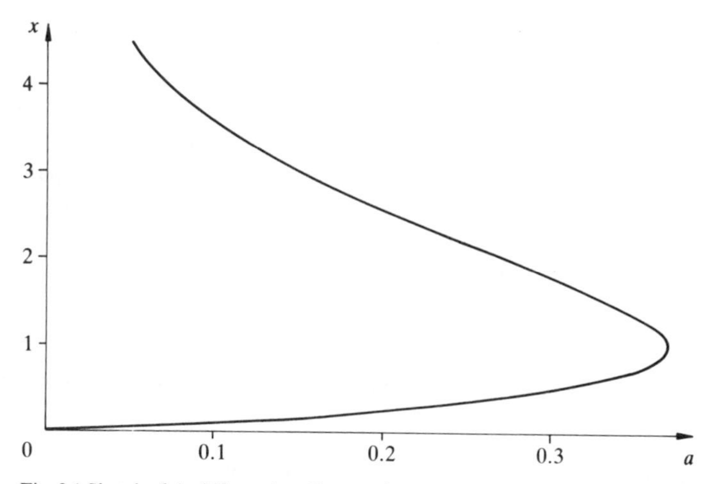

Fig. 2.1 Sketch of the bifurcation diagram for $F(a, x) = ae^x - x$, i.e. of the curve with equation $ae^x = x$ in the (a, x)-plane.

Therefore

$$(X-1)^2 = 2(1-ae) + 2(X-1)(1-ae) - \tfrac{1}{3}(X-1)^3 + \cdots.$$

Therefore

$$X = 1 \pm \{2(1-ae)\}^{1/2}[1 \pm \{\tfrac{1}{2}(1-ae)\}^{1/2} + O(1-ae)] \qquad \text{as } a \uparrow 1/\mathrm{e}.$$

Observe that $F_x(1/\mathrm{e}, 1) = 0$ and there is a turning point at $X = 1, a = 1/\mathrm{e}$ in the bifurcation diagram in the (a, x)-plane, as sketched in Fig. 2.1.

*Note that an expansion of the form $X(a) = 1 + (a - 1/\mathrm{e})X_1 + (a - 1/\mathrm{e})^2 X_2 + \cdots$ is invalid as $a \to 1/\mathrm{e}$. However, we may expand

$$a = 1/\mathrm{e} + \epsilon a_1 + \epsilon^2 a_2 + \cdots \quad \text{and} \quad X(a) = 1 + \epsilon X_1 + \epsilon^2 X_2 + \cdots$$
$$\text{as } \epsilon \to 0, \tag{6}$$

where the new small parameter ϵ is introduced instead of $a - 1/\mathrm{e}$. This gives

$$\begin{aligned} 0 &= a\mathrm{e}^X - X \\ &= (1/\mathrm{e} + \epsilon a_1 + \cdots)\exp(1 + \epsilon X_1 + \cdots) - (1 + \epsilon X_1 + \cdots) \\ &= (1 + \epsilon \mathrm{e} a_1 + \cdots)\{1 + (\epsilon X_1 + \cdots) + \tfrac{1}{2}(\epsilon X_1 + \cdots)^2 + \cdots\} \\ &\quad - 1 - \epsilon X_1 - \cdots \\ &= \epsilon \mathrm{e} a_1 + \epsilon^2(\mathrm{e} a_2 + \mathrm{e} a_1 X_1 + \tfrac{1}{2}X_1^2) + O(\epsilon^2) \qquad \text{as } \epsilon \to 0. \end{aligned}$$

Equating coefficients of ϵ and ϵ^2, we deduce that

$$a_1 = 0, \qquad X_1^2 = -2ea_2$$

respectively. There is some freedom of choice of the coefficients $a_2, a_3, \ldots$, because the same solution X may be expressed differently by the pair (6) of expansions. Requiring X to be real, we may take a_2 as any negative number. It is convenient to choose $a_2 = -1$ and $a_3 = a_4 = \ldots = 0$ as a means of defining ϵ in terms of $a - 1/\mathrm{e}$. Then $X_1 = \pm(2\mathrm{e})^{1/2}$, and

$$X = 1 \pm (2\mathrm{e})^{1/2}\epsilon + O(\epsilon^2) \qquad \text{as } \epsilon \to 0,$$

where $a = 1/\mathrm{e} - \epsilon^2$, in agreement with the solution found otherwise above. □

When the first several terms of the Taylor series (3) vanish, the local behaviour of the equilibrium curve near (a_0, X_0) is determined by the first terms which do not vanish. The main special cases may be classified as follows.

(i) *A regular point* (a_0, X_0) of the equilibrium curve is one for which $F_{x0} \neq 0$ or $F_{a0} \neq 0$. Then the implicit function theorem shows that $X(a)$ or $a(X)$ respectively exists and is continuously differentiable in a neighbourhood of (a_0, X_0). For an example (see Fig. 2.2(*a*)), take

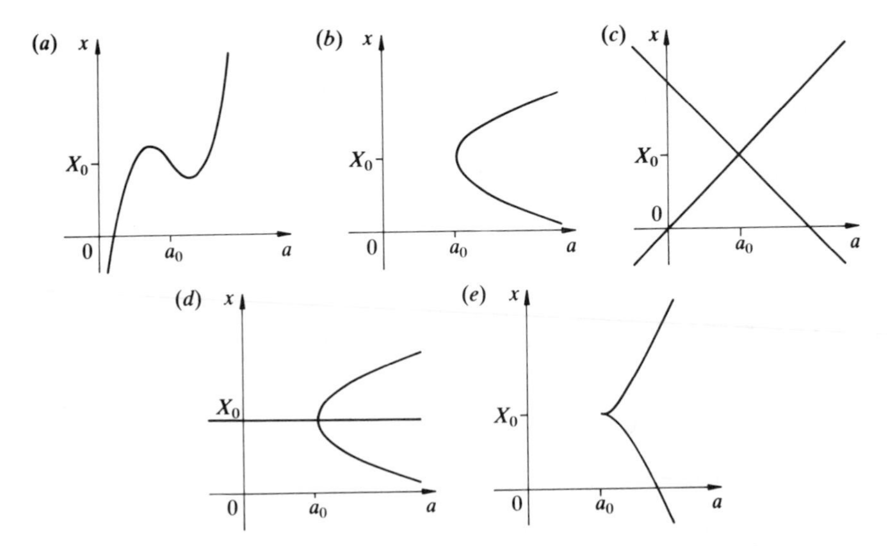

Fig. 2.2 Sketches of examples of typical kinds of bifurcation points. (*a*) A regular point: $x - X_0 + (a - a_0) - (a - a_0)^3 = 0$. (*b*) A regular turning point: $(x - X_0)^2 - (a - a_0) = 0$. (*c*) A transcritical bifurcation: $(x - X_0)^2 - (a - a_0)^2 = 0$. (*d*) A singular turning point: $\{(x - X_0)^2 - (a - a_0)\}(x - X_0) = 0$. (*e*) A cusp: $(x - X_0)^2 - (a - a_0)^3 = 0$.

$$F(a, x) = x - X_0 + a - a_0 - (a - a_0)^3.$$

(ii) A *regular turning point* or *simple turning point* (a_0, X_0) of an equilibrium curve is a regular point at which $\mathrm{d}a/\mathrm{d}X$ changes sign and $F_{a0} \neq 0$. For example (see Fig. 2.2(*b*)), take

$$F(a, x) = (x - X_0)^2 - (a - a_0).$$

Note that this is a special case of a regular point, that $X(a)$ is a 2–1 function for $a > a_0$ but does not exist for $a < a_0$, and that $F_{x0} = 0$.

(iii) A *singular point* is a point on an equilibrium curve which is not a regular point, i.e. it is a point (a_0, X_0) where $F_x = F_a = 0$.

(iv) A *double point* is a singular point (a_0, X_0) through which pass two and only two branches of the equilibrium curve with distinct tangents. This requires $F_{ax0}^2 > F_{aa0}F_{xx0}$. Typically there is a transcritical bifurcation at a double point. For example (see Fig. 2.2(*c*)), take

$$F(a, x) = (x - X_0)^2 - (a - a_0)^2.$$

(v) A *singular turning point* is a double point at which $\mathrm{d}a/\mathrm{d}X$ changes sign on one branch of the equilibrium curve with distinct tangents. We need cubic terms in the Taylor expansion of F to distinguish this special case of a double point. This gives a pitchfork bifurcation at (a_0, X_0). For example (see Fig. 2.2(*d*)), take

$$F(a, x) = \{(x - X_0)^2 - (a - a_0)\}(x - X_0).$$

(vi) A *cusp* is a point of second-order contact between the two branches of the curve. For example (see Fig. 2.2(*e*)), take

$$F(a, x) = (x - X_0)^2 - (a - a_0)^3.$$

(vii) A *conjugate point* is an isolated singular point. For example, take

$$F(a, x) = (x - X_0)^2 + (a - a_0)^2.$$

This exhausts the essentially distinct cases in which at least one of the second derivatives of F is non-zero at the equilibrium point (a_0, X_0). There are, however, other cases of higher-order singularity.

Next consider cases $l > 1$ and $m = 1$. It can be seen, as for the case $l = 1$, that, for well-behaved functions $F: \mathbb{R}^l \times \mathbb{R} \to \mathbb{R}$, a bifurcation may arise only at those values of $\mathbf{a}$ for which F has a multiple zero x, i.e. where

$$F(\mathbf{a}, x) = 0 \qquad \text{and} \qquad F_x(\mathbf{a}, x) = 0.$$

We can in principle eliminate x from these two equations to get a set, called the *bifurcation set*, of values of $\mathbf{a}$. So the bifurcation set is a subset of $\mathbb{R}^l$. (It is possible, with this informal definition, that there is a point of the bifurcation set at which there is no bifurcation; for example, if $F(\mathbf{a}, x) = (x - a)^2$ then $a = 0, x = 0$ is not a bifurcation point.)

Example 2.2: the cusp bifurcation set. Suppose that

$$F(a, b, x) = 4x^3 - 2ax + b.$$

Then we may picture $F = 0$ as a surface of height x above the (a, b)-plane, shown in Fig. 2.3. There are one, three, or, exceptionally, two, real zeros x according to the values of a and b. The number changes on the bifurcation

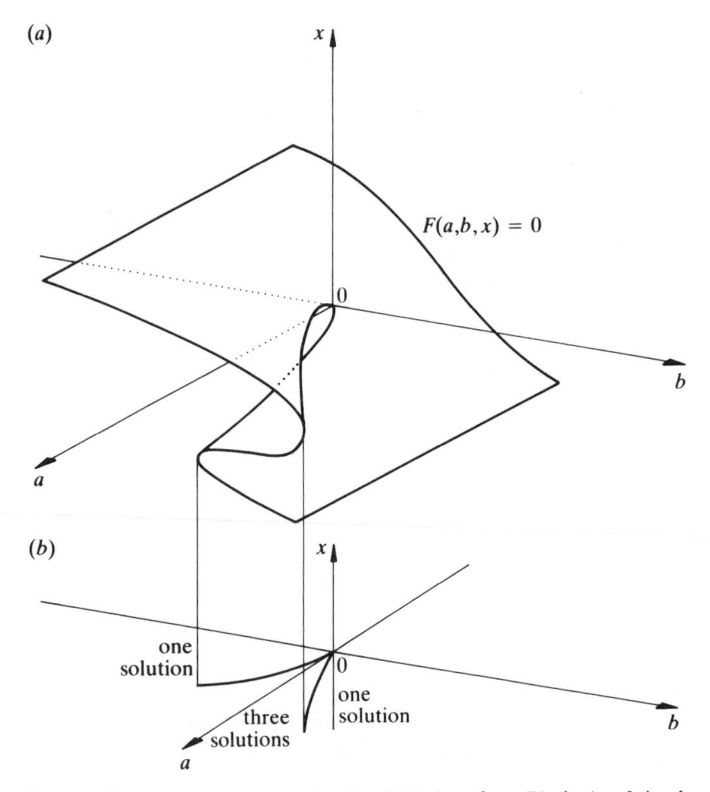

Fig. 2.3 The cusp catastrophe. (a) The folded surface $F(a, b, x) = 0$ in the (a, b, x)-space and (b) the projection in the (a, b)-plane of its points with 'vertical' tangents, and with a cusp at $(0, 0)$.

set, where there are folds, and the tangent plane to the surface is vertical, i.e. perpendicular to the (a, b)-plane. Thus to find the bifurcation set, we solve $F = F' = 0$, i.e.

$$4x^3 - 2ax + b = 0 \qquad \text{and} \qquad 12x^2 - 2a = 0$$

for a, b. Therefore

$$x = \pm(\tfrac{1}{6}a)^{1/2} \qquad \text{for } a \geqslant 0, \qquad \text{and} \qquad \pm 4(\tfrac{1}{6}a)^{3/2} \mp 2a(\tfrac{1}{6}a)^{1/2} + b = 0,$$

i.e.

$$27b^2 = 8a^3.$$

This gives a cusp at the origin in the (a, b)-plane. □

3 Imperfections

In §1.9 we introduced an example of an imperfection. In general, we use mathematical models with various idealizations, such as symmetry, steadiness, plane boundaries and unbounded domains, which are not found precisely in the 'real' world they are designed to represent. So it is important to consider the effect of small 'irregularities' on idealized models. An irregularity may be important if it changes not only the quantitative character of the solutions a little but also the qualitative character of the solutions. If the qualitative character of the set of all the solutions of a system is changed by an infinitesimal perturbation of the system (or, rather, at least one infinitesimal perturbation of a specified class) then the system is said to be *structurally unstable* (cf. Andronow & Pontryagin 1937, Andronow & Chaikin 1949, p. 337). (This is in contrast to the usual concept of instability which concerns the evolution in time of solutions of the same system but infinitesimally different initial values.) The theory of imperfections treats the change in the character of the set of solutions when an extra small parameter is introduced into the system, and thereby assesses the limitations of the simplifications in a model of a natural phenomenon, as we did in §1.9.

To substantiate these ideas, we may consider the one-dimensional evolution equation of the form

$$\frac{dx}{dt} = F(a, x, \delta), \tag{1}$$

where δ is the extra small parameter in addition to the usual parameter a. We call the system with $\delta = 0$ *perfect* and with small $\delta \neq 0$ *imperfect*. Then the equilibrium points and their bifurcations for $\delta = 0$ may be perturbed by putting

$$F(a, x, \delta) = 0 \tag{2}$$

and expanding F as a Taylor series in δ as well as $a - a_0$ and $x - X_0$.

Example 2.3: the structural stability of a turning point. Consider the perfect system with

$$F(a, x, 0) = x^2 - a.$$

Then the imperfect system may be expanded about the turning point $(0, 0)$ of the perfect system by use of a Taylor series:

$$\begin{aligned} 0 &= F(a, x, \delta) \\ &= x^2 - a + F_{\delta 0}\delta + F_{x\delta 0}x\delta + F_{a\delta 0}a\delta + \tfrac{1}{2}F_{xx\delta 0}x^2\delta + \cdots \\ &= (1 + \tfrac{1}{2}\delta F_{xx\delta 0})(x + \tfrac{1}{2}\delta F_{x\delta 0})^2 - (1 - \delta F_{a\delta 0})(a - \delta F_{\delta 0}) \\ &\quad + O(\delta^2, \delta a^2, \delta x^3) \qquad \text{as } a, x, \delta \to 0, \end{aligned}$$

on collecting the terms in a way to show that the nose of the parabola in the (a, x)-plane is preserved for small δ and neglecting very small terms. Therefore the bifurcation curve $F(a, x, \delta) = 0$ has the same *qualitative* behaviour near $(0, 0)$ in the (a, x)-plane for small δ, although the turning point is translated, rotated and magnified a little. Thus a turning point is structurally stable. The imperfection changes the position of almost all points of the curve a little but does not change the topological character of the curve locally. The perfect and imperfect systems both have a nose-shaped bifurcation curve near $(0, 0)$ in the (a, x)-plane: 'Nose is a nose is a nose is a nose' as Gentnude Stein (1922) put it. □

Example 2.4: the structural instability of a transcritical bifurcation. Similarly take

$$F(a, x, \delta) = x^2 - a^2 + \delta.$$

Then the perfect system, i.e. $F(a, x, 0) = 0$, has a transcritical bifurcation in the (a, x)-plane at $(0, 0)$. However, if $\delta \neq 0$ there are two separate solutions; both have a turning point if $\delta > 0$. We can see the bifurcation diagrams in Fig. 2.4 in the (a, x)-plane for different values of δ as sections of the bifurcation surface $F(a, x, \delta) = 0$ in (a, x, δ)-space. □

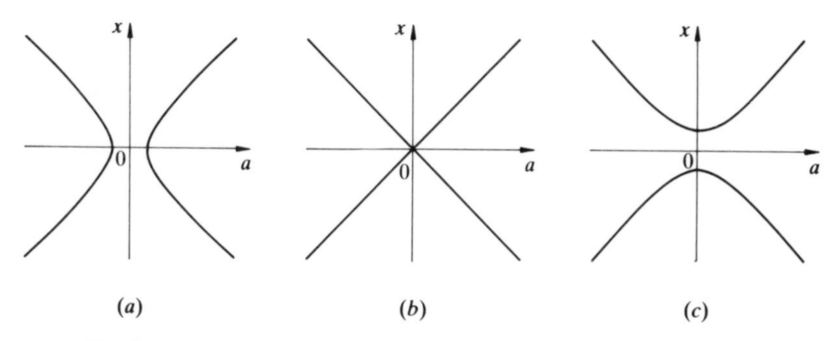

Fig. 2.4 An imperfect transcritical bifurcation: sketches of the equilibrium curves $x = \pm(a^2 - \delta)^{1/2}$ for (a) $\delta > 0$, (b) $\delta = 0$, and (c) $\delta < 0$.

Example 2.5: the structural instability of a pitchfork bifurcation. Next take

$$F(a, x, \delta) = x(x^2 - a) + \delta,$$

and consider solutions $F(a, x, \delta) = 0$. (Note that this is essentially the same function F as in Example 2.2, but now we look at it from a different point of view.) If $\delta = 0$ then there is a supercritical pitchfork bifurcation at $(0, 0)$ in the (a, x)-plane. If, however, $\delta \neq 0$ then there are two separate branches of the equilibrium curve in the (a, x)-plane, one with a turning point. Rather than solve the cubic for x, it is easier to calculate $a = x^2 + \delta/x$ as x varies for fixed δ. See the curves in Fig. 2.5, noting that the symmetry in $\pm x$ for $\delta = 0$ is broken for $\delta \neq 0$. □

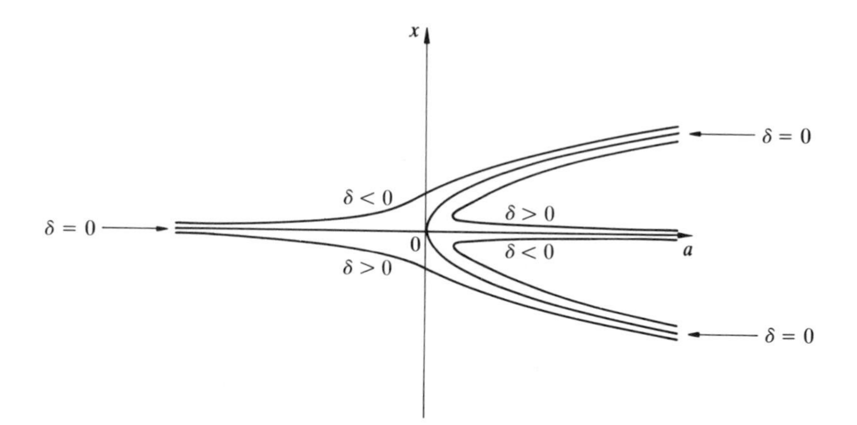

Fig. 2.5 An imperfect pitchfork bifurcation: sketches of the curves $x(x^2 - a) + \delta = 0$ in the (a, x)-plane.

Example 2.6: an isola bifurcation. The topological nature of structural stability implies at once that a conjugate point is structurally unstable, because an isolated point can be smoothly changed to a small closed curve or to nothing. The canonical example of this is given by $F(a, x, \delta) = a^2 + x^2 - \delta$. Then for $\delta > 0$ the bifurcation curves are circles with centre $(0, 0)$ and radii $\sqrt{\delta}$ in the (a, x)-plane. For $\delta = 0$ the curve is the conjugate point $(0, 0)$, and for $\delta < 0$ there is no solution of $F(a, x, \delta) = 0$. An isolated closed curve, such as one of the circles, in a bifurcation plane is called an *isola*.

Another example of an isola bifurcation is given by $F(a, x, \delta) = \frac{1}{2}x^2 + \frac{1}{3}a^3 - a - \frac{2}{3} + \delta$, shown in Fig. 2.6. (Note the *geometrical* analogy with Example 1.1, but different interpretation of the variables.) If $\delta = 0$ then there is a transcritical bifurcation at $(-1, 0)$ and a turning point at $(2, 0)$ because $F(a, x, 0) = \frac{1}{2}x^2 - \frac{1}{3}(a + 1)^2(2 - a)$. If $\delta = \frac{4}{3}$ then there is an isolated singular point at $(1, 0)$ and an unbounded branch of the bifurcation curve with a turning point at $(-2, 0)$, because $F(a, x, \frac{4}{3}) = \frac{1}{2}x^2 + \frac{1}{3}(a - 1)^2(a + 2)$. If $0 < \delta < \frac{4}{3}$ then there are two separate branches of the bifurcation curve, one an isola and the other unbounded; if $\delta < 0$ or $\delta > \frac{4}{3}$

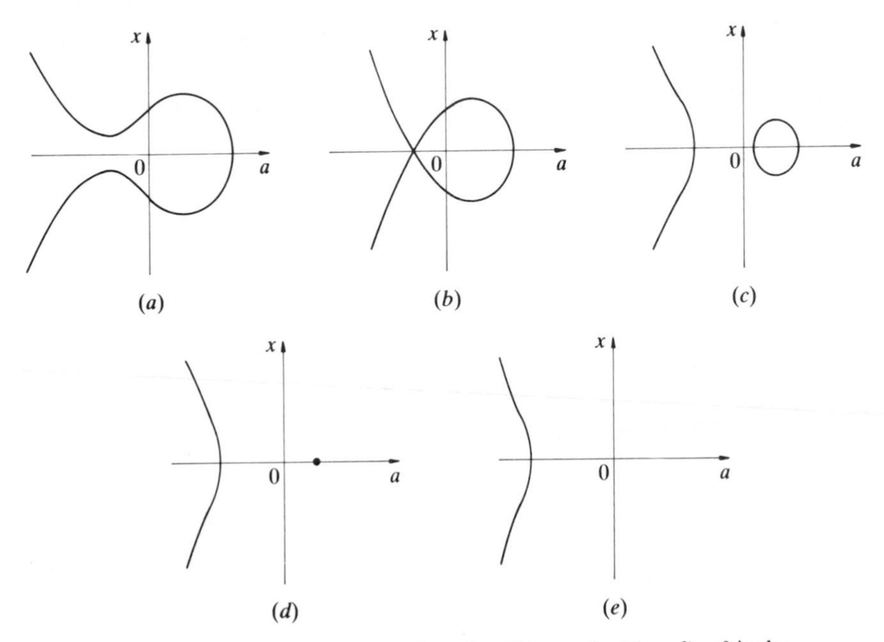

Fig. 2.6 An isola: sketches of the bifurcation diagrams for $F(a, x, \delta) = 0$ in the (a, x)-plane, where $F(a, x, \delta) = \frac{1}{2}x^2 + \frac{1}{3}a^3 - a - \frac{2}{3} + \delta$. (*a*) $\delta < 0$. (*b*) $\delta = 0$. (*c*) $0 < \delta < \frac{4}{3}$. (*d*) $\delta = \frac{4}{3}$, (*e*) $\delta > \frac{4}{3}$.

then there is a single unbounded bifurcation curve with a turning point on the a-axis. □

4 Classification of bifurcations in higher dimensions

For the general case of l *control variables*, or *parameters*, $\mathbf{a}$ and m *state variables*, or *behaviour variables*, $\mathbf{x}$ we have

$$\mathbf{F}(\mathbf{a}, \mathbf{x}) = \mathbf{0}, \tag{1}$$

for $\mathbf{F}$: $\mathbb{R}^l \times \mathbb{R}^m \to \mathbb{R}^m$. If we know one solution $(\mathbf{a}_0, \mathbf{X}_0)$ such that

$$\mathbf{F}(\mathbf{a}_0, \mathbf{X}_0) = \mathbf{0} \tag{2}$$

and $\mathbf{F}$ is infinitely differentiable at $(\mathbf{a}_0, \mathbf{X}_0)$ then we may expand equation (1) as the Taylor series

$$0 = F_{i0} + \sum_{j=1}^{l} (a_j - a_{j0})\left[\frac{\partial F_i}{\partial a_j}\right]_0 + \sum_{k=1}^{m} (X_k - X_{k0})\left[\frac{\partial F_i}{\partial x_k}\right]_0 + \cdots$$

for $i = 1, 2, \ldots, m$, where the suffix zero denotes evaluation at $(\mathbf{a}_0, \mathbf{X}_0)$, i.e.

$$\begin{aligned} 0 &= \sum_{j=1}^{l} A_{ij}(a_j - a_{j0}) + \sum_{k=1}^{m} J_{ik}(X_k - X_{k0}) + \cdots \\ &= \mathbf{A}(\mathbf{a} - \mathbf{a}_0) + \mathbf{J}(\mathbf{X} - \mathbf{X}_0) + \cdots, \end{aligned} \tag{3}$$

say, where $A_{ij} = [\partial F_i/\partial a_j]_0$ is the element of an $m \times l$ matrix $\mathbf{A}$ in its ith row and jth column, and the $m \times m$ Jacobian matrix $\mathbf{J}$ has elements $J_{ik} = [\partial F_i/\partial x_k]_0$. It follows heuristically that

$$\mathbf{X}(\mathbf{a}) = \mathbf{X}_0 - \mathbf{J}^{-1}\mathbf{A}(\mathbf{a} - \mathbf{a}_0) + o(|\mathbf{a} - \mathbf{a}_0|) \qquad \text{as } \mathbf{a} \to \mathbf{a}_0 \tag{4}$$

if $\mathbf{J}$ is non-singular. Again, the implicit function theorem may be used to prove that if (a) equations (1), (2) hold and $\mathbf{F}$ is continuously differentiable at $(\mathbf{a}_0, \mathbf{X}_0)$ and (b) $\mathbf{J}$ is non-singular, then there exists a unique solution $\mathbf{X}(\mathbf{a})$ which is continuously differentiable near $(\mathbf{a}_0, \mathbf{X}_0)$.

**Example 2.7: a pitchfork bifurcation.* Suppose that $\mathbf{F}(a, \mathbf{x}) = [\sin y - \tan x, ax - y - x^2 y]^{\mathrm{T}}$ for all $a \in \mathbb{R}$ and $\mathbf{x} = [x, y]^{\mathrm{T}} \in \mathbb{R}^2$, and seek $\mathbf{X}(a)$ such that $\mathbf{F}(a, \mathbf{X}(a)) = \mathbf{0}$. By inspection, we see that $\mathbf{X} = \mathbf{0}$ is a zero of $\mathbf{F}$ for all a. Also note the symmetry of the system whereby it is invariant under reflection in the origin in the (x, y)-plane, i.e. under the transformation $(x, y) \to (-x, -y)$.

To seek solutions near a given solution $a = a_0$, $\mathbf{X}_0(a_0) = \mathbf{0}$ such that $\mathbf{F}(a_0, \mathbf{X}_0(a_0)) = \mathbf{0}$, take the Taylor series

$$\begin{aligned}\mathbf{0} &= \mathbf{F}(a, \mathbf{X}(a)) \\ &= \mathbf{F}_0 + (a - a_0)\left[\frac{\partial \mathbf{F}}{\partial a}\right]_0 + \left[\frac{\partial \mathbf{F}}{\partial \mathbf{x}}\right]_0 \mathbf{X} + \cdots \\ &= \mathbf{JX} + \text{nonlinear terms in } (a - a_0) \text{ and } \mathbf{X},\end{aligned} \tag{5}$$

because $\mathbf{F}_0, [\partial \mathbf{F}/\partial a]_0 = \mathbf{0}$, where the Jacobian matrix is

$$\mathbf{J} = \begin{bmatrix} -1 & 1 \\ a_0 & -1 \end{bmatrix},$$

and we use the subscript zero to denote evaluation at $a = a_0$, $\mathbf{x} = \mathbf{X}(a_0)$, $= \mathbf{0}$. Therefore $\det \mathbf{J} = 1 - a_0$; so $\mathbf{J}$ is invertible if and only if $a_0 \neq 1$. Therefore the implicit function theorem gives a unique solution (in fact $\mathbf{X}(a) = \mathbf{0}$) in the neighbourhood of $(a_0, \mathbf{0})$ except when $a_0 = 1$. So we anticipate a bifurcation where $a = 1$.

To find all the neighbouring solutions $\mathbf{X}(a)$ when $a - 1$ is small, we first note that

$$\mathbf{J} = \begin{bmatrix} -1 & 1 \\ 1 & -1 \end{bmatrix}$$

when $a_0 = 1$, and that then $\mathbf{J}$ has eigenvalues $0, -2$ belonging to the respective eigenvectors $[1, 1]^{\mathrm{T}}, [-1, 1]^{\mathrm{T}}$.

Next rewrite equation (5), without approximation, in the form

$$\mathbf{JX} = \begin{bmatrix} (Y - \sin Y) - (X - \tan X) \\ -(a - 1)X + X^2 Y \end{bmatrix}. \tag{6}$$

This form is convenient because the small terms on the left-hand side are linear in $a - 1$, X, Y, but when a Taylor expansion is taken the very small terms of quadratic or higher order in $a - 1$, X, Y are all on the right-hand side.

We see that, to linear order, $\mathbf{X}$ may be any member of the null space of $\mathbf{J}$, i.e. $\mathbf{X}$ is an arbitrary multiple of the eigenvector $[1, 1]^{\mathrm{T}}$ with zero eigenvalue. We anticipate that the nonlinear terms will determine what the small multiple of the eigenvector $\mathbf{X}$ is as $a \to 1$. Accordingly define

$$\epsilon = [1, 1]\mathbf{X} = X + Y, \tag{7}$$

and assume that we may expand

$$\mathbf{X}(a) = \epsilon \mathbf{X}_1 + \epsilon^2 \mathbf{X}_2 + \cdots, \quad a = 1 + \epsilon a_1 + \epsilon^2 a_2 + \cdots \quad \text{as } \varepsilon \to 0. \tag{8}$$

(This is a generalization of the *second* method used in Example 2.1.) Now substitute the expansions (8) into equations (6) and (7), and equate coefficients of successive powers of ϵ.

The coefficients of ϵ give

$$Y_1 - X_1 = 0, \qquad X_1 - Y_1 = 0,$$

and

$$X_1 + Y_1 = 1.$$

These equations have the unique solution $\mathbf{X}_1 = \frac{1}{2}[1, 1]^{\mathrm{T}}$, thereby confirming the form of solution which we anticipated.

The coefficients of ϵ^2 give

$$Y_2 - X_2 = 0, \qquad X_2 - Y_2 = -a_1 X_1 = -\tfrac{1}{2}a_1,$$

and

$$X_2 + Y_2 = 0.$$

Therefore $a_1 = 0$ for the compatibility of the first two of these equations, and we deduce that $\mathbf{X}_2 = [0, 0]^{\mathrm{T}}$.

The coefficients of ϵ^3 give

$$Y_3 - X_3 = \tfrac{1}{6}Y_1^3 + \tfrac{1}{6}X_1^3 = \tfrac{1}{24},$$

$$X_3 - Y_3 = -a_1 X_2 - a_2 X_1 + X_1^2 Y_1 = -\tfrac{1}{2}a_2 + \tfrac{1}{8}$$

and

$$X_3 + Y_3 = 0.$$

The solvability condition, i.e. compatibility condition, of the first two of these equations gives $a_2 = \frac{1}{3}$, and then we deduce that $\mathbf{X}_3 = \frac{1}{48}[-1, 1]^{\mathrm{T}}$.

We may proceed in this way to find as many terms in the expansions of a and $\mathbf{X}$ as we wish. So far we have found that

$$\mathbf{X}(a) = \tfrac{1}{2}\epsilon[1, 1]^{\mathrm{T}} + \tfrac{1}{48}\epsilon^3[-1, 1]^{\mathrm{T}} + O(\epsilon^4), \qquad a = 1 + \tfrac{1}{3}\epsilon^2 + O(\epsilon^3)$$

as $\varepsilon \to 0$. This is enough to give the most important result, namely that in addition to the null solution there are two solutions such that

$$\mathbf{X}(a) \sim \pm\tfrac{1}{2}\{3(a-1)\}^{1/2}[1, 1]^{\mathrm{T}} \qquad \text{as } a \downarrow 1.$$

This describes a supercritical pitchfork bifurcation at $a = 1$, $\mathbf{X} = \mathbf{0}$. Note that the definition (7) of ϵ normalizes the solution and renders its expression unique. □

When the Jacobian matrix $\mathbf{J}$ is singular there are various forms of bifurcation according to its co-rank as well as the higher terms in the Taylor series. This is the subject of *singularity theory*, at one time called catastrophe theory. We shall merely summarize it briefly here. The co-rank of $\mathbf{J}$, rather than its rank, describes the topological behaviour of the solutions as $\mathbf{a}$ varies near $\mathbf{a}_0$, because many components of $\mathbf{x}$ may be 'passive', i.e. the bifurcations may occur in a subspace of $\mathbb{R}^m$ with dimension much less than m. (Also the nullity of $\mathbf{J}$ is more easily generalizable in the case $m = \infty$ when $\mathbb{R}^m$ is replaced by an infinite-dimensional function space.) We seek to retain only the 'active' components of $\mathbf{x}$ in identifying the essential topological character of the bifurcations, and these lie in the null space of $\mathbf{J}$. Similarly we define the *co-dimension* to be the dimension of the subspace of $\mathbb{R}^l$ in which $\mathbf{a}$ varies to give the essential character of a bifurcation. For example, the turning point or fold of §1.3 has co-dimension one even though it may be embedded in a space with more than one parameter (cf. Example 2.2). We also seek to make smooth invertible transformations of the coordinates to reduce the bifurcations to simple canonical forms. Thus we are interested in equivalence classes of bifurcations with the same topological character.

Thom (1975) considered $\mathbf{F} = -\text{grad}\, V$, i.e. $\mathbf{F} = -\nabla V$, for a potential function $V(\mathbf{a}, \mathbf{x})$ and showed that a structurally stable bifurcation is topologically equivalent to one of a few canonical forms. On taking $\mathbf{a} = \mathbf{0}$ and $\mathbf{X} = \mathbf{0}$ without loss of generality, the forms are as listed in Table 2.1, it being more compact to give the potential than all the components of $\mathbf{F}$. Some familiar bifurcations may be seen in this table. Transcritical and pitchfork bifurcations do not appear in their own right because they are not structurally stable. However, they do appear. A pitchfork bifurcation,

Table 2.1. *Canonical forms of structurally stable potential singularities of co-dimension* $\leqslant 4$: *the seven elementary catastrophes.*

co-rank	co-dimension	V	name
1	1	$\frac{1}{3}x^3 + ax$	*fold* or *turning point*
1	2	$\frac{1}{4}x^4 + \frac{1}{2}ax^2 + bx$	*cusp*
1	3	$\frac{1}{5}x^5 + \frac{1}{3}ax^3 + \frac{1}{2}bx^2 + cx$	*swallow's tail*
2	3	$x^3 + y^3 + cxy - ax - by$	*hyperbolic umbilic*
2	3	$x^3 - 3xy^2 + c(x^2 + y^2) - ax - by$	*elliptic umbilic*
1	4	$\frac{1}{6}x^6 + \frac{1}{4}ax^4 + \frac{1}{3}bx^3 + \frac{1}{2}cx^2 + dx$	*butterfly*
2	4	$x^2y + \frac{1}{4}y^4 + cx^2 + dy^2 - ax - by$	*parabolic umbilic*

for example, arises as a special case ($b = 0$) of the cusp; as we saw in detail in Examples 2.2 and 2.5, a small change in b destroys the pitchfork; but each cusp catastrophe has one special case which gives a pitchfork.

In this chapter we have examined the behaviour of fixed points, i.e. steady solutions, as parameters vary. In the remainder of the book we shall examine the behaviour of unsteady solutions. So, having studied statics, we shall now go on to study dynamics.

Further reading

§2.2 Courant (1936, pp. 127–9) gives the elementary theory of singular points on curves.

§2.3 Iooss & Joseph (1990) describe systematically the analytic theory, including the theory of imperfections, with many examples and exercises.

§2.4 Thom (1975) describes the elementary bifurcations, emphasizing geometrical aspects of the theory and their applications. Golubitsky & Schaeffer (1985) is a valuable treatise on bifurcations.

Problems

Q2.1 *A turning point.* Consider the difference equation

$$x_{n+1} = G(a, x_n) \qquad \text{for } n = 0, 1, \ldots,$$

where $G: \mathbb{R} \times \mathbb{R} \to \mathbb{R}$ is a well-behaved function of both its arguments. Using subscripts x and a to denote partial derivatives, show that if $G(0, X_0) = X_0$, $G_x(0, X_0) = 1$, $G_a(0, X_0) \neq 0$, $G_{xx}(0, X_0) \neq 0$ then there is a turning point at $a = 0$, $x = X_0$, and that if, moreover, $G_a(0, X_0) < 0$, $G_{xx}(0, X_0) > 0$ then there are two fixed points for $0 < a \ll 1$ but none for $0 < -a \ll 1$.

Q2.2 *A flip bifurcation.* Consider the difference equation

$$x_{n+1} = G(a, x_n) \qquad \text{for } n = 0, 1, \ldots,$$

where $G: \mathbb{R} \times \mathbb{R} \to \mathbb{R}$ is a well-behaved function of both its arguments. Show that if $G(0, 0) = 0$, $G_x(0, 0) \neq 1$, $G_a(0, 0) \neq 0$ then there is a unique fixed point near to $x = 0$ when a is near to zero.

Defining H by $H(a, x) = G(a, G(a, x)) - x$, show that $H_x(a, x) = G_x(a, G(a, x))G_x(a, x) - 1$, $H_a(a, x) = G_a(a, G(a, x)) + G_x(a, G(a, x))G_a(a, x)$. Deduce that if, moreover, $G_{x0} = -1$ then $H_0 = H_{a0} = H_{x0} = H_{xx0} = 0$, on using the subscript zero to denote evaluation at $a = 0$, $x = 0$. *Hence show that in general there is a flip bifurcation of solutions of the difference equation with map G at $a = 0$, $x = 0$.

Q2.3 *An integro-differential equation.* Given that

$$\frac{d^2y}{dx^2} + \left[a - 2\int_0^1 \{y(x)\}^2\,dx\right]y = 0 \qquad \text{and} \qquad y(0) = y(1) = 0,$$

show that all solutions are of the form $y(x) = A\sin n\pi x$, where A is a constant and n a positive integer. Hence find all the solutions, and sketch the bifurcation diagram in the (a, A)-plane.

[Griffel (1981, §11.6).]

Q2.4 *Imperfect bifurcations.* Sketch the curves $F(a, x, \delta) = 0$ in the (a, x)-plane for all 'typical' values of δ in the cases

(i) $F(a, x, \delta) = x(x - a) + \delta$,

(ii) $F(a, x, \delta) = x(a - x - x^2) + \delta$,

(iii) $F(a, x, \delta) = x^2 + ax + a^2 - \delta$.

Q2.5 *The structural instability of a pitchfork bifurcation.* Consider the roots of $F(a, x, \delta) = 0$, where F is defined by $F(a, x, \delta) = x^3 + \delta x^2 - ax$. Show that the pitchfork bifurcation at $(0, 0)$ in the (a, x)-plane for $\delta = 0$ becomes a transcritical bifurcation for small δ, and that there is a turning point at $(-\frac{1}{4}\delta^2, -\frac{1}{2}\delta)$. Sketch the bifurcation curves in the (a, x)-plane for $\delta > 0$.

Q2.6 *A model of bifurcation of some steady flows.* Consider the bifurcation diagrams of the equation $F(1, b, c, x) = 0$ in the (b, x)-plane for various values of c, where F is defined by

$$F(a, b, c, x) = x^3 - 2ax^2 - (b - 3)x + c$$

for all real a, b, c, x. Show that if $c = 0$ then there is a transcritical bifurcation but if $c \neq 0$ then there is a 'primary' and a 'secondary' branch of the equilibrium curve such that the primary branch has a fold only when $-\frac{8}{27} < c < 0$. Sketch the curves for $c = 0.4$, 0, -0.1 and $-\frac{8}{27}$. Show that the bifurcation set, i.e. the set of values of b and c such that $F(1, b, c, x) = F_x(1, b, c, x) = 0$, is a curve in the (b, c)-plane with equation

$$(27c - 18b + 38)^2 = 4(3b - 5)^3.$$

Show that this curve has a cusp at $b = \frac{5}{3}$, $c = -\frac{8}{27}$. Sketch the curve.

Show that, for all values of a, the cusp is at $b = 3 - \frac{4}{3}a^2$, $c = -\frac{8}{27}a^3$, and hence that it lies on the curve with equation $(3 - b)^3 = 27c^2$.

[Benjamin (1978), p. 17) suggested this as a model of the bifurcation of toroidal vortices in a liquid between coaxial rotating cylinders.]

Q2.7 *A problem of particle dynamics which exhibits a structurally unstable bifurcation.* A particle moves along a smooth wire which is fixed in a vertical plane. Gravity acts on the particle, and the plane rotates with constant angular velocity ω about a vertical axis in the plane. Take Cartesian coordinates (x, z), where $0z$ is the upward vertical, so that the wire has an equation of the form $z = f(x)$ and the axis of rotation has equation $x = \delta a > 0$. You are given that the equation of motion of the particle then is

$$(1 + f'^2)\frac{d^2x}{dt^2} + ff'\left(\frac{dx}{dt}\right)^2 - \omega^2(x - \delta a) + gf' = 0.$$

Show that there may be equilibrium at the point $x = X$, where X is any root of $gf'(X) = \omega^2(X - \delta a)$, and that this point is stable if $\omega^2 < gf''(X)$ and unstable if $\omega^2 > gf''(X)$.

Show that if $f(x) = \frac{1}{2}a(x/a)^2\{1 + (x/a)^2\}$ and b is defined by $b = a\omega^2/g$, then there is supercritical pitchfork bifurcation at $b = 1$ and $X = 0$ when $\delta = 0$, but that the bifurcation has a different character when $0 < \delta \ll 1$. Sketch the bifurcation curves in the (b, x)-plane for both cases. Further, show that for all real δ there is marginal stability and bifurcation where $27\delta^2b^2 = 2(b - 1)^3$, the equation of a curve with a cusp at $(0, 1)$ in the (δ, b)-plane.

[Drazin & Reid (1981, p. 463).]

Q2.8 *The van der Waals equation and the cusp catastrophe.* The pressure P, volume v and temperature T of a sample of a gas are given to satisfy the equation of state

$$(P + a/v^2)(v - b) = RT$$

for positive constants a, b, R. Regarding this as a cubic which determines v as a function of P for a fixed value of T, show that the three roots of the cubic coincide with value v_c, when $P = P_c$, $T = T_c$, where the *critical temperature* is $T_c = 8a/27bR$, the *critical pressure* is $P_c = a/27b^2$ and the *critical volume* is $v_c = 3b$. Sketch the graphs of the function $P(v)$ in the first quadrant of the (v, P)-plane for $T > T_c$, $T = T_c$, $T < T_c$, showing that the function is single-valued if $T \geqslant T_c$ but not if $T < T_c$.

Defining $\pi = P/P_c - 1$, $\phi = v_c/v - 1$ and $\theta = T/T_c - 1$, rewrite the van der Waals equation as $F(\theta, \pi, \phi) = 0$, where F is defined by

$$F(\theta, \pi, \phi) = \phi^3 + \tfrac{1}{3}(8\theta + \pi)\phi + \tfrac{2}{3}(4\theta - \pi).$$

Hence show that three is a double zero v of F if

$$81(4\theta - \pi)^2 = -(8\theta + \pi)^3,$$

which gives a cusp at $(0, 0)$ in the (θ, π)-plane.

[In the derivation of the van der Waals (1873) equation it is assumed that the distribution of the molecules of the gas in space is uniform. This assumption is invalid for a real gas near its critical point, so the equation is a poor physical model where its mathematical properties are most interesting.]

Q2.9 *Landau's theory of second-order phase transitions.* It is given that the thermodynamic potential Φ of a uniform sample of a pure substance at constant pressure is modelled by an analytic function of temperature T and the square of a measure η of the order of the sample, so that

$$\Phi(T, \eta^2) = \Phi_0(T) + A(T)\eta^2 + B(T)\eta^4 + \cdots,$$

where $A(T) \sim a(T - T_c)$ as $T \to T_c$, $B > 0$, $a > 0$, and T_c is the transition temperature of the substance. Equilibrium may occur at a stationary point of Φ with respect to variations of η, and is stable at a minimum of Φ.

Show that there is stable equilibrium with $\eta = 0$ as $T \downarrow T_c$ and $\eta \sim \{a(T_c - T)/2B(T_c)\}^{1/2}$ as $T \uparrow T_c$. Deduce plausibly that near transition the entropy $S = -(\partial\Phi/\partial T)_p$ and the specific heat $C_p = T(\partial S/\partial T)_p$ are given approximately by the forms

$$S(T) = \begin{cases} S_0(T) \\ S_0(T) - a^2(T_c - T)/2B \end{cases}, \quad C_p(T) = \begin{cases} C_0(T) \\ C_0(T) + a^2 T/2B \end{cases}$$

$$\text{for} \begin{cases} T > T_c \\ T < T_c \end{cases}.$$

[This theory, with a discontinuous specific heat, is useful, although the assumption that the thermodynamic potential is an analytic function of η is invalid physically, and leads to results which are not very accurate. Landau & Lifshitz (1980, Chap. XIV).]

Q2.10 *Weiss's theory of ferromagnetism.* The magnetic moment M per unit volume of a uniform sample of material in equilibrium is given to be

$$M = N\mu \tanh(\lambda\mu M/k_B T),$$

where N is the number of unpaired electrons per unit volume, μ is the magnetic moment of an electron, λ is a measure of the self-interaction of the magnetic moments of the electrons, k_B is Boltzmann's constant, and T is the temperature of the material. Show that

$$m = \tanh(m/t),$$

where $m = M/N\mu$, $t = T/T_c$ and the *Curie temperature* $T_c = N\lambda\mu^2/k_B$. Show that $m = 0$ is the unique solution for $t \geqslant 1$ but that there are three solutions m for $t < 1$. Sketch the bifurcation diagram in the (t, m)-plane. Deduce that $M = 0$ is the unique solution for $T > T_c$ but that there are two extra solutions M for $T < T_c$ such that

$$M(T) \sim \pm\{(3k_B N/\lambda)(T_c - T)\}^{1/2} \qquad \text{as } T \uparrow T_c.$$

[Weiss (1907) explained the ferromagnetism of a material below its Curie temperature, although his original work, in the absence of a knowledge of quantum mechanics, was oversimplified. Physical arguments show that the solution $M = 0$ is stable for $T > T_c$, and that the other two solutions are stable for $T < T_c$. See Kittel (1986, Chap. 15).]

Q2.11 *Phase instability.* Consider the equation

$$\frac{dz}{dt} = (a - |z|^2)z + \epsilon,$$

where z is a complex function of the real variable t but a and ϵ are real constants. Expressing $z = re^{i\theta}$ for non-negative modulus r and real phase θ, show that

$$\frac{dr}{dt} = (a - r^2)r + \epsilon \cos\theta, \qquad \frac{d\theta}{dt} = -\frac{\epsilon \sin\theta}{r}.$$

Investigate the steady solutions and their stability. Sketch the bifurcation curves in the (a, r)-plane for fixed $\epsilon > 0$, $\epsilon = 0$ and $\epsilon < 0$.

[Note that at a turning point an eigenvalue in general has a real part which changes sign, but that *both* branches of the bifurcation curve may represent unstable solutions if another eigenvalue has positive real part at the turning point.]

Q2.12 *Generalized Newton–Raphson method.* Given that $\mathbf{f}: \mathbb{R}^m \to \mathbb{R}^m$ is continuously differentiable, that its Jacobian matrix $\mathbf{J}(\mathbf{x})$ at $\mathbf{x}$ is invertible for all $\mathbf{x}$, and that $\mathbf{X}$ is a zero of $\mathbf{f}$; show that a linear approximation to $\mathbf{f}$ near $\mathbf{X}$ gives rise to the difference equation

$$\mathbf{x}_{n+1} = \mathbf{F}(\mathbf{x}_n) \qquad \text{for } n = 0, 1, \ldots,$$

in order to find successive approximations to $\mathbf{X}$, where $\mathbf{F}(\mathbf{x}) = \mathbf{x} - \mathbf{J}^{-1}(\mathbf{x})\mathbf{f}(\mathbf{x})$; and deduce that $\mathbf{x}_n \to \mathbf{X}$ as $n \to \infty$ if $\mathbf{x}_0$ is sufficiently close to $\mathbf{X}$.

Q2.13 *The swallow's tail.* Take $V = \frac{1}{5}x^5 + \frac{1}{3}ax^3 + \frac{1}{2}bx^2 + cx$ and find $F = -dV/dx$. Consider the bifurcation set in (a, b, c)-space defined by $F = dF/dx = 0$. Sketch the intersections of the set and three planes $a =$ constant, for $a < 0$, $a = 0$, and $a > 0$. Hence sketch a perspective view of the bifurcation set in (a, b, c)-space.

[Thom (1975, §5.3).]

3

Difference equations

Where do we come from? What are we? Where are we going?

Paul Gaugin ('*D'où venons-nous? Que sommes-nous? Où allons-nous?*', painting of 1897, now in Museum of Fine Arts, Boston, MA)

1 The stability of fixed points

To continue our detailed examination of the ideas introduced in Chapter 1, we shall next consider the dynamics of difference equations, i.e. of iterated maps, because they are in most respects the simplest of nonlinear systems. We noted that the methods of solution of difference equations are similar to the methods of solution of differential equations, and we related how a fixed point is the analogue of an equilibrium point of a system of ordinary differential equations. In this section we shall elaborate the analogy and then examine the stability of fixed points, much as we examined the stability of equilibrium points. Periodic solutions and their stability are examined in §2, and some fundamental analysis of difference equations is given in §3. The logistic difference equation is treated at length in §4, not only because it is a fascinating equation whose solutions have a rich structure but also because it is used as a prototype of many nonlinear one-dimensional difference equations. Some relevant numerical methods are introduced in §5. Iterated two-dimensional maps are described in §6 and iterated maps of the complex plane in §7.

First note that it is sufficient to consider a system of first-order difference equations rather than an equation or equations of higher order. For an mth-order difference equation,

$$x_{n+m} = F(x_n, x_{n+1}, \ldots, x_{n+m-1}, n) \qquad \text{for } n = 0, 1, \ldots,$$

where $F\colon \mathbb{R}^m \times \mathbb{Z} \to \mathbb{R}$, can be expressed as m simultaneous first-order equations $\mathbf{y}_{n+1} = \mathbf{G}(\mathbf{y}_n, n)$ for $\mathbf{y}_n \in \mathbb{R}^m$ and $\mathbf{G}\colon \mathbb{R}^m \times \mathbb{Z} \to \mathbb{R}^m$ by defining $\mathbf{y}_n = [x_n, x_{n+1}, \ldots, x_{n+m-1}]^{\mathrm{T}}$ and $\mathbf{G}(\mathbf{y}, n) = [y_2, y_3, \ldots, y_{n+m-1}, F(y_1, \ldots, y_m, n)]^{\mathrm{T}}$, where $\mathbf{y} = [y_1, \ldots, y_m]^{\mathrm{T}}$. The converse is false, i.e. m first-order equations are, in general, not equivalent to one mth-order equation.

Example 3.1: reduction of a second-order equation to two first-order equations. It is easier to understand this idea by looking at a simple example. So consider the second-order linear difference equation,

$$x_{n+2} + 2bx_{n+1} + cx_n = 0,$$

and define

$$\mathbf{y}_n = \begin{bmatrix} x_n \\ x_{n+1} \end{bmatrix} = \begin{bmatrix} y_{n1} \\ y_{n2} \end{bmatrix}.$$

Then

$$\mathbf{y}_{n+1} = \begin{bmatrix} x_{n+1} \\ x_{n+2} \end{bmatrix} = \begin{bmatrix} y_{n2} \\ -2by_{n2} - cy_{n1} \end{bmatrix} = \mathbf{G}(\mathbf{y}_n),$$

say, where we define $\mathbf{G}$ by $\mathbf{G}([y_1, y_2]^{\mathrm{T}}) = [y_2, -2by_2 - cy_1]^{\mathrm{T}}$. □

Next note that we may consider *autonomous systems*, i.e. those that do not depend explicitly on the independent variable n, without loss of generality. To show this consider the system

$$\mathbf{x}_{n+1} = \mathbf{F}(\mathbf{x}_n, n)$$

for $\mathbf{F}\colon \mathbb{R}^m \times \mathbb{Z} \to \mathbb{R}^m$. It may be expressed in the autonomous form,

$$\mathbf{y}_{n+1} = \mathbf{G}(\mathbf{y}_n),$$

where $\mathbf{y}_n = [\mathbf{x}_n, y_{n,m+1}]^{\mathrm{T}} \in \mathbb{R}^{m+1}$ and $\mathbf{G}(\mathbf{y}) = [\mathbf{F}(\mathbf{x}, y_{m+1}), 1 + y_{m+1}]^{\mathrm{T}}$ and $y_{0,m+1} = 0$. This gives $\mathbf{G}\colon \mathbb{R}^{m+1} \to \mathbb{R}^{m+1}$. The essential idea here is to extend the definition of $\mathbf{x}_n \in \mathbb{R}^m$ to $\mathbf{y}_n \in \mathbb{R}^{m+1}$ by the addition of an extra component which we arrange to be n itself.

Example 3.2: reduction to an autonomous system. Again a simple example shows the idea more clearly. To put the equation

$$x_{n+1} = x_n^2 - n$$

into autonomous form, we define $\mathbf{y}_n = [y_{n1}, y_{n2}]^{\mathrm{T}}$ and take $\mathbf{y}_{n+1} = [y_{n1}^2 - y_{n2}, 1 + y_{n2}]^{\mathrm{T}}$ for $n = 0, 1, \ldots$, where $\mathbf{y}_0 = [x_0, 0]^{\mathrm{T}}$. Then it follows that $\mathbf{y}_n = [x_n, n]^{\mathrm{T}}$ and that $x_{n+1} = y_{n+1,1} = x_n^2 - n$, as required. □

Now, to consider the stability of a fixed point, it is sufficient to confine our attention to an autonomous system of first-order difference equations. First it will help to recall the theory for the case $m = 1$ in §1.8. There we tacitly assumed the intuitive notion of stability of a fixed point of a difference equation, namely that if the initial solution is sufficiently close to the fixed point then the solution remains close to the fixed point thereafter. This notion can be formalized mathematically in many ways which are similar but not the same. Here we shall adopt a definition which is essentially due to Liapounov (1892): a fixed point $\mathbf{X}$ of $\mathbf{F}: \mathbb{R}^m \to \mathbb{R}^m$ is *stable* if for all $\epsilon > 0$ there exists $\delta(\epsilon)$ such that

$$|\mathbf{F}^n(\mathbf{x}_0) - \mathbf{X}| < \epsilon \qquad \text{for } n = 1, 2, \ldots \tag{1}$$

for all $\mathbf{x}_0$ such that $|\mathbf{x}_0 - \mathbf{X}| < \delta$, where $\mathbf{x}_n = \mathbf{F}^n(\mathbf{x}_0)$ is defined iteratively by $\mathbf{x}_{n+1} = \mathbf{F}(\mathbf{x}_n)$. The definition means that the set of solutions $\{\mathbf{x}_n\}$ is uniformly continuous at $\mathbf{X}$ for all $n > 0$ with respect to variations of $\mathbf{x}_0$. Thus, for example, when $m = 3$, 'the fixed point $\mathbf{X}$ is stable' means that 'given any small sphere with centre $\mathbf{X}$ and radius ϵ, there exists a second sphere with centre $\mathbf{X}$ and radius δ such that if $\mathbf{x}_0$ lies in the second sphere then $\mathbf{x}_1, \mathbf{x}_2, \ldots$ *all* lie in the first sphere' (in exceptional cases the second sphere is the same as the first, but in general it is smaller). The fixed point $\mathbf{X}$ is *asymptotically stable* if it is stable and $\mathbf{x}_n \to \mathbf{X}$ as $n \to \infty$. It is *globally stable* if it is stable and if for all $\epsilon > 0$ and for all $\mathbf{x}_0$ there exists $N(\epsilon)$ such that $|\mathbf{F}^n(\mathbf{x}_0) - \mathbf{X}| < \epsilon$ for $n = N, N + 1, \ldots$. It is *metastable* if it is stable but not globally stable, i.e. if it is stable to all small initial perturbations $\mathbf{x}_0 - \mathbf{X}$ but not to some which are not small.

Generalizing the theory of §1.8, take the fixed point $\mathbf{X}$ of the difference equation $\mathbf{x}_{n+1} = \mathbf{F}(\mathbf{x}_n)$ and define its perturbation as

$$\mathbf{x}'_n = \mathbf{x}_n - \mathbf{X}. \tag{2}$$

Therefore the difference equation becomes

$$\begin{aligned} \mathbf{X} + \mathbf{x}'_{n+1} &= \mathbf{F}(\mathbf{X} + \mathbf{x}'_n) \\ &= \mathbf{F}(\mathbf{X}) + \mathbf{J}\mathbf{x}'_n + o(|\mathbf{x}'_n|) \qquad \text{as } |\mathbf{x}'_n| \to 0, \end{aligned}$$

if $\mathbf{F}$ is continuously twice differentiable, where $\mathbf{J} = [\text{grad}\,\mathbf{F}]_{\mathbf{X}}$ is the $m \times m$ Jacobian matrix of $\mathbf{F}$ evaluated at $\mathbf{X}$, i.e. the matrix with element $[\partial F_i/\partial x_j]_{\mathbf{X}}$ in the ith row and jth column. Now $\mathbf{X} = \mathbf{F}(\mathbf{X})$. Therefore

$$\mathbf{x}'_{n+1} = \mathbf{J}\mathbf{x}'_n, \tag{3}$$

the linearized system, arises in the limit as $|\mathbf{x}'_n| \to 0$. It is plausible that the linearized system will govern the behaviour of the perturbation $\mathbf{x}'_n$ when it is small, and so determine the stability of the fixed point of the nonlinear equation. In any event, equation (3) has solutions of the form

$$\mathbf{x}'_n = q^n \mathbf{u}, \tag{4}$$

where

$$\mathbf{Ju} = q\mathbf{u}, \tag{5}$$

i.e. where $\mathbf{u}$ is an eigenvector and the multiplier q is the corresponding eigenvalue of the real $m \times m$ Jacobian matrix $\mathbf{J}$. More generally, we find m eigenvalues q_j belonging to eigenvectors $\mathbf{u}_j$, where the m eigenvectors form a complete set in $\mathbb{R}^m$, although they may occur in complex conjugate pairs. It follows that, given $\mathbf{x}'_0 \in \mathbb{R}^m$, there exist numbers $\xi_1, \xi_2, \ldots, \xi_m$ (each is either real or one of a complex conjugate pair) such that

$$\mathbf{x}'_0 = \sum_{j=1}^{m} \xi_j \mathbf{u}_j. \tag{6}$$

(For the special case where $\mathbf{J}$ does not have m linearly independent eigenvectors, this expansion is not possible for all $\mathbf{x}'_0$, and special treatment is needed, cf. Q3.4.) Therefore

$$\mathbf{x}'_n = \sum_{j=1}^{m} \xi_j q_j^n \mathbf{u}_j \qquad \text{for } n = 0, 1, \ldots. \tag{7}$$

We deduce that the fixed point $\mathbf{X}$ is linearly stable if and only if $\mathbf{x}'_n$ is bounded as $n \to \infty$ for all $\mathbf{x}'_0$, and therefore is linearly stable if $|q_j| < 1$ for $j = 1, 2, \ldots, m$ and unstable if $|q_j| > 1$ for at least one value of j. It can be shown under quite general conditions that the solution is nonlinearly stable when it is linearly stable except possibly at the margin of linear stability.

Example 3.3: the logistic map. Cf. Example 1.5. For the map $F(x) = ax(1 - x)$ we have $F: \mathbb{R} \to \mathbb{R}$ and the Jacobian matrix $\mathbf{J}$ is a 1×1 matrix which is simply the derivative $[\mathrm{d}F/\mathrm{d}x]_X = a(1 - 2X)$. The eigenvalue is $q = a(1 - 2X)$. Therefore $q = a$ for the fixed point $X = 0$. Therefore the point is stable for $-1 < a < 1$. Similarly, the fixed point $X = (a - 1)/a$ has $q = 2 - a$ and is stable for $1 < a < 3$. Note (cf. Example 1.5) that $q = 1$ and there is a transcritical bifurcation when $a = 1$, but $q = -1$ and there is a flip bifurcation when $a = 3$. □

Example 3.4: the two-dimensional case. If $m = 2$ then

$$\mathbf{x}'_n = \xi_1 q_1^n \mathbf{u}_1 + \xi_2 q_2^n \mathbf{u}_2.$$

If q_1 and q_2 are real and distinct then the eigenvectors $\mathbf{u}_1$ and $\mathbf{u}_2$ are real and independent, so we may use Cartesian components $x'_n = \xi_1 q_1^n$ and $y'_n = \xi_2 q_2^n$ to express the linear perturbation as

$$\mathbf{x}'_n = x'_n \mathbf{u}_1 + y'_n \mathbf{u}_2. \, \square$$

In the special case of marginal stability we may order the multipliers so that $|q_1| = |q_2| = \cdots = |q_k| = 1$ and $1 > |q_{k+1}| \geqslant \cdots \geqslant |q_m|$ for some integer k such that $1 \leqslant k \leqslant m$. Then, in the limit as $n \to \infty$, $\mathbf{x}'_n = \mathbf{x}_n - \mathbf{X} \in \mathrm{E}$, where E is the k-dimensional subspace of $\mathbb{R}^m$ spanned by the eigenvectors $\mathbf{u}_1, \mathbf{u}_2, \ldots, \mathbf{u}_k$. Note that E is independent of $\mathbf{x}'_0$, although $\mathbf{x}'_n$ does depend on $\mathbf{x}'_0$.

Bifurcation often occurs at the onset of instability as a parameter varies. There is a close relationship between the bifurcation of a fixed point as a parameter changes (§2.4) and the stability of a fixed point at a given value of a parameter. The same Jacobian matrix plays a role in each of the local analyses, because each involves a Taylor expansion about the same fixed point at the same value of the parameter.

Suppose that a nonlinear system depends on a parameter a such that the fixed point $\mathbf{X}(a)$ is stable for $a < a_c$ and unstable for $a > a_c$. Then there is marginal stability when $a = a_c$, as described for the linearized system above. There are *centre manifold* theorems which, under a variety of quite general hypotheses, show that for sufficiently small values of $a - a_c$ solutions of the nonlinear system are such that, in the limit as $n \to \infty$, $\mathbf{x}_n - \mathbf{X}$ belongs to a k-dimensional manifold. For real systems, an eigenvalue is either real or one of a complex conjugate pair. So, for $k = 1$ and $q = 1$, we expect, in general, a turning point or a transcritical bifurcation at $a = a_c$, with $\mathbf{x}_n - \mathbf{X} \to \xi_1 \mathbf{u}_1$ as $n \to \infty$ and $a \to a_c$. For $k = 1$ and $q = -1$ we expect a flip bifurcation at $a = a_c$, with $\mathbf{x}_n - \mathbf{X} \sim (-1)^n \xi_1 \mathbf{u}_1$ as $n \to \infty$ and $a \to a_c$. For $k = 2$, $q_1 = \mathrm{e}^{\mathrm{i}\theta}$ for real θ and $q_2 = \bar{q}_1$, we except a Hopf-like bifurcation, with $\mathbf{x}_n - \mathbf{X} \sim \xi_1 \mathrm{e}^{\mathrm{i}n\theta}\mathbf{u}_1 + \bar{\xi}_1 \mathrm{e}^{-\mathrm{i}n\theta}\bar{\mathbf{u}}_1$ as $n \to \infty$. A Hopf bifurcation, supercritical or subcritical, of the nonlinear system leads to an orbit along a closed curve in a two-dimensional submanifold of $\mathbb{R}^m$.

2 Periodic solutions and their stability

In §1.8 we met a solution of period two as well as fixed points. More generally, we say that if there exist a map $\mathbf{F}: \mathbb{R}^m \to \mathbb{R}^m$ and distinct points $\mathbf{X}_r$ for $r = 1, 2, \ldots, p$ such that $\mathbf{X}_{r+1} = \mathbf{F}(\mathbf{X}_r)$ for $r = 1, \ldots, p-1$ and $\mathbf{X}_1 = \mathbf{F}(\mathbf{X}_p)$ then the set $\mathrm{S} = \{\mathbf{X}_1, \mathbf{X}_2, \ldots, \mathbf{X}_p\}$ of p points is a solution of the difference equation $\mathbf{x}_{n+1} = \mathbf{F}(\mathbf{x}_n)$ with period p or a *p-cycle*. Thus, for a p-cycle, p is the least positive integer such that $\mathbf{x}_{n+p} = \mathbf{x}_n$ for all $n > 0$. Note that $\mathbf{F}(\mathrm{S}) = \{\mathbf{F}(\mathbf{X}_1), \mathbf{F}(\mathbf{X}_2), \ldots, \mathbf{F}(\mathbf{X}_p)\} = \{\mathbf{X}_2, \mathbf{X}_3, \ldots, \mathbf{X}_1\} = \mathrm{S}$, so we say that S is an *invariant set* under the operation of the map $\mathbf{F}$, just as a fixed point is.

The stability of a p-cycle S may be defined by replacing $\mathbf{X}$ by each point of S in the above definition of stability of a fixed point, so that S is stable if $\mathbf{F}^n(\mathbf{x}_0)$ is in one of the p hyperspheres with radius ϵ and centre $\mathbf{X}_j$ for $j = 1, \ldots, p$ if $\mathbf{x}_0$ is in one of the p hyperspheres with radius δ and centre $\mathbf{X}_j$.

To consider the stability of a p-cycle it is helpful to consider the *pth-generation map*, i.e. the map $\mathbf{F}$ composed with itself p times. For the simplest example, consider a two-cycle $\{\mathbf{X}, \mathbf{Y}\}$ such that

$$\mathbf{Y} = \mathbf{F}(\mathbf{X}), \qquad \mathbf{X} = \mathbf{F}(\mathbf{Y}), \tag{1}$$

where $\mathbf{F}\colon \mathbb{R}^m \to \mathbb{R}^m$, and define the *second-generation map* $\mathbf{G}$ of $\mathbf{F}$ by $\mathbf{G}(\mathbf{x}) = \mathbf{F}^2(\mathbf{x}) = \mathbf{F} \circ \mathbf{F}(\mathbf{x}) = \mathbf{F}(\mathbf{F}(\mathbf{x}))$ for all $\mathbf{x} \in \mathbb{R}^m$. Note that a fixed point of $\mathbf{G}$ must be either a fixed point of $\mathbf{F}$ (when $\mathbf{Y} = \mathbf{X}$) or a point of a two-cycle of $\mathbf{F}$ (when $\mathbf{Y} \neq \mathbf{X}$).

A two-cycle $\{\mathbf{X}, \mathbf{Y}\}$ of a continuous map $\mathbf{F}$ is stable if and only if the fixed point $\mathbf{X}$ (or $\mathbf{Y}$) of the second-generation map $\mathbf{G}$ is stable. To understand this note that if $\mathbf{x}_0$ lies within a distance δ of $\mathbf{X}$ implies that $\mathbf{G}^n(\mathbf{x}_0)$ lies within a distance ϵ of $\mathbf{X}$ for all $n > 0$ then it follows by continuity that there exists δ' such that $\mathbf{F}^n(\mathbf{x}_0)$ lies within a distance ϵ of either $\mathbf{X}$ or $\mathbf{Y}$ for all $n > 0$ if $\mathbf{x}_0$ lies within a distance δ' of either $\mathbf{X}$ or $\mathbf{Y}$.

We shall suppose that $\mathbf{F}$ is continuously twice differentiable. Then the linearized system for $\mathbf{G}$ at $\mathbf{X}$ is

$$\mathbf{x}'_{n+1} = \mathbf{K}(\mathbf{X})\mathbf{x}'_n,$$

where $\mathbf{K}(\mathbf{X})$ is the Jacobian matrix of $\mathbf{G}$, i.e. the matrix with element $[\partial G_i/\partial x_j]_{\mathbf{X}}$ in the ith row and jth column. Now

$$\begin{aligned}\frac{\partial G_i(x_j)}{\partial x_j} &= \frac{\partial F_i(F_k(x_j))}{\partial x_j} \\ &= \sum_{k=1}^{m} \left[\frac{\partial F_i}{\partial x_k}\right]_{\mathbf{F}(\mathbf{X})} \left[\frac{\partial F_k}{\partial x_j}\right]_{\mathbf{X}},\end{aligned}$$

on differentiating a function of functions, so

$$\begin{aligned}\mathbf{K}(\mathbf{X}) &= \mathbf{J}(\mathbf{F}(\mathbf{X}))\mathbf{J}(\mathbf{X}) \\ &= \mathbf{J}(\mathbf{Y})\mathbf{J}(\mathbf{X}),\end{aligned}$$

where $\mathbf{J}$ is the Jacobian matrix of $\mathbf{F}$. This gives stability of the two fixed points, $\mathbf{X}$ and $\mathbf{Y}$, of $\mathbf{G}$ if all the eigenvalues of the $m \times m$ product matrix $\mathbf{K}(\mathbf{X})$ have moduli less than one, and instability if the modulus of at least one eigenvalue is greater than one. It follows that the two-cycle of the first-generation map $\mathbf{F}$ is stable or unstable respectively.

Example 3.5: the logistic map. In §1.8 we found that the logistic map $F(x) = ax(1-x)$ has the two-cycle $X, Y = [a + 1 \pm \{(a+1)(a-3)\}^{1/2}]/2a$ for $a > 3$. This gives $F: \mathbb{R} \to \mathbb{R}$ with $\mathrm{d}F/\mathrm{d}x$ as the 1×1 Jacobian matrix. The multiplier of the second-generation map is the product

$$\begin{aligned} q &= \left[\frac{\mathrm{d}F}{\mathrm{d}x}\right]_X \left[\frac{\mathrm{d}F}{\mathrm{d}x}\right]_Y \\ &= a(1-2X) \times a(1-2Y) \\ &= a^2\{1 - 2(X+Y) + 4XY\} \\ &= a^2\{1 - 2(a+1)/a + 4(a+1)/a^2\} \\ &= 4 + 2a - a^2. \end{aligned}$$

Therefore the two-cycle is stable for $1 > |q| = |4 + 2a - a^2|$, i.e. for $3 < a < 1 + \sqrt{6} = 3.4495$, and unstable for $a > 1 + \sqrt{6}$. Note further that the second-generation map has $q = -1$ at $a = 1 + \sqrt{6}$, where it has a flip bifurcation (and so there is a bifurcation of the first-generation map from a stable two-cycle to a stable four-cycle). □

In summary, we note that, for a given map $\mathbf{F}$ and initial point $\mathbf{x}_0$, either (a) $\mathbf{x}_0$ is a fixed point, (b) $\mathbf{F}^n(\mathbf{x}_0)$ is a fixed point for some positive integer n, i.e. $\mathbf{x}_0$ is an *eventually fixed point* of $\mathbf{F}$, (c) $\mathbf{F}^n(\mathbf{x}_0)$ tends to a fixed point as $n \to \infty$, i.e. $\mathbf{x}_0$ is an *asymptotically fixed point*, (d) $\mathbf{x}_0$ is a point of period p for some integer $p \geqslant 2$, (e) $\mathbf{F}^n(\mathbf{x}_0)$ is a point of period p for some positive integer n, i.e. $\mathbf{x}_0$ is an *eventually periodic point*, (f) $\mathbf{F}^n(\mathbf{x}_0)$ tends to a solution of period p, so $\mathbf{F}^n(\mathbf{x}_0)$ has a subsequence which converges to a point of period p, i.e. $\mathbf{x}_0$ is an *asymptotically periodic point*, or (g) $\mathbf{F}^n(\mathbf{x}_0)$ is aperiodic and does not approach a p-cycle, i.e. none of the previous cases is applicable. In the last case we sometimes say that the orbit is *chaotic*.

3 Attractors and volume

3.1 *Attractors*

In studying difference equations of the general form $\mathbf{x}_{n+1} = \mathbf{F}(\mathbf{x}_n)$ for $n = 0, 1, \ldots$, where $\mathbf{F}: \mathbb{R}^m \to \mathbb{R}^m$, we are often interested in sequences $\{\mathbf{x}_0, \mathbf{x}_1, \ldots\}$ for given $\mathbf{F}$ for all $\mathbf{x}_0$, i.e. in the orbits $\{\mathbf{x}_n\} = \{\mathbf{F}^n(\mathbf{x}_0)\}$. In particular, we are interested in the behaviour of $\mathbf{x}_n$ as $n \to \infty$, just as we are interested in solutions $\mathbf{x}(t)$ of differential equations as the independent variable $t \to \infty$. If there exists $\mathbf{x}_\infty$ and an infinite subsequence $\{\mathbf{x}_{n_r}\}$ for integers n_r such that $0 < n_1 < n_2 < n_3 \ldots$ and $\mathbf{x}_{n_r} \to \mathbf{x}_\infty$ as $r \to \infty$ then we say that $\mathbf{x}_\infty$ is a *cluster point*, a *point of accumulation* or a *limit point*, of the sequence $\{\mathbf{x}_n\}$. It follows, for example, that if $\mathbf{x}_n \to \mathbf{X}$ as $n \to \infty$ then $\mathbf{X}$ is the cluster point of $\{\mathbf{x}_n\}$, and if $\{x_n\} = \{\frac{1}{2}, -\frac{1}{2}, \frac{2}{3}, -\frac{2}{3}, \frac{3}{4}, -\frac{3}{4}, \frac{4}{5}, -\frac{4}{5}, \ldots\}$ then 1 and -1 are the cluster points.

Let C be the set of cluster points of the orbit $\{\mathbf{F}^n(\mathbf{x}_0)\}$ for given $\mathbf{F}$ and $\mathbf{x}_0$. The set C is called the *ω-limit set* of the orbit, because it is essentially the limit of the orbit for large positive n, and ω is the last letter of the Greek alphabet. Then $\mathbf{F}(\mathrm{C}) = \mathrm{C}$, because $\{\mathbf{F}^{n+1}(\mathbf{x}_0)\}$ has the same set of cluster points as $\{\mathbf{F}^n(\mathbf{x}_0)\}$, i.e. C is an invariant set of the map $\mathbf{F}$. Thus the concept of an ω-limit set embraces both a fixed point and a p-cycle. The ω-limit set C is closed, because the limit of a sequence of points of the set is a limit of some subsequence of points of the orbit and so belongs to C.

We define the *attractor* A of $\mathbf{F}$ and $\mathbf{x}_0$ as the set C when all orbits near C have the same set C of cluster points. Thus all attractors are ω-limit sets but not all ω-limit sets are attractors. The definition of attractor may be made precise in various slightly different ways, but our informal definition will suffice here. It follows that A is a closed invariant set of $\mathbf{F}$. We may define an *α-limit set* and a *repeller* of the orbit $\{\mathbf{F}^n(\mathbf{x}_0)\}$ similarly by taking cluster points in the limit as $n \to -\infty$.

*A less informal definition is that the set $\mathrm{A} \subset \mathbb{R}^m$ is an *attractor* of $\mathbf{F}$ if $\exists$ an open set $\mathrm{N} \subset \mathbb{R}^m$ such that $\forall\ \mathbf{x} \in \mathrm{N}$ the distance between $\mathbf{F}^n(\mathbf{x})$ and A tends to zero as n tends to infinity and there is no proper subset of A satisfying these conditions. Thus $\mathbf{F}^n(\mathbf{x})$ approaches *all* points of A and no proper subset of A is an attractor.

Different authorities define an attractor in slightly different ways, but an attractor is essentially what we would find after a long time, by starting at any point near it and then computing the iterations of the map, whereas an ω-limit set is an ideal property of a single orbit. Note that an attractor may be a single point, i.e. an asymptotically stable fixed point, or p points, i.e. a cycle of period p. We have met examples of fixed points and of two-cycles, i.e. cycles of period two, above. In §4 we shall show that some limits of orbits are aperiodic and so an attractor may be an infinity of points.

Example 3.6: some attractors of the logistic map. The previous examples give the following. The fixed point $X = 0$ of the logistic map $F(a, x) = ax(1 - x)$ is an attractor for $-1 < a < 1$; it is in fact a cluster point of $\{x_n\}$ for all $x_0 \in [0, 1]$. For $1 < a < 3$, the point $X = 0$ is a cluster point of $\{x_n\}$, but only for $x_0 = 0$ or 1; the fixed point $X = (a - 1)/a$ is the attractor and

$X = 0$ the repeller. For $3 < a < 1 + \sqrt{6}$, the two-cycle is an attractor and 0 and $(a-1)/a$ are repellers. □

For a given map $\mathbf{F}$ and attractor A, we define the *domain of attraction*, or *basin of attraction*, D(A) as the set such that if $\mathbf{x}_0 \in$ D then A is the set of cluster points of the sequence $\{\mathbf{x}_n\}$, i.e.

$$\text{D(A)} = \{\mathbf{x}\text{: the set of cluster points of } \{\mathbf{F}^n(\mathbf{x})\} \text{ is contained in A}\}.$$

It follows that D(A) is an invariant set of $\mathbf{F}$, i.e. $\mathbf{F}(\text{D(A)}) = \text{D(A)}$, because if $\mathbf{x}$ is in D(A) then $\mathbf{F}^n(\mathbf{x})$ is in D(A). Also if $\mathbf{F}$ is continuous then D(A) is an open set; because an orbit tending to A lies in D(A) and an orbit in D(A) tends to A, and so if $\mathbf{F}^n(\mathbf{x}) \to \text{A}$ as $n \to \infty$ then $\mathbf{F}^n(\mathbf{x}+\mathbf{y}) \to \text{A}$ for all sufficiently small $|\mathbf{y}|$, by continuity.

Suppose that $\mathbf{x}_{n+1} = \mathbf{F}(\mathbf{a}, \mathbf{x}_n)$ for $n = 0, 1, \ldots,$ where $\mathbf{F}\colon \mathbb{R}^l \times \mathbb{R}^m \to \mathbb{R}^m$. Then $\mathbf{a}_0 \in \mathbb{R}^l$ is said to be a *bifurcation value* if the topological character of the set of attractors or repellers of $\mathbf{F}$ changes when $\mathbf{a}$ varies in a small neighbourhood of $\mathbf{a}$. This concept of bifurcation embraces a change in the number or the stability of the fixed points of $\mathbf{F}$ as $\mathbf{a}$ varies through $\mathbf{a}_0$.

Example 3.7: some domains of attraction of the logistic map. Given $F(a,x) = ax(1-x)$, consider what points are mapped where; in particular it is helpful to find what are the maps and inverse maps (i.e. the images and pre-images) of the fixed points and to remember that the map is continuous. Then careful algebra shows that D(0) is the open interval $((a-1)/a, 1/a)$ when $0 < a < 1$, and $\text{D}((a-1)/a)$ is the open interval $(0, 1)$ when $1 < a < 3$. When $3 < a < 1 + \sqrt{6}$ the domain of attraction of the two-cycle is the union of the two open intervals $(0, (a-1)/a)$ and $((a-1)/a, 1)$.

These results may be more easily understood geometrically, by sketching the curve $y = F(a,x)$ and the line $y = x$ in the (x,y)-plane for fixed a, than algebraically. In the construction of Fig. 3.1 (see also Fig. 3.3(b)) the intersections of the curve $y = f(x)$ and the line $y = x$ give the points (X, X), where $f(X) = X$, and thence the fixed points. The construction can be seen to determine geometrically the point $(f(x), 0)$ when given the point $(x, 0)$. This enables us to see qualitatively the fixed points, the pre-images of the fixed points, the two-cycles etc. and thence the domains of attraction. □

Example 3.8: the Bernoulli shift (cf. von Neumann 1951). Let θ_{n+1} be the *fractional part* of $2\theta_n$ for $n = 0, 1, \ldots,$ i.e. consider the difference equation $\theta_{n+1} = \sigma(\theta_n)$, where the function σ is defined by

$$\sigma(x) \equiv 2x \text{ modulo } 1 \qquad \text{and } 0 \leqslant \sigma(x) < 1. \tag{1}$$

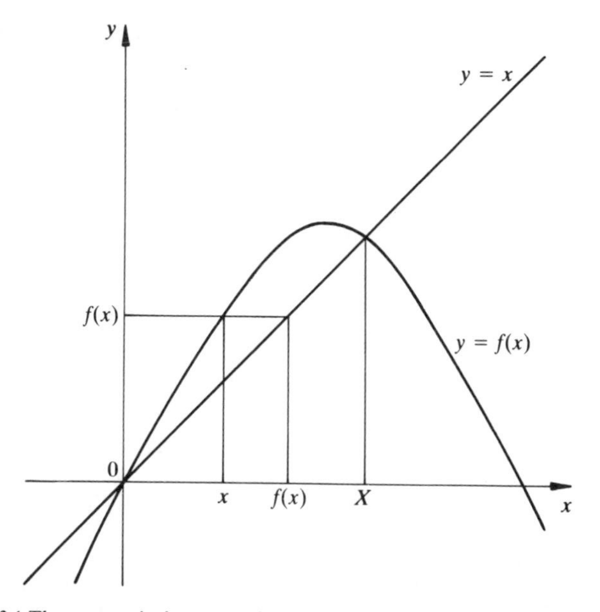

Fig. 3.1 The geometrical construction of the fixed points of a map f of the real line, and of the map $f(x)$ of a given value x.

Therefore $\sigma(x) = 2x$ if $0 \leqslant x < \frac{1}{2}$ and $\sigma(x) = 2x - 1$ if $\frac{1}{2} \leqslant x < 1$. The relationship of θ_{n+1} to θ_n by this 'sawtooth' map is illustrated in Fig. 3.2.

We see at once that 0 is the unique fixed point. Rather than setting out methodically to find the p-cycles, and then their stability, in turn, we can find all the dynamics of the difference equation as follows. We shall meet our first example of chaos.

Note that $0 \leqslant \theta_1 < 1$ so that we may take $0 \leqslant \theta_0 < 1$ with little loss of generality. Then we can express $\theta_0 = d_1/2 + d_2/2^2 + d_3/2^3 + \cdots$ as a binary number with digits $d_j = 0$ or 1 for $j = 1, 2, \ldots$. Now it can be seen that

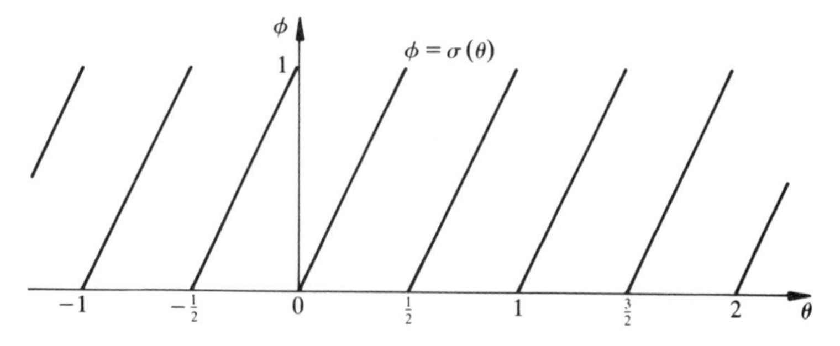

Fig. 3.2 The sawtooth map, $\phi = \sigma(\theta)$, where $\sigma(\theta)$ is the fractional part of 2θ.

the binary expression of θ_{n+1} is the binary expression of θ_n after the removal of the first digit (zero or one) to the right of the binary point. (This is called the *Bernoulli shift* after Jakob Bernoulli's (1713) posthumous study of independent random events with only two possible outcomes, e.g. coin tossing.) So if θ_0 has a binary expression of finite length, i.e. if $\theta_0 = q/2^s$ for some positive integers s and odd q, then $\theta_n = 0$ for all $n > s$. Also if θ_0 is rational and has a binary expression of infinite length then $\theta_0 = q/2^s r$ for some integers $q, s \geqslant 0, r \geqslant 3$ such that q and $2^s r$ are coprime and r is odd; therefore $\theta_n \equiv q/2^{s-n}r$ modulo 1; therefore $\theta_{n+p} = \theta_n$ for all $n \geqslant s$, where the period p is such that $1 \leqslant p \leqslant r$, by the pigeon-hole principle. If θ_0 is irrational then $\{\theta_n\}$ is an aperiodic sequence.

For an example, first take $\theta_0 = \frac{1}{3}$; then $\{\theta_n\} = \{\frac{1}{3}, \frac{2}{3}, \frac{1}{3}, \frac{2}{3}, \ldots\}$, a two-cycle. Again, if $\theta_0 = \frac{1}{4}$ then $\{\theta_n\} = \{\frac{1}{4}, \frac{1}{2}, 0, 0, \ldots\}$, with an eventually fixed point. If $\theta_0 = \frac{1}{5}$ then $\{\theta_n\} = \{\frac{1}{5}, \frac{2}{5}, \frac{4}{5}, \frac{3}{5}, \frac{1}{5}, \frac{2}{5}, \ldots\}$, a four-cycle. If $\theta_0 = \frac{1}{6}$ then $\theta_1 = \frac{1}{3}$ and the eventually periodic sequence proceeds as above. If $\theta_0 = \frac{1}{7}$ or $\frac{3}{7}$ then the sequence has period three, and so forth. All the periodic sequences are unstable, because each rational value of θ_0 is arbitrarily close to an irrational value of θ_0 which generates an aperiodic sequence $\{\theta_n\}$. Each aperiodic sequence in fact goes either through or arbitrarily close to each point of the interval $[0, 1]$.

It follows for this map σ that there is an infinity of different ω-limit sets C according to the values of θ_0; C is finite if θ_0 is rational and infinite if θ_0 is irrational. With the definition of an attractor which we have chosen, no set C is an attractor, because an initial point θ_0 in each neighbourhood of C may give a sequence $\{\theta_n\}$ with a different set of cluster points.

Let us consider a wider issue next. Two infinite sequences $\{\theta'_n\}$ and $\{\theta''_n\}$ can be generated by iterating the sawtooth map (1) with irrational initial values θ'_0 and θ''_0 respectively. You can see that, however small $\theta''_0 - \theta'_0$ is, $\theta''_n - \theta'_n$ may have almost any value between zero and one for sufficiently large values of n, because the value of $\theta''_n - \theta'_n$ depends only on the digits in the binary expressions of θ''_0 and θ'_0 which lie more than n places to the right of the binary points. When the difference $\theta''_n - \theta'_n$ is small it in general doubles each time n increases by one, because $\theta''_n - \theta'_n \equiv 2^n(\theta''_0 - \theta'_0)$ modulo 1. Thus the value of $\sigma^n(\theta_0)$ for a large value of n depends very sensitively on the initial value θ_0. This property, with exponentially growing separation of neighbouring orbits, is called *sensitive dependence on initial conditions* (after Ruelle 1979). Indeed, there is a periodic solution for a rational value of θ_0 and an aperiodic one for an irrational value of θ_0, and, of course, the sets of rationals and irrationals are each dense in the interval $[0, 1]$. We may say that an aperiodic solution rapidly 'forgets' the initial condition as n increases. Sensitive dependence on initial conditions is sometimes used as a characteristic to define chaos, although a satisfactory precise definition is elusive. This issue is elaborated in §4.4.

There is an important geometrical application of the sawtooth map. Each point on the circumference S^1 of a circle may be represented by its polar angle ϕ and so uniquely by $\theta \equiv \phi/2\pi$ modulo 1, where $0 \leqslant \theta < 1$. Then the map represents a doubling of the polar angle ϕ of a point on the circle, and in this representation is a continuous map. □

Example 3.9: the rotation map. Another simple, but instructive, map is the piecewise linear map, $F: [0, 1) \times [0, 1) \to [0, 1)$ defined by

$$F(a, x) \equiv x + a \text{ modulo } 1 \qquad \text{and } 0 \leqslant F < 1.$$

We take $0 \leqslant a < 1$ without loss of generality. We may again regard this as a continuous map of the circle S^1 in which a point of the circumference is rotated by a constant angle $2\pi a$, because a point with angle $2\pi x_n$ is mapped to the point with angle $2\pi x_{n+1} = 2\pi x_n + 2\pi a$.

Note that $x_n = F^n(a, x_0) \equiv x_0 + na$ modulo 1.

If $a = 0$ then F is the identity map. Therefore all $x \in [0, 1)$ are fixed points and stable. However, none is an attractor because they are not asymptotically stable.

If $a > 0$ is rational then $a = q/p$ for coprime integers $0 < q < p$. Therefore $x_n - x_0 \equiv nq/p$ modulo 1. Therefore $x_n \neq x_0$ for $0 < n < p$ and $x_p = x_0$. Therefore $\{x_n\}$ has period p, and the p-cycle is $\{x_0, x_0 + 1/p, x_0 + 2/p, \ldots, x_0 + (p-1)/p\}$ in some order. This cycle is stable, but not asymptotically stable, because a small change of x_0 merely translates the cycle a little.

Conversely, if $\{x_n\}$ has period p, then $0 = x_p - x_0 \equiv pa$ modulo 1, i.e. pa is an integer. Therefore a is rational.

It may be intuitively sensed that as an irrational a is approximated by a rational q/p with large p the sequence $\{x_n\}$ is approximated by a large p-cycle which covers the interval $[0, 1)$ uniformly. This can be stated more carefully as follows.

*Suppose next that a is irrational and $0 < \xi < 1$. Then x_n goes arbitrarily close to ξ. We shall show that *either* there exist non-negative integers m, n such that $x_0 + na - m = \xi$, in which case $x_n = \xi$; *or* $x_0 + na - m \neq \xi$ for all $m, n \geqslant 0$. In the former case, one and only one member of $\{x_n\}$ equals ξ. In the latter case, it may be proved that there exists an infinite subsequence $\{n_r\}$ of the integers such that $0 \leqslant n_1 < n_2 < \cdots$ and $x_{n_r} \to \xi$ as $r \to \infty$, i.e. that ξ is a cluster point of $\{x_n\}$. To prove this, first choose any integer $r > 0$ and note that at least two of the $r + 1$ points $x_0, x_1, \ldots, x_r$ must lie in one of the r equal subintervals $[0, 1/r), [1/r, 2/r), \ldots, [(r-1)/r, 1)$, by the pigeon-hole principle. Denote these two points by x_l, x_m for $l < m$, and define $\epsilon = x_m - x_l$, so $|\epsilon| < 1/r$. Therefore $x_{m-l} \equiv x_0 + (m-l)a \equiv x_0 + x_m - x_l \equiv x_0 + \epsilon$ modulo 1. Similarly, $x_{k(m-l)} \equiv x_0 + k\epsilon$. Therefore there exists k such that the distance of $x_{k(m-l)}$ from ξ is less than $|\epsilon|$. Therefore $|x_{n_r} - \xi| < 1/r$, where $n_r = k(m-l)$. So we can construct a subsequence $\{x_{n_1}, x_{n_2}, \ldots\}$ such that $|x_{n_r} - \xi| < 1/r$ for $r = 1, 2, \ldots$. The result follows. Thus in this case the ω-limit set is the interval $[0, 1]$, but is not an attractor. □

3.2 *Volume*

Another important, but rather different, concept is that of how the volume of a set of points is mapped. Consider a set $S_n \subset \mathbb{R}^m$ and a given continuously differentiable map $\mathbf{F}: \mathbb{R}^m \to \mathbb{R}^m$. Then we may define S_{n+1} as the set $\mathbf{F}(S_n)$, i.e the points $\mathbf{x}_{n+1} = \mathbf{F}(\mathbf{x}_n)$ for all $\mathbf{x}_n \in S_n$. Also define μ_n as the hypervolume, i.e. the measure, of the set S_n of points; so μ_n is the total length of the points of the set on the real line if $m = 1$, the total area of the points of the set in the plane if $m = 2$, or the total volume of the points of the set if $m = 3$. Now the local ratio of the hypervolume μ_{n+1} to μ_n is the modulus of the Jacobian $\det \mathbf{J}(\mathbf{x}_n)$, where the $m \times m$ Jacobian matrix $\mathbf{J}$ has element $\partial F_i/\partial x_j$ in its ith row and jth column. Therefore if the modulus is everywhere less than one then μ_n decreases monotonically as n increases; in this case the hypervolume μ_n shrinks as n increases, and the dimension of the attractor must be less than m.

If $|\det \mathbf{J}| = 1$ everywhere then $\mu_n = \mu_0$ for all n, i.e. the hypervolume μ_n is independent of n, and we say that $\mathbf{F}$ is a *volume-preserving map* or an *area-preserving map* as is appropriate, or simply a *measure-preserving map*. More generally, $\mathbf{F}$ is a measure-preserving map if the measure of all sets S is the measure of the set of all points mapped to S, i.e. $\mu(\mathrm{S}) = \mu(\mathbf{F}^{-1}(\mathrm{S}))$ for all $\mathrm{S} \subset \mathbb{R}^m$. The definition of attractor we have taken, with the condition that all neighbouring orbits are 'drawn' into the attractor, implies that a measure-preserving map does not have an attractor.

If $\mathbf{F}$ preserves volume and orientation, e.g. if $\mathbf{F}$ is a rotation about an axis, then $\det \mathbf{J} = 1$; if $\mathbf{F}$ preserves volume but reverses orientation, e.g. it is a reflection, then $\det \mathbf{J} = -1$.

Example 3.10: an area-preserving map. Take $\mathbf{x}_n = (x_n, y_n) \in \mathbb{R}^2$, $\mathbf{F}: \mathbb{R}^2 \to \mathbb{R}^2$, where the rotation $\mathbf{R}(\mathbf{x}) = (y, -x)$, the shear $\mathbf{S}(\mathbf{x}) = (x, y + f(x))$,

$$\mathbf{F}(\mathbf{x}) = \mathbf{S}(\mathbf{R}(\mathbf{S}(\mathbf{x}))) = (y + f(x), -x + f(y + f(x)))$$

and f is differentiable. Therefore the Jacobian matrix is

$$\mathbf{J}(\mathbf{x}) = \begin{bmatrix} f'(x) & 1 \\ -1 + f'(x)f'(y + f(x)) & f'(y + f(x)) \end{bmatrix}$$

and hence the Jacobian is $\det \mathbf{J} = 1$ for all x and y. Therefore $\mathbf{F}$ is an area-preserving map. □

Example 3.11: the product of the eigenvalues. If the eigenvalues of the Jacobian matrix of a map $\mathbf{F}: \mathbb{R}^m \to \mathbb{R}^m$ at $\mathbf{x}$ are $q_1, q_2, \ldots, q_m$ then $q_1 q_2 \ldots q_m = \det(\partial F_i/\partial x_j)$. It follows that if the map is area-preserving then the product of the eigenvalues is ± 1. □

4 The logistic equation

Many of the important properties of bifurcation of nonlinear systems are exhibited by the logistic difference equation, so we shall examine it in detail, finding immensely greater richness and depth of results than might be anticipated from so simple an equation. Some of the concepts which arise in this way will be elaborated in later sections. We shall see that the logistic equation not only is a prototype of a large class of one-dimensional nonlinear difference equations but also shares many properties of higher-dimensional difference equations and other nonlinear systems. It may be said that the logistic difference equation is used in the theory of iterated maps as the fruit fly, *Drosophila*, is used in genetics.

We again take

$$x_{n+1} = F(a, x_n), \qquad \text{for } n = 0, 1, \ldots, \tag{1}$$

where

$$F(a, x) = ax(1 - x) \qquad \text{for all } a \geqslant 0. \tag{2}$$

If $a \neq 0$ and $x > 1$ or $x < 0$ then $F(a, x) < 0$, so let us here take $0 \leqslant x_0 \leqslant 1$ and $1 \geqslant \max_{0 \leqslant x \leqslant 1} F(a, x) = \frac{1}{4}a$, i.e. $a \leqslant 4$, in order that F maps the interval $[0, 1]$ into itself. We seek to examine the sequences $\{x_n\}$, mostly for given values of $a \in [0, 4]$ and $x_0 \in [0, 1]$. We have found in §1.8 that the fixed points of F are $X = 0$ for all a and $X = (a - 1)/a$ for $a \neq 0$. The fixed point $X = 0$ is stable if $0 \leqslant a \leqslant 1$, and $X = (a - 1)/a$ is stable if and only if $1 < a \leqslant 3$. Also there is a two-cycle with $X_1, X_2 = [a + 1 \pm \{(a + 1)(a - 3)\}^{1/2}]/2a$ for $3 < a$ and it is stable if and only if $3 < a \leqslant 1 + \sqrt{6}$.

It is helpful to examine graphically how the fixed points and the periodic solutions arise, and also see why they become unstable. The graphical method will show that the qualitative behaviour of the logistic map is shared by all maps of the form $F(a, x) = af(x)$ for a smooth and convex

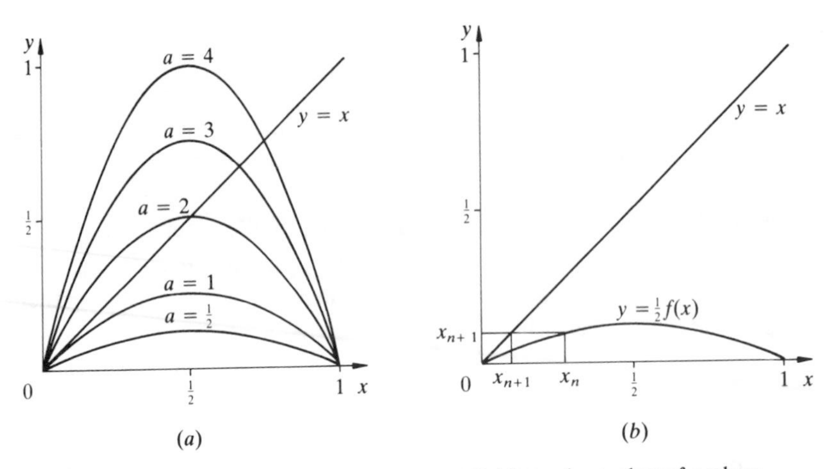

Fig. 3.3 The line $y = x$ and some curves $y = af(x)$ for various values of a, where $f(x) = x(1 - x)$. (a) $a = \frac{1}{2}, 1, 2, 3, 4$. (b) The graphical construction of x_{n+1} from x_n, for the case $a = \frac{1}{2}$.

function f such that $f(0) = 0$, $f(b) = 0$ for some $b > 0$, and $f(x) > 0$ for $0 < x < b$.

First examine the graphs of $y = F(a, x)$ for various values of a illustrated in Fig. 3.3(a). Each cuts the line $y = x$ in the fixed points. It can be seen at once that there is one fixed point $X = 0$ for all a and also another for $a > 1$, because the convex curve lies below its tangent $y = ax$ at the origin if and only if $a \leqslant f'(0) = 1$.

Secondly consider the graphical construction of x_{n+1} from x_n illustrated in Fig. 3.3(b). When the tangent $y = ax$ to the curve $y = F(a, x)$ at the fixed point $X = 0$ lies below the line $y = x$, it can be seen that $x_{n+1} > x_n$ and x_n approaches the fixed point as $n \to \infty$. Similarly, if the tangent lies above the line $y = x$, i.e. if $a > 1$, then the fixed point is unstable. Also it can be seen that the slope of the tangent to the curve at the other fixed point $X = (a - 1)/a$ determines the stability likewise.

The geometrical effect of the map can be seen to be equivalent to non-uniform stretching (when $a > 2$) of the interval $[0, 1]$ followed by folding back onto the subinterval $[0, \frac{1}{4}a]$ to give the curve $y = F(a, x)$. This stretching and folding is characteristic of maps with infinite attractor sets (cf. §6).

Next look at the graphs of some of the curves with equations $y = G(a, x)$ for various values of a, where G is the second-generation map defined by $G(a, x) = F(a, F(a, x)) = F^2(a, x)$ for all x, as illustrated in Fig. 3.4(a). Note that G is a quartic. Each curve cuts the line $y = x$ in the fixed points of the second-generation map, and thereby gives the two-cycle of F as well as the fixed points of F. When $a < 1$ the curve cuts the line only at the fixed

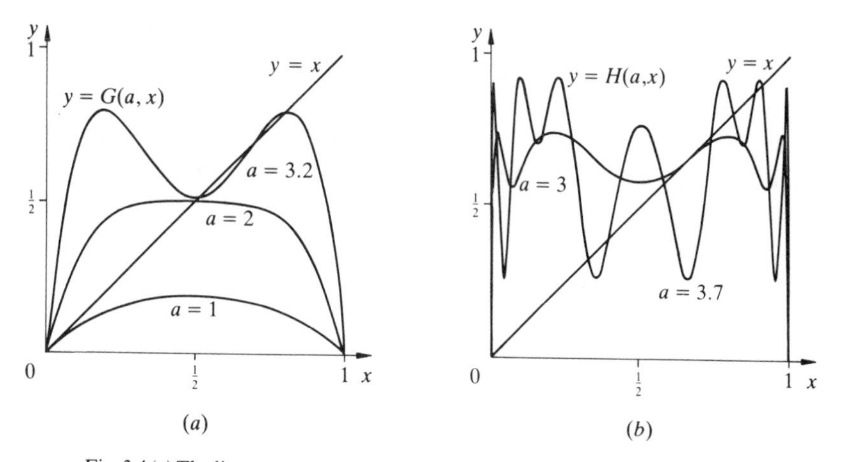

Fig. 3.4 (a) The line $y = x$ and curves $y = G(a, x)$ for the second-generation map $G = F^2$ of the logistic map F, with $a = 1, 2, 3.2$. (b) The line $y = x$ and some curves $y = H(a, x)$ for the fourth-generation map $H = F^4$ of the logistic map F, with $a = 3, 3.7$.

point $X = 0$. When $a > 1$ it also cuts the line in the other fixed point, $X = (a - 1)/a$. The curve $y = G(3, x)$ touches the line at the fixed point $X = \frac{2}{3}$. When $a > 3$ the curve $y = G(a, x)$ cuts the line at the origin and three other points, the middle of which is a fixed point of F and the rest of which are the two-cycle of F, so the two-cycle exists if and only if $a > 3$. This illustrates the geometrical essence of a flip bifurcation and period doubling. If the slope of the curve $y = G(a, x)$ is less than one in modulus where it intersects $y = x$, then the intersection represents a stable fixed point of G, and therefore a stable fixed point or two-cycle of F.

It is also instructive to look at the graphs of some higher-generation maps F^{2^r}. Note that F^q is a polynomial of degree 2^q because F is a polynomial of degree two. Two graphs of the fourth-generation map H are illustrated in Fig. 3.4(b), where H is defined by $H(a, x) = G(a, G(a, x)) = F^4(a, x)$. This suggests that four-cycles arise as a increases. Indeed, the graph of the curve with equation $y = F^q(a, x)$ oscillates more rapidly as q increases, so we anticipate repeated period doubling as a increases.

These and other properties are illustrated by the special cases in which we can solve the logistic difference equation more-or-less explicitly by analysis: $a = -2$ (Q3.19), $a = 0$ (trivial), $a = 1$ (Q1.32), $a = 2$ (Q1.33), and $a = 4$ (later in this section).

**Example 3.12: local analysis of a flip bifurcation.* We have just seen the flip bifurcation from a general geometrical point of view, having seen it in explicit analytic terms for the logistic map in Example 1.6. Another point of view is that of perturbation theory. We shall describe here quite a general theory of the flip bifurcation, and apply it to the logistic map.

The linear theory of stability of the fixed point $X = (a-1)/a$ of the logistic map $F(a,x) = ax(1-x)$ shows (Example 3.3) that X becomes unstable, the multiplier q decreasing through -1, as a increases through $a_c = 3$. It gives $x_n = X + (-1)^n x_0' + O(x_0'^2)$ as $x_0' \to 0$, where $a = a_c$; so there is a solution of period two, with $x_{2n} = x_0 = X + x_0'$ and $x_{2n+1} = X - x_0'$ for $n = 0, 1, \ldots$.

These results of the linear theory guide our choice of the ansatz (or assumed form) of the perturbation theory of the flip bifurcation as $a \to a_c$, for the ansatz must allow for the appropriate behaviour. The start of the problem is that there is a fixed point $X(a)$ with a multiplier $q(a)$ such that $q(a_c) = -1$, i.e.

$$F(a,X) = X \qquad \text{for } a \text{ near } a_c \text{ and} \qquad F_x(a_c, X_c) = -1,$$

where $X_c = X(a_c)$ and a subscript x denotes a partial derivative with respect to x. We assume that there is also a two-cycle $\{X_1, X_2\}$ as $a \to a_c$ (either from above or from below, but not both) such that

$$F(a, X_k) = X_j, \qquad \text{where } j, k = 1, 2 \text{ or } 2, 1, \tag{3}$$

$$X_j = X_c + \epsilon X_{j1} + \epsilon^2 X_{j2} + \cdots \qquad \text{for } j = 1, 2 \tag{4}$$

and

$$a = a_c + \epsilon a_1 + \epsilon^2 a_2 + \cdots \qquad \text{as } \epsilon \to 0. \tag{5}$$

It only remains to substitute the series (4) and (5) into equation (3), equate coefficients of successive powers of ϵ, and find the coefficients a_1, $X_{11}, X_{21}, a_2, X_{12}, \ldots$ in turn. The form of the series (4) and (5), with the introduction of the new small parameter ϵ, rather than series in powers of $a - a_c$, is motivated in Example 2.1: the more flexible form (5) allows the expansion (4) to be effectively either in powers of $(a - a_c)^{1/2}$ or in powers of $(a_c - a)^{1/2}$ rather than in powers of $a - a_c$, and so to match the properties of the solution. The expansion (5) will be seen not to be unique, although the choice of $a_2, a_3, \ldots$ does not affect the solution itself, of course.

Now substitution of series (4), (5) into equation (3) gives

$$F_c + \epsilon(a_1 F_{ac} + X_{k1} F_{xc}) + \epsilon^2(a_2 F_{ac} + X_{k2} F_{xc} + \tfrac{1}{2} a_1^2 F_{aac} + a_1 X_{k1} F_{axc} + \tfrac{1}{2} X_{k1}^2 F_{xxc}) + \cdots = X_c + \epsilon X_{j1} + \epsilon^2 X_{j2} + \cdots,$$

where the subscript c denotes evaluation of a partial derivative at $a = a_c$ and $X = X_c$. But $F_c = X_c$ and $F_{xc} = -1$. Therefore

$$\epsilon(a_1 F_{ac} - X_{k1}) + \epsilon^2(a_2 F_{ac} - X_{k2} + \tfrac{1}{2}a_1^2 F_{aac} + a_1 X_{k1} F_{axc} + \tfrac{1}{2}X_{k1}^2 F_{xxc}) + \cdots$$
$$= \epsilon X_{j1} + \epsilon^2 X_{j2} + \cdots.$$

Equating coefficients of ϵ, we find

$$X_{11} + X_{21} = a_1 F_{ac} \tag{6}$$

twice.

Equating coefficients of ϵ^2, we find

$$a_2 F_{ac} - X_{k2} + \tfrac{1}{2}a_1^2 F_{aac} + a_1 X_{k1} F_{axc} + \tfrac{1}{2}X_{k1}^2 F_{xxc} = X_{j2}.$$

Therefore

$$X_{12} + X_{22} = a_2 F_{ac} + \tfrac{1}{2}a_1^2 F_{aac} + a_1 X_{11} F_{axc} + \tfrac{1}{2}X_{11}^2 F_{xxc}$$

and

$$X_{12} + X_{22} = a_2 F_{ac} + \tfrac{1}{2}a_1^2 F_{aac} + a_1 X_{21} F_{axc} + \tfrac{1}{2}X_{21}^2 F_{xxc}.$$

Therefore

$$a_1 X_{21} F_{axc} + \tfrac{1}{2}X_{21}^2 F_{xxc} = a_1 X_{11} F_{axc} + \tfrac{1}{2}X_{11}^2 F_{xxc}.$$

In general $X_{21} \neq X_{11}$ because $X_2 \neq X_1$. Therefore

$$X_{11} + X_{21} = -2a_1 F_{axc}/F_{xxc} \tag{7}$$

unless $F_{xxc} = 0$. It follows from equations (6) and (7) that either $a_1 = 0$ or $-2F_{axc}/F_{xxc} = F_{ac}$. Assuming the general case, in which the latter equality is not satisfied, we deduce that

$$a_1 = 0.$$

Therefore

$$X_{21} = -X_{11} \text{ and } X_{12} + X_{22} = a_2 F_{ac} + \tfrac{1}{2}X_{21}^2 F_{xxc}.$$

Equating coefficients of ϵ^3 in equation (3) (and assuming that $F_{aa} = 0$ in order to simplify the calculation a little), we deduce that

$$a_3 F_{ac} - X_{k3} + a_2 X_{k1} F_{axc} + X_{k1} X_{k2} F_{xxc} + \tfrac{1}{6}X_{k1}^3 F_{xxxc} = X_{j3}.$$

Therefore

$$X_{13} + X_{23} = a_3 F_{ac} + a_2 X_{k1} F_{axc} + X_{k1} X_{k2} F_{xxc} + \tfrac{1}{6}X_{k1}^3 F_{xxxc} \quad \text{for } k = 1, 2.$$

Therefore

$$a_2 X_{21} F_{axc} + X_{21} X_{22} F_{xxc} + \tfrac{1}{6}X_{21}^3 F_{xxxc} = a_2 X_{11} F_{axc} + X_{11} X_{12} F_{xxc}$$
$$+ \tfrac{1}{6}X_{11}^3 F_{xxxc}.$$

Therefore

$$2a_2F_{axc} + (X_{12} + X_{22})F_{xxc} + \tfrac{1}{3}X_{11}^2F_{xxxc} = 0$$

because $X_{21} = -X_{11}, \neq 0$. Therefore

$$a_2(2F_{axc} + F_{ac}F_{xxc}) + \tfrac{1}{2}X_{11}^2F_{xxc}^2 + \tfrac{1}{3}X_{11}^2F_{xxxc} = 0, \tag{8}$$

a formula sufficient to give the leading approximation to the behaviour of the two-cycle as $a \to a_c$.

To verify this, we apply the results to the logistic map, for which $F(a,x) = ax(1-x)$, $a_c = 3$ and $X_c = \frac{2}{3}$, and so $F_{ac} = \frac{2}{9}$, $F_{axc} = -\frac{1}{3}$, $F_{xxc} = -6$ and $F_{xxxc} = 0$. Therefore equation (8) gives $X_{11}^2 = \frac{1}{9}a_2$. Therefore

$$X_j - \tfrac{2}{3} \sim \tfrac{1}{3}(-1)^j(a-3)^{1/2} \qquad \text{as } a \downarrow 3 \text{ for } j = 1, 2,$$

in agreement with Example 1.6. □

Another important approach to the problem is the numerical one. We can calculate the periodic solutions of a difference equation in various ways. A simple program to calculate the periodic solutions and many other properties of the logistic equation with a computer is described in the next section. You can easily confirm for yourself the results in Table 3.1 of attractors appropriate to various intervals of the parameter a. You may infer the occurrence of an infinite sequence of period doubling with values a_r at bifurcations such that

$$\frac{a_{r-1} - a_r}{a_r - a_{r+1}} \to \delta \qquad \text{as } r \to \infty, \tag{9}$$

and therefore

Table 3.1. *Some attractors of the logistic map for $a \geqslant 0$*

Interval	Attractor
$0 \leqslant a \leqslant 1$	$X = 0$
$1 < a \leqslant a_1 = 3$	$X = (a-1)/a$
$a_1 < a \leqslant a_2 = 3.449$	2^1-cycle
$a_2 < a \leqslant a_3 = 3.544$	2^2-cycle
$a_3 < a \leqslant a_4 = 3.564$	2^3-cycle
...	...
$a_r < a \leqslant a_{r+1}$	2^r-cycle
...	...

$$a_r = a_\infty - A\delta^{-r} + o(\delta^{-r}),$$

where $a_\infty = 3.5700\ldots$ and $\delta = 4.6692\ldots$. A sequence with an asymptotic property of the form (9) for the same value of δ is called a *Feigenbaum sequence*, after Feigenbaum (1978), who showed that the qualitative character of the period doubling and the exact value of the constant δ in relation (9) are independent of the particular map F chosen, although the constants a_∞ and A do depend on F. So a Feigenbaum sequence is characteristic of a wide class of maps with the same form as F and not just of the logistic map. In this sense δ is a universal constant. Indeed, period doubling and Feigenbaum sequences will be seen to occur for difference equations which are not measure-preserving, whether they are one- or higher-dimensional, and also for many nonlinear ordinary and partial differential equations. We shall show some details of the period doubling in the limit as $r \to \infty$ and demonstrate the universality of the limit of the period doubling in §4.3.

The properties are even more interesting when a is increased above a_∞. All the solutions of periods 2^r which originated for $a < a_\infty$ persist for $a > a_\infty$, but they are unstable and in addition there is an attractor with an infinite number of points x for most values of $a > a_\infty$. Such an attractor gives an aperiodic solution, and is called a *strange attractor* (after Ruelle & Takens 1971). It is often called a *chaotic solution*, or simply *chaos* (after Li & Yorke 1975), because, as we shall see, the solution may be regarded as a sample of a random variable. It may help to imagine the chaos as being due to the infinity of repellers, each of which repels the solution in its orbit in the manner in which the ball is repelled by the pins in a pin-ball machine; this model is crude but has some essence of the truth. The attractor is called a *Cantor set* because it is topologically equivalent to a famous set defined by Cantor (1883). In the next chapter we shall elaborate the aspects of set theory which arise. For the present we add that a Cantor set of points is uncountable and its points are disconnected, in the sense of there being no interval contained in the set, although to each point of the set there is another point arbitrarily close, and each convergent subsequence of the set converges to a member of the set, i.e. a Cantor set is closed. You will find the nature of the attractor easier to understand if you use the computer program described in the next section and you read the set theory in the next chapter.

Although chaos is the rule for $a > a_\infty$, there are 'windows' i.e. small intervals of a, for which a stable periodic attractor exists. In particular you may like to use the computer program of the next section to demonstrate

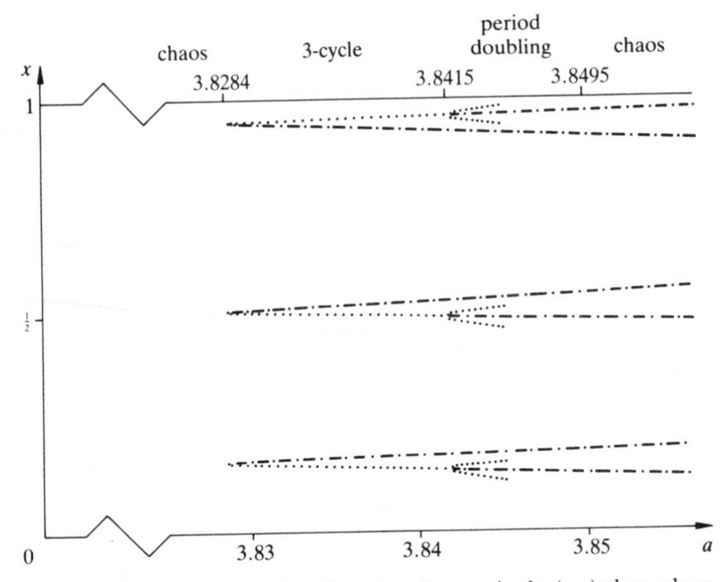

Fig. 3.5 A sketch of part of the bifurcation diagram in the (a, x)-plane where there are orbits of periods 3×2^r for $r = 0, 1, \ldots$. Dotted curves denote the stable three- and six-cycles, and dot-dashed curves the unstable three-cycles.

that there is a stable three-cycle or (3×2^r)-cycle for $3.8284 < a < 3.8495$. Look at Fig. 3.5, and note the origin of the stable and unstable three-cycles due to a turning point of the third-generation map F^3. In fact, the solution undergoes period doubling so that solutions of periods 3×2^r arise as a increases, and the critical values of a at which there is bifurcation from another Feigenbaum sequence! These three-cycles originate because F^3 is a polynomial of degree eight, and so F^3 has eight fixed points, which are real or occur in complex conjugate pairs; two are the fixed points of F for all values of a; and three complex conjugate pairs become real as a increases through $1 + \sqrt{8} = 3.8284$ (Q3.17). Further, there is an infinity of windows in which p-cycles are stable for $3 < a < 4$, such that for each integer p there is some interval of a for which an attractor is a p-cycle. For example, a stable five-cycle arises when a increases through 3.7382. Of course, some of the windows in which there exists a stable periodic solution are very narrow indeed.

The bifurcation diagram for $0 \leqslant a \leqslant 4$ is summarized in Fig. 3.6. Note the many windows for $a > 3.6$.

The case $a = 4$ is rather special (von Neumann 1951), because then $\max_{0<x<1} F(4, x) = 1$ and F maps the interval $[0, 1]$ onto itself. This property helps us to 'solve' difference equation (1). We find that according to the

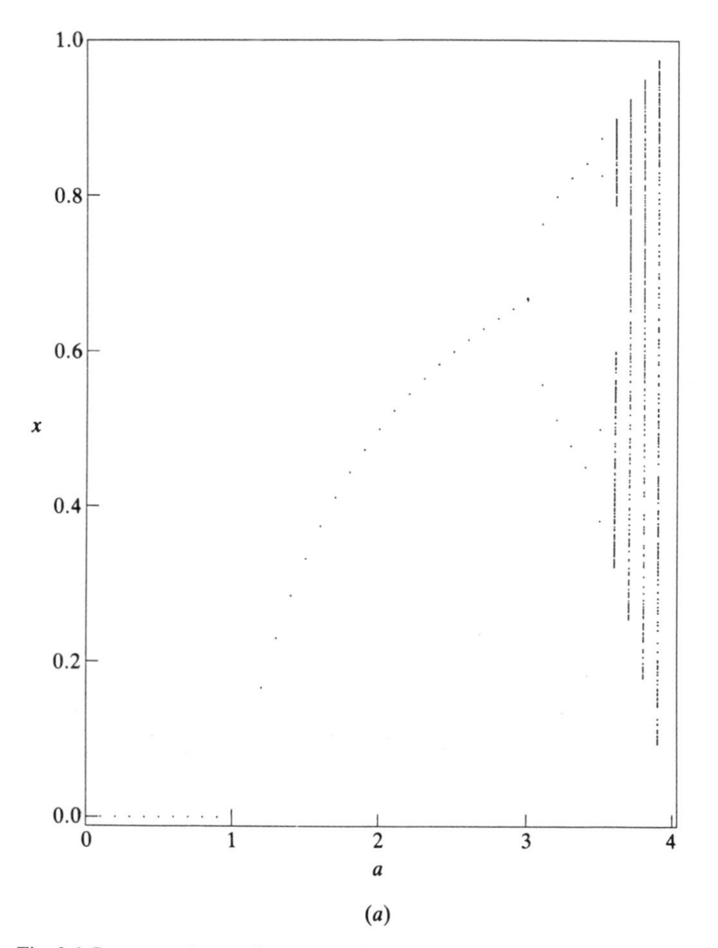

Fig. 3.6 Computer-drawn diagrams of the attractors of $F(a, x) = ax(1 - x)$ in the (a, x)-plane for (a) $0 \leqslant a \leqslant 3.9$ and (b) $3.5 \leqslant a \leqslant 4$.

initial value x_0 there may be an unstable orbit of period p for all positive integers p. Thus there is a countable infinity of periodic orbits. These unstable periodic orbits are the continuation of all the periodic orbits which arose for $3 < a < 4$. However, almost all initial values x_0 give orbits of points x_n with a set of cluster points which is the interval $[0, 1]$ itself. These results may be shown as follows.

Suppose that

$$x_n = P(\theta_n), \tag{10}$$

where

$$P(\theta) = \sin^2 \pi\theta \tag{11}$$

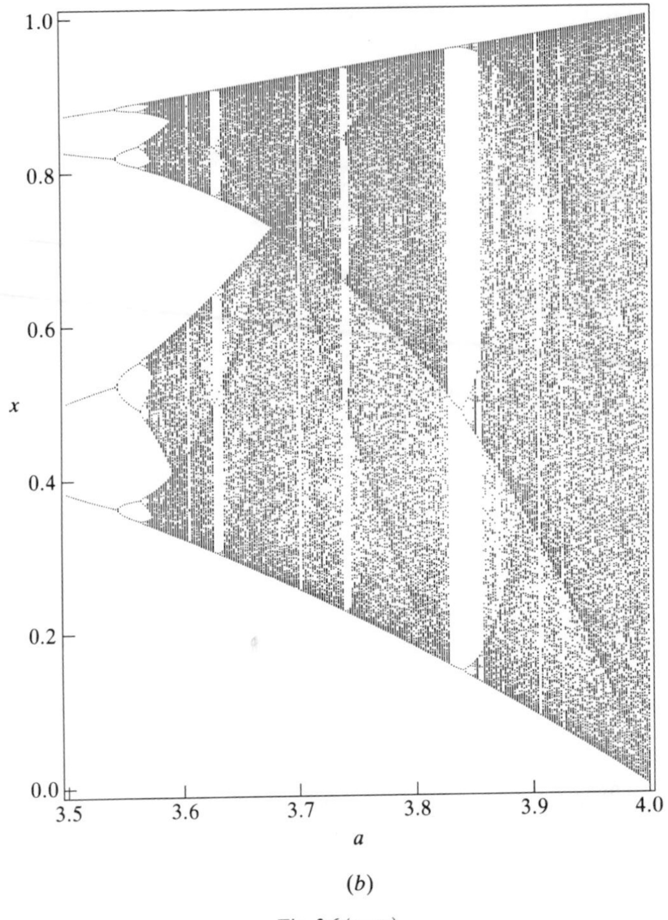

(b)

Fig. 3.6 (*cont.*)

Note that P has period one and is a $2-1$ mapping of the interval $[0, 1]$ onto itself, as shown in Fig. 3.7, so in general there are two possible values of θ_n for a given value of $x_n \in [0, 1]$. Next suppose that θ_{n+1} is the fractional part of $2\theta_n$, i.e. that

$$\theta_{n+1} \equiv 2\theta_n \text{ modulo } 1 \qquad \text{and} \qquad 0 \leqslant \theta_{n+1} < 1, \tag{12}$$

and $0 \leqslant \theta_0 < 1$. Now relations (10) and (11) give

$$\begin{aligned} x_{n+1} &= \sin^2 \pi\theta_{n+1} \\ &= \sin^2 2\pi\theta_n, \end{aligned}$$

because θ_{n+1} is given by equation (12), and P has period one,

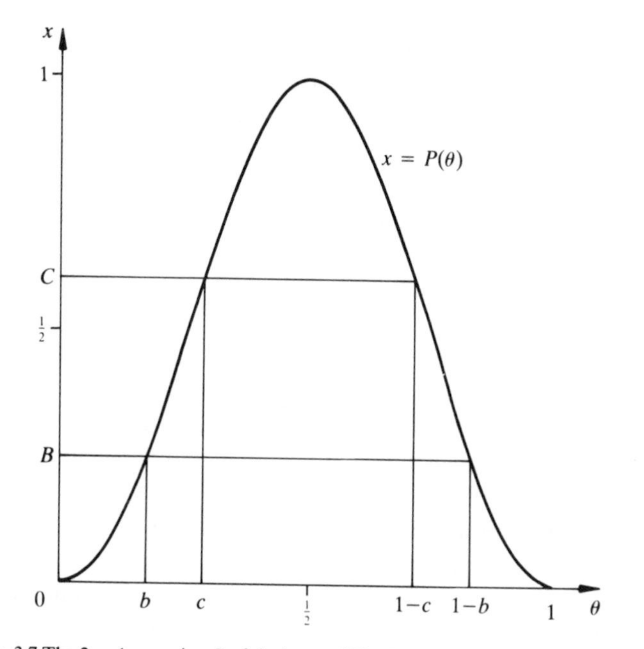

Fig. 3.7 The $2-1$ mapping P of the interval $[0, 1]$ onto itself, for $P(\theta) = \sin^2 \pi\theta$.

$$
\begin{aligned}
&= 4\sin^2 \pi\theta_n \cos^2 \pi\theta_n \\
&= 4x_n(1 - x_n) \\
&= F(4, x_n). \qquad (13)
\end{aligned}
$$

We can now regard the original nonlinear difference equation (13) as equivalent to the simpler piecewise-linear equation (12) by use of the transformation (10). So we may in principle solve equation (13) for a given value of x_0 by first inverting equation (10) to find a value of θ_0, then solving equation (12) as in §3 to find θ_n, and finally finding x_n from equation (10), so that $x_n = \sin^2(2^n\pi\theta_0)$.

We have met the sawtooth map (12) in Example 3.8. Here we note that a fixed point X of the logistic equation (13) is the map (11) either of the fixed point $\theta = 0$ of equation (10) or of a two-cycle of equation (12), because P is a $2-1$ map and $P(1-\theta) = P(\theta)$ for all θ. In fact $X = 0$ corresponds to $\theta = 0$ and $X = \frac{3}{4}$ to the two-cycle, $\theta_{2m-1} = \frac{1}{3}$ and $\theta_{2m} = \frac{2}{3}$ for $m = 1, 2, \ldots$.

The two-cycle of equation (13) must similarly correspond to a two-cycle or a four-cycle of equation (12). In fact the two-cycle $X_1, X_2 = \frac{1}{8}(5 \pm \sqrt{5}) = \sin^2\{(3 \pm 1)\pi/10] = \sin^2[(7 \mp 1)\pi/10]$ corresponds to the four-cycle $\theta_{4m-3} = \frac{2}{5}, \theta_{4m-2} = \frac{4}{5}, \theta_{4m-1} = \frac{3}{5}$ and $\theta_{4m} = \frac{1}{5}$.

Let us consider two wider issues next. Although they are of general importance, they can be simply illustrated by the logistic equation with $a = 4$. First recall from Example 3.8 that if there are two infinite sequences $\{\theta'_n\}$ and $\{\theta''_n\}$ generated by relation (12), with irrational initial values θ'_0 and θ''_0 respectively, then $\theta''_n - \theta'_n \equiv 2^n(\theta''_0 - \theta'_0)$ modulo 1. There is a corresponding property of the sequences $\{x_n\}$ that the value of x_n for large values of n depends sensitively on the initial value x_0, with exponentially growing separation of neighbouring orbits. Indeed, there is a periodic solution if x_0 corresponds to a rational value of θ_0 and an aperiodic one if x_0 corresponds to an irrational value of θ_0, and, of course, the sets of rationals and irrationals are each dense in the interval $[0, 1]$. We have demonstrated the sensitivity to initial conditions for the special value $a = 4$ because the sensitivity can be made explicit, but a similar sensitivity occurs for other values of a for which there is an attractor which is infinite. The attractor is the same set for all initial values within the domain of attraction, but the position of x_n within the attractor for a given large value of n depends sensitively upon x_0. Also many other one-dimensional and higher-dimensional nonlinear ordinary differential equations and other nonlinear systems show sensitive dependence on initial conditions.

Secondly, we see how ideas of probability are useful. Each difference equation determines x_n precisely in terms of x_0, yet in some respects a sequence $\{x_n\}$ which has an infinite set of cluster points may be regarded as a sequence of random numbers. Indeed, piecewise-linear difference equations of the form $\phi_{n+1} \equiv p\phi_n$ modulo q for integers p and q have been used in computers to generate so-called random numbers. We can illustrate this with equation (12). It can be proved that if θ_0 is irrational then the infinite sequence $\{\theta_n\}$ is uniformly distributed, i.e. there exists

$$\lim_{N\to\infty} \{(\text{number of times } \theta_n \in [b, c] \text{ for } n = 1, 2, \ldots, N)/N\} = c - b$$

for all b, c such that $0 \leqslant b \leqslant c \leqslant 1$. In the language of probability theory we may say that the probability that θ_n lies in the interval $[b, c]$ is $c - b$, i.e. θ_n is a sample of a random variable Θ which satisfies

$$\mathbf{P}(b \leqslant \Theta \leqslant c) = c - b. \tag{14}$$

Now $P(1 - \theta) = P(\theta)$ and P is a $2 - 1$ function, as shown in Fig. 3.7, so relation (10) gives x_n as a sample of a random variable X such that

$$\begin{aligned}\mathbf{P}(\sin^2 \pi b \leqslant \mathrm{X} \leqslant \sin^2 \pi c) &= \mathbf{P}(b \leqslant \Theta \leqslant c) + \mathbf{P}(1 - c \leqslant \Theta \leqslant 1 - b) \\ &= 2(c - b),\end{aligned}$$

when b and c lie on the same side of $\frac{1}{2}$, i.e.

$$\mathbf{P}(B \leqslant \mathrm{X} \leqslant C) = 2\pi^{-1}(\arcsin\sqrt{C} - \arcsin\sqrt{B}), \tag{15}$$

on putting $\pi b = \arcsin\sqrt{B}$ and $\pi c = \arcsin\sqrt{C}$. In particular, this gives the *probability distribution function* of the random variable X as

$$\begin{aligned} F(x) &= \mathbf{P}(\mathrm{X} \leqslant x) \\ &= 2\pi^{-1} \arcsin\sqrt{x}. \end{aligned} \tag{16}$$

Also the *probability density function* of X is given by

$$\begin{aligned} f(x) &= F'(x) \\ &= \frac{1}{\pi\{x(1-x)\}^{1/2}}. \end{aligned} \tag{17}$$

This function gives the probability $f(x)\,\mathrm{d}x$ that X lies in the elemental 'interval' $[x, x + \mathrm{d}x]$, and so shows that x_n is found much more often near the ends than it is at the middle of the interval $[0, 1]$. (Note that the variable X and these functions F and f are different from others which have been denoted above by the same letters.) We see that, although a particular value x_n may depend sensitively on x_0, the *statistical* properties of the sequence $\{x_n\}$ are the same for all values of x_0 which correspond to irrational values of θ_0. The Cantor nature of the attractor for other values of a complicates the probability distribution function, because it may be discontinuous, and *a fortiori* not differentiable, at an infinity of points. Nonetheless probabilistic ideas are still useful.

The philosophical implications of these ideas deserve consideration. For a given value of x_0 the sequence $\{x_n\}$ is *in principle* determined completely and precisely by an algorithm, namely iteration of the map F. Yet in practice the sequence may be indistinguishable from a sequence of random numbers. In laboratory experiments, however careful, observations are not absolutely precise, and there is no possibility of distinguishing an irrational from some rational value of a datum. Also numerical errors, such as round-off error, are inevitable in the practice of processing the data to find x_{n+1} from x_n. So we could not predict all the future of a chaotic system, even if we had an exact model on which to base our predictions. The errors in prediction grow exponentially in time (or, rather, in n) because of our lack of exact knowledge of the present, so that doubling the accuracy of our measurements and data processing will avail only a little: we may predict with confidence only the near future of chaotic solutions and their statistical properties over a long time. It is because of this, perhaps, that such chaos is sometimes called *deterministic chaos.*

In summary of this section, we state that each ω-limit set for the logistic difference equation is one of four types. (a) It may be a finite attractor as a stable periodic solution or fixed point, for example when $a = 3.2$. (b) It may be an unstable fixed point or p-cycle, for example the two-cycle when $a = 3.5$. (c) The ω-limit set may be an infinite attractor as a Cantor set with chaos, for example when $a = 3.6$. (d) It may be an interval with no attractor, for example when $a = 4$. For a given value of a, the ω-limit sets may be of more than one type according to the value of x_0 (see Q3.18 for an example). The qualitative behaviour of the solutions of the logistic equation can be discerned to be independent of the symmetry and of the simple parabolic form of the map, because most of the arguments we have used do not depend on these special properties of the map, although the quantitative behaviour does. Therefore the ideas and qualitative results of this section are widely applicable to all one-dimensional maps of similar form.

5 Numerical and computational methods

Computers are invaluable tools with which to investigate nonlinear systems. Numerical results, especially when they are presented graphically, are useful aids to learning. Also they may suggest conjectures and stimulate new lines of mathematical enquiry, just as observations of natural phenomena have done for millennia. The conjectures may be proved mathematically afterwards. The aim of this section is to encourage readers to make their own numerical experiments and use them.

First, however, a warning should be given. Sometimes numerical experiments may suggest false results, because they usually model a mathematical ideal only approximately, and small errors may be important when a solution is unstable or chaotic. We have seen how a differential equation may be modelled by a difference equation in numerical calculations. This leads to truncation errors. Most computers are digital rather than analogue, using discrete rather than continuous variables. So no fixed- or floating-point variable, however accurate the computer is, can represent a real variable exactly. This leads to round-off errors. We have already met sensitive dependence on initial conditions, for which it may be significant whether a number is rational or irrational, and so can see an inherent limitation of computers. Indeed, when solving the logistic equation on any digital computer, there is only a finite number of discrete values to repre-

Table 3.2. *List of the BASIC program 'LOGIS' (You may find it more convenient to replace lower by upper case letters throughout.)*

```
10 INPUT "a0, astep, ia", a0, astep, ia%
11 REM Type in initial value of a, step for values of a, number of steps
20 MODE 0
30 FOR i% = 0 TO ia%
40 a = a0 + astep*i%
50 PRINT TAB(60, 5), a
60 x = .501
61 REM Take x0 = 0.501
70 FOR n% = 1 TO 192
80 x = a*x*(1.-x)
81 REM Specify the logistic map
90 IF n% < 65 THEN SOUND 1, -10, 255*x, 2 ELSE PLOT 69, i%*16, x*1000
91 REM If n < 65 play a note of frequency representing x; if n >= 65 plot
the pixel for current values of a (horizontal coordinate), x (vertical coordinate)
100 NEXT n%: NEXT i%
110 STOP
```

sent a real variable, so *every* solution of the equation calculated must be eventually periodic! Nonetheless, a solution with a very large period usually represents well many aspects of an aperiodic solution.

Next we present a computer program to give the bifurcation diagram of the logistic equation. It will give Figs. 1.18, 3.5, 3.6 and many others which you may choose for yourself. The program 'LOGIS', listed in Table 3.2, is in BBC BASIC, but you will find it easy to translate into another dialect of BASIC or into FORTRAN if you know these computer languages.

This program calculates x_n, where $x_{n+1} = ax_n(1 - x_n)$ and $x_0 \approx \frac{1}{2}$ (line 60) for various values of a. First for a = a0, the values of $x_1, x_2, \ldots, x_{64}$ are recorded (line 90) as a succession of short notes, a higher note for a higher value. This serves both to let the value of n become 'large' before results are plotted on the screen and to demonstrate some properties audibly. You can 'hear' convergence, whether it is monotonic or oscillatory, fast or slow; you can hear the difference between a two-cycle and a four-cycle; you can hear the strange 'melody' of a Cantor set. Next the 128 values of $x_{65}, x_{66}, \ldots, x_{192}$ are plotted (line 90) in a vertical line on the screen. Of course, two values may be plotted as the same point on the screen if the values are too close together to be resolved. The scale has been chosen so that $x = 0$ at the bottom of the screen and $x = 1$ near the top. The whole procedure is next repeated ia times, starting with the value a0 + astep of a, and increasing by astep each time so that the final value is a0 + astep × ia. (Note that ia is represented by the integer variable ia% in the program.) The vertical lines, each of 128 values of x_n, are plotted according to the values of a, starting with a0 at the left-hand side of the screen and moving right as a changes. Thus a is the abscissa and x is the ordinate for the points plotted. The current value of a is shown (line 50) near the top right-hand corner of the screen.

Run the program and type the values of a0, astep and ia in response to the cue (line 10). You are recommended first to try a0 = 0, astep = 0.1 and ia = 40 to plot the whole of the bifurcation diagram of Fig. 1.18. Look at the crude attempt to plot the Cantor sets. Next you might try a0 = 3.4, astep = 0.02 and ia = 30 in order to examine the period doubling and strange attractors a little more closely. You could re-plot Fig. 3.5 by taking a0 = 3.82, astep = 0.001 and ia = 30. You might also like to change line 80 of the program to examine other difference equations.

6 Some two-dimensional difference equations

There are many interesting one-dimensional difference equations for functions $F(a, x)$ other than the logistic map. A few are given in the problems at the end of this chapter. However, in this and the next section we shall discuss some two-dimensional difference equations of the form,

$$x_{n+1} = f(x_n, y_n), \qquad y_{n+1} = g(x_n, y_n) \qquad \text{for } n = 0, 1, \ldots.$$

Sometimes we write them as

$$\mathbf{x}_{n+1} = \mathbf{F}(\mathbf{x}_n) \qquad \text{for } n = 0, 1, \ldots,$$

where $\mathbf{F}: \mathbb{R}^2 \to \mathbb{R}^2$. Higher-dimensional spaces permit a greater variety of properties than one-dimensional ones do. In particular, the occurrence of strange attractors in $\mathbb{R}^2$ can be seen geometrically by use of a *horseshoe map*. Smale (1967) showed how a class of maps $\mathbf{F}: \mathrm{S} \to \mathbb{R}^2$ give attractors which are the Cartesian product of an interval and a Cantor set, where S is the unit square $[0, 1] \times [0, 1]$. The essential idea is the qualitative one of taking $\mathbf{F}$ as a map of S which is the result of first contracting S in the vertical, then stretching it in the horizontal direction, and finally folding it over back into S. Then $\mathbf{F}(\mathrm{S})$ resembles a horseshoe, as shown in Fig. 3.8. Consideration of $\bigcap_{n=0}^{\infty} \mathbf{F}^n(\mathrm{S})$ shows that it converges to the attractor of $\mathbf{F}$ for almost all initial points of S, and that the attractor is like a Cantor set in a vertical section but like an interval in a transverse section.

Example 3.13: the baker's transformation. This is a simple example of a map similar to a horseshoe, although it is a discontinuous map. Consider

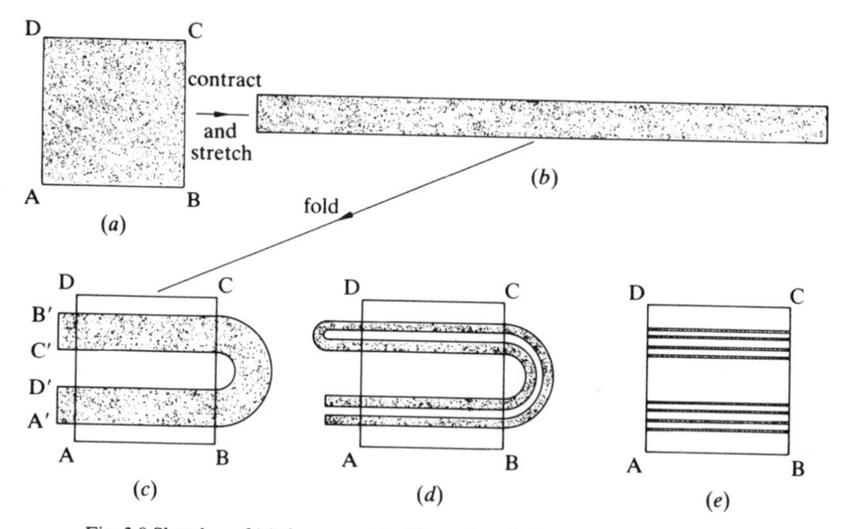

Fig. 3.8 Sketches of (*a*) the square S with vertices A, B, C and D; (*b*) contracting, stretching and folding; (*c*) the horseshoe $\mathbf{F}(S)$, where $A' = \mathbf{F}(A)$ etc.; (*d*) $\mathbf{F}^2(S)$; (*e*) $\mathbf{F}^3(S)$ in S.

the map $\mathbf{F}$ of the half-open square $[0, 1) \times [0, 1)$ onto itself where $\mathbf{F}(\mathbf{x}) = (\sigma(x), g(a, x, y))$,

$$\sigma(x) \equiv 2x \text{ modulo } 1 \qquad \text{and} \qquad 0 \leqslant \sigma(x) < 1,$$

$$g(a, x, y) \equiv \left.\begin{cases} \frac{1}{2}ay & \text{for } 0 \leqslant x \leqslant \frac{1}{2} \\ \frac{1}{2}(ay + 1) & \text{for } \frac{1}{2} \leqslant x < 1 \end{cases}\right\} \text{modulo } 1 \qquad \text{and} \qquad 0 \leqslant g < 1.$$

As in Example 3.8 of the Bernoulli shift, the sawtooth map σ maps the interval $[0, 1)$ onto itself, and leads to chaotic orbits $\sigma^n(x)$ independently of y when x is irrational.

Note that $(0, 0)$ is the only fixed point of $\mathbf{F}$, and it is unstable, as for the Bernoulli shift. There is, in fact, an infinity of unstable p-cycles.

The Jacobian matrix of $\mathbf{F}$ is $\mathbf{J} = \begin{bmatrix} 2 & 0 \\ 0 & \frac{1}{2}a \end{bmatrix}$ for all $\mathbf{x}$, so $\det \mathbf{J} = a$. Therefore $\mathbf{F}$ is area preserving if and only if $a = \pm 1$.

Consider first the case $a = 1$. Then the geometrical nature of the map is illustrated in Fig. 3.9(*a*)–(*c*), where it is seen that the map is equivalent to a horizontal stretching and vertical contraction, followed by a vertical cutting and stacking. This resembles the preparation of dough, so $\mathbf{F}$ is often called the baker's transformation.

When $0 < a < 1$, areas are contracted by $\mathbf{F}$. We show $\mathbf{F}(S)$ and $\mathbf{F}^2(S)$ in Figs. 3.9(*d*), (*e*), where S is the unit square. Iterating, we see plausibly that $\mathbf{F}^n(S)$ has 2^n slices of thickness $(\frac{1}{2}a)^n$ each, like *mille feuilles* (or filo) pastry.

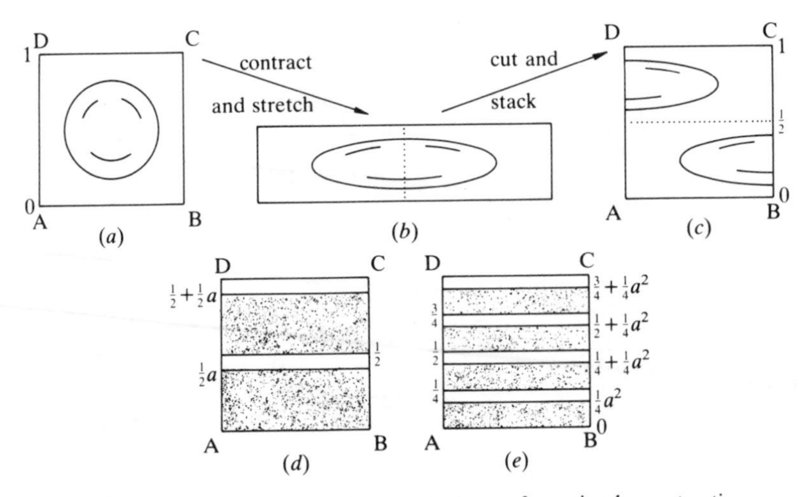

Fig. 3.9 (*a*) The unit square S; (*b*) the baker's transformation by contracting, stretching, cutting and stacking; and (*c*) $\mathbf{F}(\mathrm{S})$ for $a = 1$. (*d*) $\mathbf{F}(\mathrm{S})$; and (*e*) $\mathbf{F}^2(\mathrm{S})$ for $0 < a < 1$.

In the limit as $n \to \infty$, this gives an attractor with the topological character of the interval $[0, 1)$ in the x-direction and of a Cantor set in the y-direction.
□

Example 3.14: the Hénon map. Next we shall follow Hénon (1976) and Hénon & Pomeau (1976), who considered the pair of difference equations with

$$f(x, y) = 1 + y - ax^2, \qquad g(x, y) = bx \tag{1}$$

for given parameters a and b. This is not a horseshoe map, but it is somewhat similar geometrically, and it allows us to follow the details closely. It is easy to show that there are two fixed points $\mathbf{X} = (X, Y)$ if $a > a_0 = -\frac{1}{4}(1 - b)^2$ and $a \neq 0$, namely

$$X = [b - 1 \pm \{(1 - b)^2 + 4a\}^{1/2}]/2a, \qquad Y = bX. \tag{2}$$

We see that there is a turning point at $(a_0, -(1 - b)/2a_0)$ in the bifurcation diagram in the (a, x)-plane for fixed $b \neq 1$.

A fixed point is stable if the 2×2 Jacobian matrix,

$$\mathbf{J}(\mathbf{X}) = \begin{bmatrix} -2aX & 1 \\ b & 0 \end{bmatrix},$$

has eigenvalues, q_1 and q_2, such that $|q_1| < 1$ and $|q_2| < 1$. Now

$$q_1, q_2 = -aX \pm (a^2X^2 + b)^{1/2}.$$

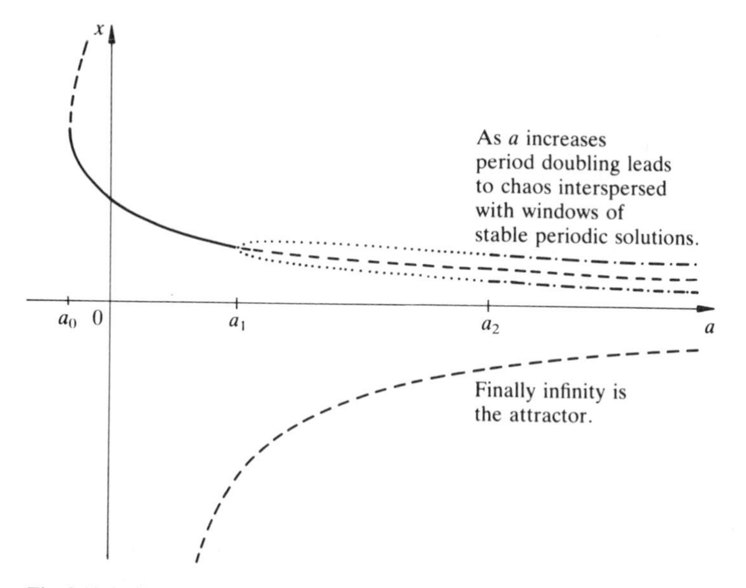

Fig. 3.10 A sketch of some of the bifurcation diagram of the Hénon map in the (a, x)-plane for $-1 < b < 1$.

It follows that the fixed point with $X = -(1-b)/2a + (a-a_0)^{1/2}/a$ is stable in the limit as $a \downarrow a_0$ when $-1 < b < 1$, but the other fixed point $X = -(1-b)/2a - (a-a_0)^{1/2}/a$ is unstable as $a \downarrow a_0$. It can be further shown that the former fixed point is stable for $a_0 < a < a_1$ and $-1 < b < 1$, where $a_1 = \frac{3}{4}(1-b)^2$, there being a flip bifurcation as a increases through a_1. The other fixed point is always unstable. As a increases beyond a_1 there are two-cycles (Q3.34), period doubling with a Feigenbaum sequence, strange attractors interspersed with windows of stable periodic solutions; and finally infinity becomes the attractor. This is illustrated in Fig. 3.10. In fact the structure of the windows of stable p-cycles amid chaos is similar to that seen in Fig. 3.6(b) for the logistic map, but is complicated by the coexistence of other attractors and their 'collision' as a increases for fixed b.

The Jacobian is $\det \mathbf{J}(\mathbf{x}) = -b$ for all $\mathbf{x}$. Therefore the map is a uniform contraction for $-1 < b < 1$, area-preserving for $b = \pm 1$, and a uniform expansion for $|b| > 1$.

Hénon & Pomeau (1976) interpreted the map geometrically as the product of successively a folding, a contraction (when $-1 < b < 1$), and a reflection in the line $y = x$, and showed that the map is a canonical form to which any quadratic map with a constant Jacobian may be reduced by a linear transformation. If $a < a_0$ or if a is sufficiently large then $x_n \to \infty$ as $n \to \infty$. Otherwise $x_n \to \infty$ or A according to the value of x_0, where A is an attractor. (Remember that there may be more than one attractor for the same pair of values of a, b.) Hénon & Pomeau showed that, for some values of a and b, a certain domain of the (a, x)-plane is mapped into itself and therefore that the points of the attractor are bounded within the domian, although they may be uncountable. Computations indicate that it is a strange attractor. A characteristic of strange attractors in dimensions higher than one is that the average, over an attractor, of the modulus of the Jacobian is less than one and the average of the modulus of at least one eigenvalue is greater than one.

The numerical results of Hénon & Pomeau are very revealing. They worked mostly with $a = 1.4$ and $b = 0.3$, for which there is a strange attractor. They took $\mathbf{x}_0$ near the unstable fixed point (0.63135448..., 0.18940634...), though this initial point is little better than many others, and plotted several points $\mathbf{x}_n$, as illustrated in Fig. 3.11(a). The other fixed point (the one which would be unstable for all b) is (−1.13135..., −0.339406...), a short distance from the attractor. The point $\mathbf{x}_n$ wanders over the attractor in an apparently random fashion. They next enlarged a region near the first fixed point and plotted those of a large number of calculated points $\mathbf{x}_n$ which happen to lie in the region, as shown in Fig. 3.11(b). They then repeated this process, enlarging a region within the first region and plotting those of a very large number of calculated points which lie in the new region, as shown in Fig. 3.11(c). They repeated this process once more, calculating even more points, as shown in Fig. 3.11(d). Note the self-similar pattern which they discovered: the successive enlargements look like one another, and it appears that the process of enlargement may be continued *ad infinitum* (as for Cantor's middle-thirds set, which is discussed in §4.1). The set appears to have a similar structure on all small length scales. Also the set is dense like a curve along one of the 'braids', but is like a Cantor set in a transverse direction, so that the attractor is the Cartesian product of a line and a Cantor set.

You can prepare these and other plots for yourself by use of the simple program 'HENON' listed in Table 3.3. It is in BBC BASIC, but may be translated easily. First the parameters are read (line 10). The value of SCALE is to be chosen to determine the magnification of the plot. A fixed point is calculated in line 40. The initial values, $x_0 = 0$ and $y_0 = 0$, are assigned in line 50. The first 300 points x_n are computed but not plotted in order that the initial values may be 'forgotten'. Thereafter the points (x_n, y_n) are plotted on the video screen, with the x-axis horizontal, the y-axis vertical, and the fixed point at the middle of the screen. As SCALE is increased both the number of points computed and the magnification are increased.

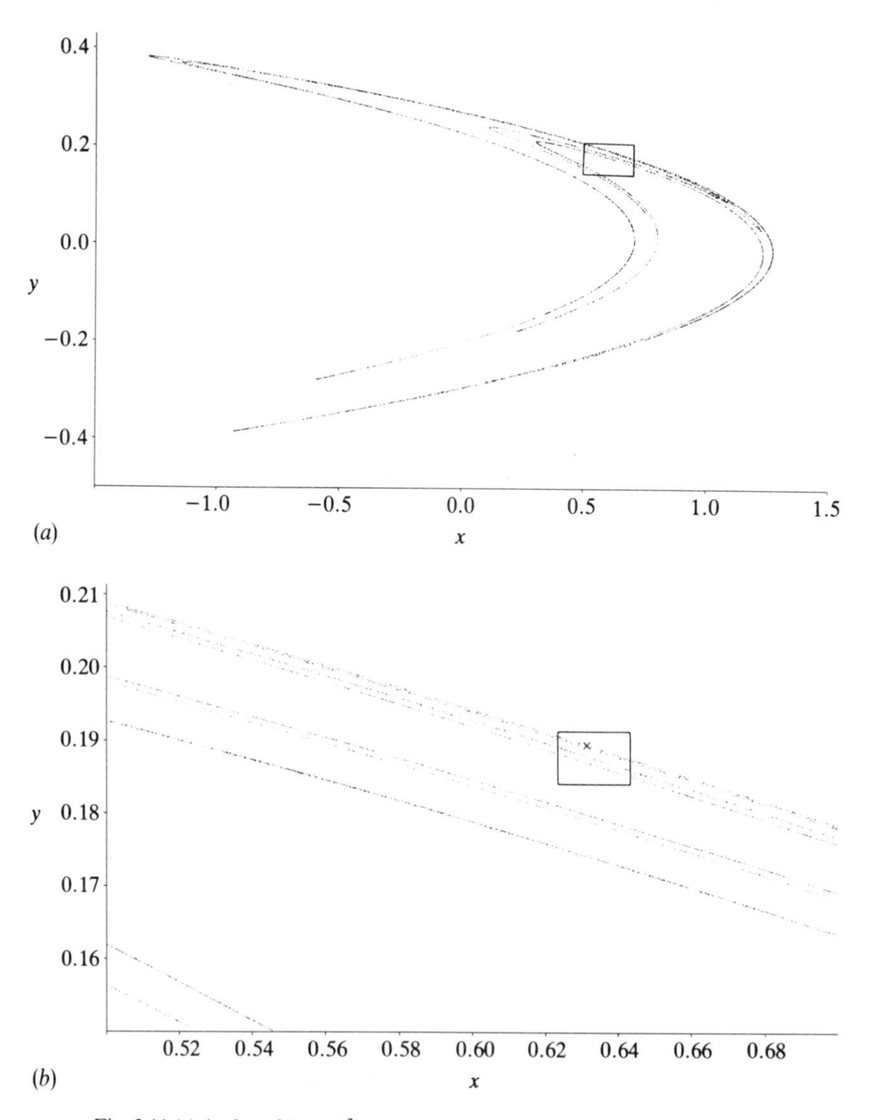

Fig. 3.11 (*a*) A plot of 4×10^3 successive points $\mathbf{x}_n$ of the Hénon map for $a = 1.4$, $b = 0.3$, with $x_0 = 0.63135448$, $y_0 = 0.18940634$. (*b*) Enlargement of the rectangle shown in (*a*) when 7.5×10^4 points are computed. The cross denotes a fixed point. (*c*) Enlargement of the rectangle shown in (*b*) when 1.5×10^6 points are computed. (*d*) Enlargement of the rectangle shown in (*c*) when 10^7 points are computed.

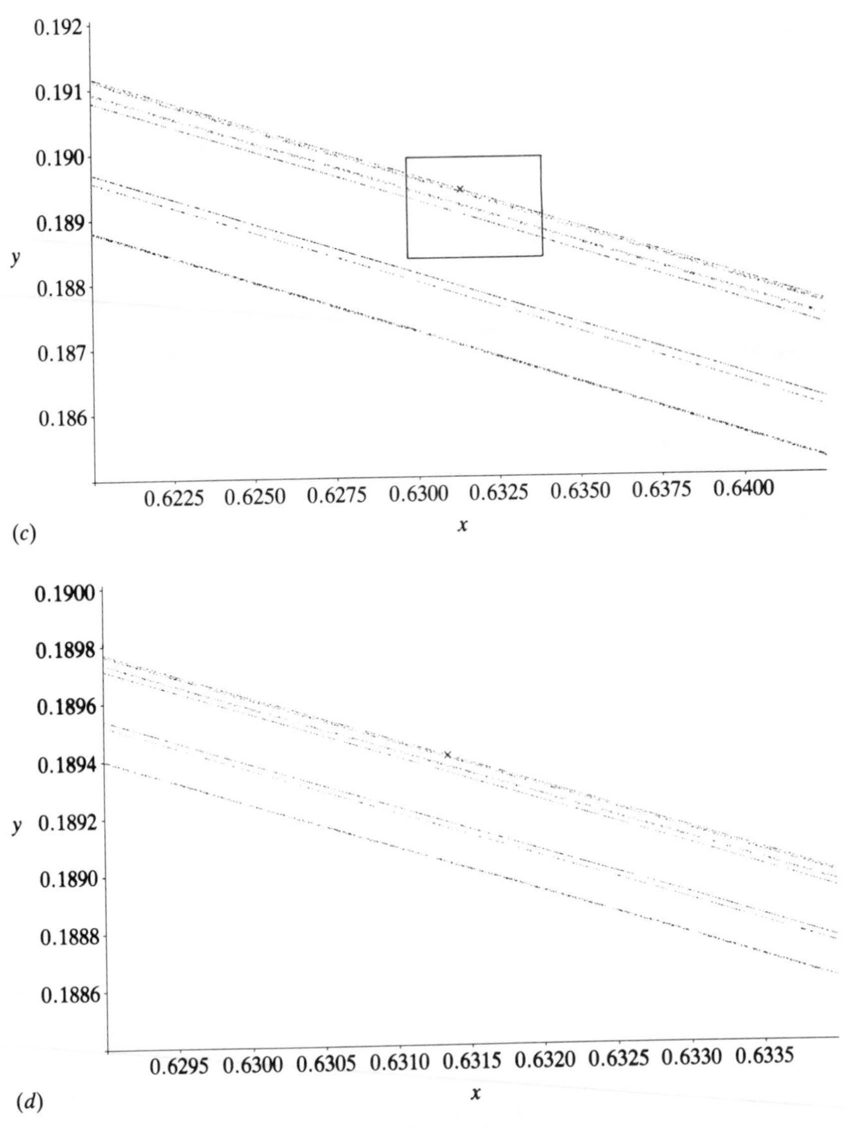

Fig. 3.11 (*cont.*)

You might start with a = 1.4, b = 0.3 and SCALE = 200. Then, in following runs, fix these values of a and b but increase SCALE bit by bit to 10 00000, say. Again, you might fix b = 0.3 and SCALE = 200, but decrease a bit by bit from 1.4. You might like to experiment for yourself afterwards. In particular, you might like to change lines 80 and 90 in order to study a different two-dimensional difference equation. □

Table 3.3. *List of the BASIC program 'HENON' (You may find it more convenient to use capital letters throughout.)*

```
 10 INPUT "a, b, SCALE", a, b, SCALE
 11 REM Type in values of a, b and the scale of the phase plane to be plotted on
line 120
 20 MODE 0
 30 C = b - 1.
 40 X = .5*(C + SQR(C*C + 4.*a))/a: Y = b*X
 41 REM (X, Y) is an unstable fixed point to centre plot
 50 xn = 0: yn = 0
 60 NN% = 300 + SCALE*2.
 70 FOR n% = 1 TO NN%
 80 xn1 = xn
 90 xn = yn + 1. -a*xn*xn: yn = b*xn1
 91 REM The Henon map
 100 IF n% < 301 GOTO 130
 110 U = (xn - X)*SCALE + 512: V = (yn - Y)*SCALE + 512
 111 REM The horizontal and vertical coordinates of the pixel to be plotted to
represent (xn, yn) in the (x, y)-plane
 120 PLOT 69, U, V
 130 NEXT n%
 140 PRINT a, b, xn, yn, NN%
 150 STOP
```

7 Iterated maps of the complex plane

Iterated analytic maps of the complex plane into itself are a special and important class of two-dimensional difference equations. They have been studied for over a century. The difference equation

$$z_{n+1} = F(z_n) \qquad \text{for } n = 0, 1, \ldots,$$

where $F: \mathbb{C} \to \mathbb{C}$ is analytic, represents the iteration of the map F of the complex plane into itself. We may write $z_n = x_n + iy_n$ and take the real and imaginary parts of the complex difference equation to get a real pair of difference equations for x_n and y_n. In considering fixed points and their stability, it is, however, more convenient to use the fact that F is an analytic function of the complex variable z. Thus Z is defined as a fixed point of F if it is a zero of $F(Z) - Z$, and it can be shown, by the argument of §1.8, to be stable if $|F'(Z)| < 1$ and to be unstable if $|F'(Z)| > 1$.

Example 3.15: the logistic map. Iterate the map

$$F(a, z) = az(1 - z),$$

where $F: \mathbb{C} \times \mathbb{C} \to \mathbb{C}$. The fixed points 0 and $(a-1)/a$ have the same forms as for the real case of §3, and they are stable if $|a| < 1$ and $|2-a| < 1$ respectively. The periodic solutions likewise have the same forms as for the real logistic equation, but here they exist for all complex $a \neq 0$ because we have removed the restriction that all the solutions are real. For some purposes it is more convenient to transform z and a so that we replace the logistic difference equation by

$$z_{n+1} = G(b, z_n),$$

where $G(b, z) = z^2 - b$ and $b = \frac{1}{4}a^2 - \frac{1}{2}a$ (Q1.31). It is found that G may map certain sets into themselves for a given value of b, i.e. that G has invariant sets. These sets may be finite, when composed of fixed points and periodic solutions, or infinite. The infinite sets may be curves or have a more complicated structure. □

Example 3.16: the square map. In the special case $b = 0$, we find $z_{n+1} = G(z_n)$, where $G(z) = z^2$, i.e.

$$z_{n+1} = z_n^2 \qquad \text{for } n = 0, 1, \ldots.$$

It can be seen that the fixed points are $Z = 0, 1, \infty$, and that

$$z_n = z_{n-1}^2 = z_{n-2}^4 = z_{n-3}^8 = \cdots = z_0^{2^n}.$$

Therefore $z_n \to 0$ as $n \to \infty$ if $|z_0| < 1$, $z_n \to \infty$ if $|z_0| > 1$, and $z_n \in \mathrm{S}^1$ if $z_0 \in \mathrm{S}^1$, where S^1 is the unit circle with equation $|z| = 1$. Thus the origin and infinity are stable fixed points with domains of attraction $\mathrm{D}(0) = \{z: |z| < 1\}$ and $\mathrm{D}(\infty) = \{z: |z| > 1\}$, and $\partial\mathrm{D}(0) = \mathrm{S}^1 = \partial\mathrm{D}(\infty)$. Also 1 is an unstable fixed point. Further, $G(\mathrm{S}^1) = \mathrm{S}^1 = G^{-1}(\mathrm{S}^1)$, i.e. all the images and pre-images of all the points of S^1 lie in S^1. If $z_0 = \exp(2\pi i\theta_0)$ then $z_n = \exp(2\pi i\theta_n) = \exp(2^n\pi i\theta_0)$, because θ_n is mapped by the Bernoulli shift (of Example 3.8). Therefore there are unstable p-cycles of all periods p in S^1. In addition there are aperiodic solutions dense in S^1. □

Let $R: \mathbb{C} \to \mathbb{C}$ be a quadratic map with two attractors, A_1 and A_2, say, each a finite set; then $\partial\mathrm{D}(\mathrm{A}_2) = \partial\mathrm{D}(\mathrm{A}_1)$, and we call $\partial\mathrm{D}(\mathrm{A}_1)$ the *Julia set* J of R. Thus, if $z \in \mathrm{J}$ then $R^n(z)$ does not tend to a limit point unless z is an unstable fixed point of R or a pre-image of the fixed point. The notion of a Julia set can be generalized for all rational maps R. It can be shown that (a) $R(\mathrm{J}) = \mathrm{J} = R^{-1}(\mathrm{J})$, i.e. the set of images and pre-images of all points of J lie in J; (b) $\mathrm{J} \neq \varnothing$, i.e. J is not empty, and J is closed; (c) unstable p-cycles are dense in J; and (d) J contains no point which belongs to an attractor.

Example 3.17: some Julia sets. The following two examples have been chosen for their simplicity to illustrate the fundamental ideas, but they are atypical; Julia sets usually have a complicated structure. (i) Suppose that $R(z) = \frac{1}{2}(z + 1/z)$. Then $R(Z) = Z$ gives the fixed points $Z = \pm 1$ (or ∞). Now $R'(z) = \frac{1}{2}(1 - 1/z^2)$. Therefore $R'(\pm 1) = 0$. Therefore $A_1 = \{1\}$ and $A_2 = \{-1\}$ are attractors. It can be shown that there are no others, because $D(1) = \{z: \text{Re}(z) > 0\}$ and $D(-1) = \{z: \text{Re}(z) < 0\}$. Then J is the imaginary axis. The four properties of J listed above may be verified for this example. (ii) Take $R(z) = z^2$. Then the fixed points are $Z = 0, 1$. Now $R'(z) = 2z$, so that $Z = 0$ is stable and $Z = 1$ unstable. Also infinity is a stable fixed point. So we may take $A_1 = \{0\}$ and $A_2 = \{\infty\}$. Now $D(0) = \{z: |z| < 1\}$ and $D(\infty) = \{z: |z| > 1\}$, so $J = S^1$, where S^1 is the set of points $\{z: |z| = 1\}$, i.e. the unit circle with centre 0. If $R(z) = z^2 + a$ for $a \neq 0$, then J has a complicated structure, but is close to the circle S^1 when a is small. In fact part (i) is equivalent to (ii) in the sense that if we define G by $w = (z - 1)/(z + 1)$ and $G(w) = \{F(z) - 1\}/\{F(z) + 1\}$, where $F(z) = \frac{1}{2}(z + 1/z)$, then $G(w) = w^2$. □

Consider next the map $F: \mathbb{C} \times \mathbb{C} \to \mathbb{C}$ defined by $F(a, z) = z^2 - a$. Then $F_z(a, 0) = 0$ and there is an attractor of $F(a, z)$ at $z = \infty$ for all $a \in \mathbb{C}$. Accordingly, the *Mandelbrot set* of F is defined as

$$M = \{a: a \in \mathbb{C}, \lim_{n \to \infty} F^n(a, 0) \not\to \infty\}.$$

Thus M is the set of values of the complex parameter a for which the given initial point $z = 0$ does *not* lie in the domain of attraction of infinity. (The initial point $z = 0$ is chosen because $F_z(a, 0) = 0$.) Mandelbrot (1982, pp. 188–9) shows computer-drawn diagrams of M for $F(a, z) = z^2 - a$ and initial point $z = 0$, and for $F(a, z) = az(1 - z)$ and $z = \frac{1}{2}$. The boundaries ∂M in these two cases appear to be complicated sets with a self-similar character (see §4.1).

We shall end by mentioning an important theorem. Julia (1918) and Fatou (1919) showed that the Julia set J of $F(a, z)$ is connected if and only if $\lim_{n \to \infty} F^n(a, 0) \not\to \infty$, i.e. if and only if a belongs to the Mandelbrot set of $F(a, z)$ for $z = 0$. So if a belongs to the Mandelbrot set then J, although a complicated set with many apparently isolated parts, is in fact joined together. Look for this in Fig. 3.12. Also see the Mandelbrot set in Fig. 3.13, and the beautiful coloured pictures widely available in the literature. Note the finer scales appearing as the picture is enlarged; there is, in fact, an infinite regression of finer and finer scales. The set itself is shown in black; an explanation of the significance of the colouring is indicated below.

Fig. 3.12 Tableau of the Julia set of $F(a, z) = z^2 + a$, in the rectangle $-0.3194417 \leqslant \operatorname{Re} z \leqslant -0.3193553$, $-0.4452514 \leqslant \operatorname{Im} z \leqslant -0.4451650$ for $a = -0.2232 - 0.7296i$. The alternation of black and white denotes different rates of divergence to infinity, but many different rates of divergence are denoted by black and many other rates by white; so the set itself is *not* any black or white region.

Perhaps the best way to enjoy the pictures aesthetically as well as to understand them is to make your own.

Tables 3.4 and 3.5 list some programs in BBC BASIC to compute Julia and Mandelbrot sets. They have been written to be simple and effective for many purposes rather than fast or good for all purposes. The program 'JULIA' gives a plot in colour of the rates of divergence rather than the Julia sets themselves for the iterated map $F(a, z) = z^2 + a$. To run the pro-

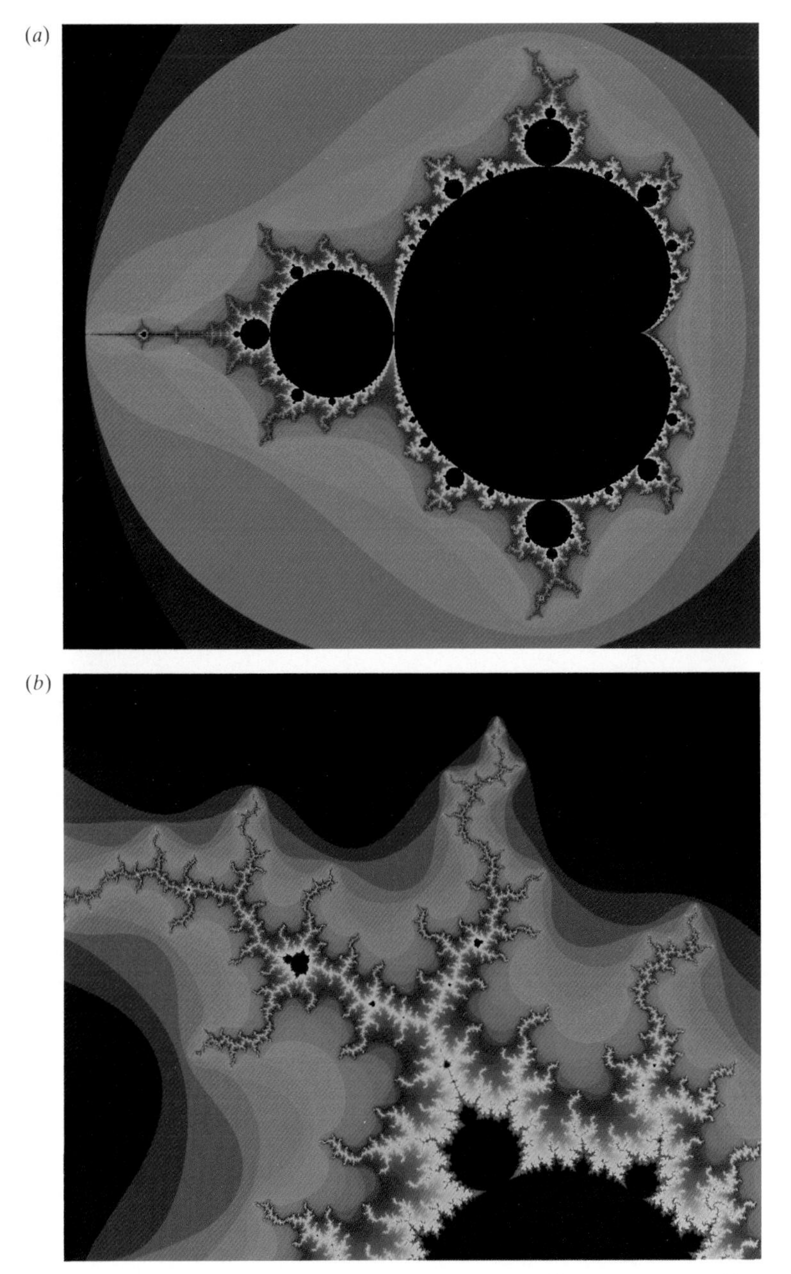

Fig. 3.13 Parts of the Mandelbrot set for $F(a, z) = z^2 + a$. The set itself is coloured black; the meaning of the other colours is indicated in the program 'MANDEL' of Table 3.5. (*a*) In the rectangle, $-2.25 \leqslant \mathrm{Re}a \leqslant 2.25$, $-1.25 \leqslant \mathrm{Im}a \leqslant 1.25$. (*b*) In the rectangle, $-1.23 \leqslant \mathrm{Re}a \leqslant -1.1$, $0.25 \leqslant \mathrm{Im}a \leqslant 0.358$.

Table 3.4. *List of the BASIC program 'JULIA'*

```
10 XS% = 80: YS% = 128: NIT% = 200: M = 4
11 REM These specify the numbers of steps for x, y; the maximum number of
iterations; and the effective size of infinity
20 INPUT "AR, AI", AR, AI
21 REM These are the real and imaginary parts of a
30 INPUT "XMIN, XMAX, YMIN, YMAX", XN, XX, YN, YX
31 REM These specify the rectangle of the complex plane for z = x + iy
40 MODE 0
50 GAPX = (XX - XN)/XS%: GAPY = (YX - YN)/YS%
51 REM These evaluate the steps for x and y
60 FOR NY% = 0 TO YS% - 1
70 FOR NX% = 0 TO XS% - 1
80 X = XN + GAPX*NX%: Y = YN + GAPY*NY%: COUN% = 0
81 REM This specifies the coordinates of the pixel (in the complex z-plane) whose
colour must be found next
90 COUN% = COUN% + 1
100 X2 = X*X: Y2 = Y*Y
110 Y = 2.*X*Y + AI: X = X2 - Y2 + AR
111 REM Lines 100, 110 replace z = x + iy by z^2 + a
120 IF X2 + Y2 < M AND COUN% < NIT% THEN GOTO 90
121 REM X2 + Y2 = |z|^2
130 C% = 7*COUN%/NIT%
140 GCOL 0, 7 - C%
141 REM Lines 120-140 determine crudely the colour of the pixel according to
how many iterations COUN% it has taken for |z|^2 to exceed M, a rough estimate
of infinity.
150 PLOT 69, NX%*8, NY%*4
160 NEXT NX%: NEXT NY%
170 STOP
```

gram you need to answer the cue (line 20) by typing in the real and the imaginary parts of a to specify the set. The next four numbers specify the Cartesian coordinates of the four vertices of the rectangle in the complex z-plane in which the program will represent the Julia set. You might try, for your first run, to type in 0.32, 0.043 and thereafter $-2.$, $2.$, -1.5, 1.5. For your second run try typing in -0.12375, 0.56508 and thereafter $-2.$, $2.$, -1.5, 1.5. Then you might care to experiment for yourself.

Beware of the slowness of the program. The program may take a long time to run completely, although it has been written to fill only a quarter of the screen in order to reduce the running time. You may alter the size of the picture or change the resolution according to whether you prefer a good big picture or a rapid program. You may be able to improve the colour

Table 3.5. *List of the BASIC program 'MANDEL'*

```
10 RS% = 80: IS% = 128: NIT% = 100: M = 4
11 REM These specify the numbers of steps for ar, ai; the maximum number of
iterations; and the effective size of infinity
20 INPUT "ARMIN, ARMAX, AIMIN, AIMAX", ARN, ARX, AIN, AIX
21 REM These specify the rectangle of the complex plane for a = ar + iai
30 MODE 0
40 GAPR = (ARX - ARN)/RS%: GAPI = (AIX - AIN)/IS%
41 REM These evaluate the steps for ar and ai
50 FOR NI% = 0 TO IS% - 1
60 AI = AIN + GAPI*NI%
70 FOR NR% = 0 TO RS% - 1
80 AR = ARN + GAPR*NR%
81 REM This specifies the coordinates of the pixel (in the complex a-plane) whose
colour must be found next
90 X = 0: Y = 0: COUN% = 0
100 COUN% = COUN% + 1
110 X2 = X*X: Y2 = Y*Y
120 Y = 2.*X*Y + AI: X = X2 - Y2 + AR
121 REM Lines 110, 120 replace z = x + iy by z^2 + a
130 IF X2 + Y2 < M AND COUN% < NIT% THEN GOTO 100
131 REM X2 + Y2 = |z|^2
140 C% = COUN%/10
150 IF C% = 5 THEN C% = 4
160 IF C% = 6 OR C% = 7 THEN C% = 5
170 IF C% = 8 OR C% = 9 THEN C% = 6
180 IF C% > 9 THEN C% = 7
190 GCOL 0, 7 - C%
191 REM Lines 140-190 determine the pixel colour from the value of COUN%
200 PLOT 69, NR%*8, NI%*4
210 NEXT NR%: NEXT NI%
220 STOP
```

scheme, which has been chosen more for simplicity of the program than for aesthetics. You may use the program for different maps by changing lines 100–120.

The program 'MANDEL' computes and plots the Mandelbrot set for the map $F(a, z) = z^2 + a$. To run the program you need to answer the cue (line 20) by typing in four numbers to specify the Cartesian coordinates of the four vertices of the rectangle in the complex a-plane in which the program will plot the Mandelbrot set. You might try the input -2.25, 2.25, -1.25, 1.25 for a start, and for your next run -1.23, -1.1, 0.25, 0.358 to emulate Fig. 3.13.

Again, you may wish to improve the program. In particular you may plot sets for other maps by changing lines 110, 120.

Further reading

§3.1 Devaney (1989) gives an excellent account of difference equations.

§3.3 The properties of the sawtooth map may be found from the properties, largely due to Borel, of the binary representation of numbers. These are described by Hardy & Wright (1979, Chap. 9), who, in particular, prove the statistical properties used in §4.

§3.4 A fuller account of the iteration of one-dimensional maps is the rigorous treatment by Collet & Eckmann (1980).

§3.7 The literature of iterated maps of the complex plane includes classic papers by Cayley (1879), Fatou (1906, 1919) and Julia (1918) as well as recent analytical and numerical work. Accounts by Mandelbrot (1982) and Falconer (1990) are recommended in order to learn more than is in the short account here. Some beautiful coloured pictures of Julia and Mandelbrot sets are shown by Peitgen & Richter (1986) and Peitgen *et al.* (V1990), and some more-advanced computer programs to make your own pictures of Julia and Mandelbrot sets are given by Peitgen & Saupe (1988).

Problems

Q3.1 *Instability of a fixed point.* Suppose that $\mathbf{F}(\mathbf{X}) = \mathbf{X}$, where $\mathbf{F}: \mathbb{R}^m \to \mathbb{R}^m$. Then show that $\mathbf{X}$ is unstable if and only if $\exists\, \epsilon > 0$ such that $\forall\, \delta > 0\ \exists\, n > 0$ and $\mathbf{x}_0 \in \mathbb{R}^m$ such that $|\mathbf{x}_0 - \mathbf{X}| < \delta$ and $|\mathbf{F}^n(\mathbf{x}_0) - \mathbf{X}| \geqslant \epsilon$.

Q3.2 *An unstable fixed point which might seem to be asymptotically stable.* With plane polar coordinates (r, θ) such that $x = r\cos\theta$, $y = r\sin\theta$, $r \geqslant 0$, and $-\pi < \theta \leqslant \pi$, a map $\mathbf{F}: \mathbb{R}^2 \to \mathbb{R}^2$ is defined by

$$\mathbf{F}(r, \theta) = (r \sin\tfrac{1}{2}\theta / \sin\theta, \tfrac{1}{2}\theta) \qquad \text{if } \theta \neq 0, \pi,$$

$$\mathbf{F}(r, 0) = (\tfrac{1}{2}r, 0)$$

$$\mathbf{F}(r, \pi) = \left\{\begin{matrix} (2r, \pi) & \text{if } r < 1 \\ (2r, 0) & \text{if } r \geqslant 1 \end{matrix}\right\}.$$

Show that if $\theta_0 \neq 0$, π then $\mathbf{x}_n = \mathbf{F}^n(r_0, \theta_0)$ lies on the circle with centre $x = 0$, $y = r_0/2\sin\theta_0$ and radius $r_0/2\sin\theta_0$ for all n. Deduce that $\mathbf{x}_n \to \mathbf{0}$ as $n \to \infty$ for all θ_0. Show that nonetheless there are points $\mathbf{x}_0$ near $\mathbf{0}$ such that $|\mathbf{x}_n|$ may be arbitrarily large (when θ_0 is close enough to π and $n = 1$). Deduce that $\mathbf{0}$ is an unstable fixed point of $\mathbf{F}$ *in the sense of Liapounov.*

Q3.3 *Fréchet derivatives.* If V is a normed linear vector space, T: V → V is a nonlinear operator, $\mathbf{X} \in$ V, and there exists a linear operator J: V → V such that

$$\mathrm{T}(\mathbf{X} + \epsilon \mathbf{x}) = \mathrm{T}(\mathbf{X}) + \epsilon \mathrm{J}(\mathbf{x}) + o(\epsilon) \qquad \text{as } \epsilon \to 0$$

for all $\mathbf{x} \in$ V then J is called the *Fréchet derivative of* T *at* $\mathbf{X}$.

(i) Show that if $F: \mathbb{R} \to \mathbb{R}$ is a differentiable function then the Fréchet derivative of F at $X \in \mathbb{R}$ is the multiplicative function $J: \mathbb{R} \to \mathbb{R}$ defined by $J(x) = F'(X)x \ \forall \ x \in \mathbb{R}$. (So the Fréchet derivative is a generalization of the ordinary derivative.)

(ii) Show that if $\mathbf{F}: \mathbb{R}^m \to \mathbb{R}^m$ is differentiable at $\mathbf{X} \in \mathbb{R}^m$ then J is the Jacobian matrix of $\mathbf{F}$ at $\mathbf{X}$.

(iii) Show that if $T(x) = \sin\{x'(0)\}$ for all $x \in \mathrm{C}^1[0, 1]$ then J at X is given by $J(x) = \cos\{X'(0)\}x'(0)$.

(iv) Show that if $T(x) = \int_0^1 \{t^3 x^3(t) - x^2(t)x'(t)\}\,dt$ for all $x \in \mathrm{C}^1[0, 1]$ then J at $X \in \mathrm{C}^1[0, 1]$ is given by $J(x) = 3\int_0^1 t^3 X^2(t)x(t)\,dt - [X^2(t)x(t)]_0^1$.

*(v) Use Taylor's theorem to show that if T: $\mathrm{C}(-\infty, \infty) \to \mathrm{C}(-\infty, \infty)$ is defined by

$$\mathrm{T}(x(t)) = -\alpha x(x(-t/\alpha))$$

for all $x \in \mathrm{C}(-\infty, \infty)$, where α is a given positive constant, then J at X is defined by

$$\mathrm{J}(y(t)) = -\alpha X'(X(-t/\alpha))y(-t/\alpha) - \alpha y(X(-t/\alpha))$$

for all $y \in \mathrm{C}(-\infty, \infty)$.

Q3.4 *A linear difference equation whose matrix does not have a complete set of eigenvectors.* Show that the real 2 × 2 matrix $\mathbf{A} = \begin{bmatrix} 0 & 1 \\ -a^2 & 2a \end{bmatrix}$ has a double eigenvalue $q = a$. Show further that if $a \neq 0$ then the eigenvalue belongs to a unique (except for a multiplicative constant) eigenvector $[1, a]^{\mathrm{T}}$, and there exist real ξ_1, ξ_2 such that

$$\mathbf{x}_0 = \xi_1 \begin{bmatrix} 1 \\ a \end{bmatrix} + \xi_2 \begin{bmatrix} 1 \\ 0 \end{bmatrix}$$

for all $\mathbf{x}_0 \in \mathbb{R}^2$.

Deduce that if $\mathbf{x}_{n+1} = \mathbf{A}\mathbf{x}_n$ for $n = 0, 1, \ldots$, and $\mathbf{x}_0$ is as above, then

$$\mathbf{x}_n = (\xi_1 - n\xi_2)a^n \begin{bmatrix} 1 \\ a \end{bmatrix} + \xi_2 a^n \begin{bmatrix} 1 \\ 0 \end{bmatrix}.$$

Q3.5 *Stability of a periodic solution.* Show that the p-cycle $\{X_1, X_2, \ldots, X_p\}$ of the continuously differentiable map $F: \mathbb{R} \to \mathbb{R}$ is stable if $|F'(X_1)F'(X_2)\ldots F'(X_p)| < 1$.

Q3.6 *A difference equation with a supercritical pitchfork bifurcation and flip bifurcations.* Suppose that $x_{n+1} = F(a, x_n)$ for $n = 0, 1, \ldots$, where $F(a, x) = ax - x^3$. Find the fixed points and for what values of a each point exists. Find for what values of a each is linearly stable.

Show that if $X_1 = F(a, X_2)$, $X_2 = F(a, X_1)$ then X_1, X_2 are roots of the equation

$$X(X^2 - a + 1)(X^2 - a - 1)(X^4 - aX^2 + 1) = 0.$$

Hence or otherwise find all the two-cycles of F. For what values of a is each cycle linearly stable?

Sketch the bifurcation diagram in the (a, x)-plane, indicating clearly the fixed points and the two-cycles.

Q3.7 *A difference equation with a subcritical pitchfork bifurcation and flip bifurcations.* Suppose that $x_{n+1} = F(a, x_n)$ for $n = 0, 1, \ldots$, where $F(a, x) = ax + x^3$. Find the fixed points and for what values of a each point exists. Find for what values of a each is linearly stable.

Show that if $X_1 = F(a, X_2)$, $X_2 = F(a, X_1)$ then X_1, X_2 are roots of the equation

$$X(X^2 + a + 1)(X^2 + a - 1)(X^4 + aX^2 + 1) = 0.$$

Hence or otherwise find all the two-cycles of F. For what values of a is each cycle linearly stable?

Sketch the bifurcation diagram in the (a, x)-plane, indicating clearly the fixed points and the two-cycles.

Q3.8 *Monotonic maps and their p-cycles.* Consider differentiable $F: \mathbb{R} \to \mathbb{R}$. (a) Show that if $F'(x) > 0$ for all x then F has no p-cycle for $p \geqslant 2$. (b) Show that if $F'(x) < 0$ for all x then F has a unique fixed point, and F has no p-cycle for $p \geqslant 3$.

[Hint for the last part of (b): consider the sign of the derivative of the qth-generation map. Devaney (1989, p. 59).]

Q3.9 *Cluster points of the sawtooth map.* Show that if $\sigma(\theta)$ is defined as the fractional part of 2θ and $\theta_0 = \sum_{j=1}^{\infty} 2^{-j(j+1)}$, then θ_0 is irrational and the cluster points of the sequence $\{\sigma^n(\theta_0)\}$ are 2^{-r} for $r = 1, 2, \ldots$.

Q3.10 *A pitchfork bifurcation.* Given that $F(a, x) = x[a/\{x^2 + (a - x^2)e^{-4\pi a}\}]^{1/2}$, as in Example 1.3, show that if $a < 0$ then F has a stable fixed point $X = 0$ and if $a > 0$ then F has an unstable fixed point $X = 0$ and two stable fixed points $X = \pm\sqrt{a}$. Find $F^n(a, x)$. What are the domains of attraction of the stable points?

Q3.11 *Shuffling cards.* A pack, i.e. a deck, of $2s$ playing cards is given a riffle shuffle by cutting the pack into two halves and then interleaving the halves, so that the bottom card returns to the bottom. Denoting the cards of the pack by the numbers 0, 1, ..., $2s - 1$, where 0 represents the bottom card, 1 the next card above, ..., and $2s - 1$ the top card before the shuffle; and denoting by $F(r)$ the position, above the bottom after the shuffle, of the rth card before the shuffle; show, by a plausible interpretation of the shuffle, that

$$F(r) \equiv 2r \text{ modulo } 2s - 1, \quad 0 \leqslant F(r) < 2s - 1 \quad \text{for } 0 \leqslant r < 2s - 1,$$

and $F(2s - 1) = 2s - 1$.

Deduce that if a pack of 52 cards is riffle shuffled eight times then the pack ends up in the same order as it started.

Q3.12 *Reflection of a light ray by the interior of a cylinder.* Consider the path of a light ray which is reflected repeatedly by a circular cylinder whose interior surface is silvered. Suppose that the path lies in a plane perpendicular to the axis of the cylinder, and take polar coordinates such that the ray's nth reflection is at a point with polar angle $\theta = \theta_n$, for $n = 0, 1, \ldots$.

Show that if the angle of reflection is α at first then

$$\theta_{n+1} = \theta_n + \pi - 2\alpha.$$

Deduce that the path of the ray is closed if and only if α/π is rational.

Q3.13 *Twist maps.* The map $\mathbf{F}: \mathbb{R}^2 \to \mathbb{R}^2$ is defined by $\mathbf{F}(x, y) = (f(x, y), g(x, y))$,

$$f(x, y) = x \cos \psi - y \sin \psi, \qquad g(x, y) = x \sin \psi + y \cos \psi,$$

where ψ is a differentiable function of $(x^2 + y^2)^{1/2}$.

Show that the map is area preserving.

Express the map as a transformation of the polar coordinates r and θ, where $x = r \cos \theta$, $y = r \sin \theta$, and show that a circle with centre at O is invariant under the map.

Describe geometrically the iterations of the map when ψ is a constant, according to the various values of that constant.

[Moser (1973) showed the importance of perturbations of such maps.]

Q3.14 *A logarithmic map.* Consider the difference equation $x_{n+1} = af(x_n)$ for $n = 0, 1, \ldots$, where $f(x) = \ln x$ and $a > 0$. Show, graphically or otherwise, that there are 2, 1 or 0 fixed points according as $a \gtreqless \mathrm{e}$ respectively. Suppose that $a > \mathrm{e}$ and let the fixed points be X_1 and $X_2 > X_1$; then find which is stable and find the domain of attraction of each stable fixed point.

Q3.15 *Stability of fixed points of difference equations which represent equilibrium points of an ordinary differential equation.* First just read this paragraph. There are several methods to find the equilibrium points of an ordinary differential equation. The Newton–Raphson method to find an equilibrium point of the equation

$$\frac{\mathrm{d}x}{\mathrm{d}t} = f(x)$$

gives the difference equation

$$x_{n+1} = N(x_n) \qquad \text{for } n = 0, 1, \ldots,$$

where $N(x) = x - f(x)/f'(x)$. Alternatively, although the equilibrium points are just the zeros of f, we may take forward finite differences with a chosen small positive time step h and approximate $\mathrm{d}x/\mathrm{d}t$ in the differential equation by $(x_{n+1} - x_n)/h$ to give

$$x_{n+1} = F(x_n),$$

where $F(x) = x + hf(x)$. This may give a *stable* equilibrium point by taking $\lim_{n\to\infty} x_n$, or an iterative approximation thereto.

Now find N and F for the equation

$$\frac{\mathrm{d}x}{\mathrm{d}t} = a - x^2$$

by these two methods. Show that the steady solutions given by $X = \pm\sqrt{a}$ for $a \geqslant 0$ are fixed points of the appropriate two difference equations, and find whether they are stable fixed points. What are the domains of attraction of the stable fixed points for each of the two difference equations?

Q3.16 *A qualitative difference between the set of solutions of an ordinary differential equation and of its finite-difference approximation.* Show that if

$$\frac{\mathrm{d}x}{\mathrm{d}t} = -x^3$$

then $x(t) = x(0)/\{1 + 2x^2(0)t\}^{1/2}$ for $t \geqslant 0$. Deduce that $X = 0$ is the only equilibrium point and that it is globally asymptotically stable.

Show that if

$$x_{n+1} = x_n - hx_n^3 \qquad \text{for } n = 0, 1, \ldots,$$

where $h > 0$, then $X = 0$ is the only fixed point and it is stable. What is its domain of attraction? Show further that $\{-(2/h)^{1/2}, (2/h)^{1/2}\}$ is a two-cycle and it is unstable.

[Stuart (1990).]

Q3.17 *Some three-cycles.* Show that if $\{X, Y, X\}$ is a three-cycle of a differentiable map $F: \mathbb{R} \to \mathbb{R}$, i.e. $Y = F(X)$, $Z = F(Y)$, $X = F(Z)$, and if $G = F^3$ is the third-generation map of F, then

$$G'(X) = G'(Y) = G'(Z).$$

Defining F by $F(a, x) = ax(1 - x)$ for $x \in \mathbb{R}$, show that if X belongs to a three-cycle of F then X is a root of the sextic equation,

$$a^6X^6 - (3a + 1)a^5X^5 + (3a + 1)(a + 1)a^4X^4 - (a^3 + 5a^2 + 3a + 1)a^3X^3$$

$$+ (2a + 1)(a^2 + a + 1)a^2X^2 - (a + 1)(a^2 + a + 1)aX + (a^2 + a + 1) = 0.$$

*Hence or otherwise show that two real three-cycles of the logistic map exist if $a > \sqrt{8} + 1$ or $a < -(\sqrt{8} - 1)$.

[Hint: it *may* help you to let the three-cycles be $\{X_1, Y_1, Z_1\}$ and $\{X_2, Y_2, Z_2\}$, to define $t_j = X_j Y_j Z_j$, $u_j = \sum Y_j Z_j$ and $v_j = \sum X_j$ for $j = 1, 2$, and then to find the quadratic equation whose roots are v_1, v_2. Myrberg (1958, p. 13).]

Q3.18 *The mapping of a subinterval onto itself by the logistic map.* Consider the logistic map $F(a, x) = ax(1 - x)$ and its second-generation map $G(a, x) = F(a, F(a, x))$ over the interval $0 \leqslant x \leqslant 1$.

Show that $G(a, \frac{1}{2}) = \frac{1}{16}a^2(4 - a)$ and that if $a > 2$ then $G(a, x) \geqslant G(a, \frac{1}{2})$ for all x sufficiently close to $\frac{1}{2}$, i.e. that G has a local minimum at $x = \frac{1}{2}$.

Deduce that if A is the root of $\frac{1}{16}a^2(4 - a) = 1/a$ such that $3 < a < 4$ then $G(A, \frac{1}{16}A^2(4 - A)) = (A - 1)/A$. Hence or otherwise show that $G(A, x)$ maps the subinterval $[1/A, (A - 1)/A]$ onto itself. Deduce that $F(A, x)$ maps the subinterval $[1/A, \frac{1}{4}A]$ onto itself, $[0, 1/A]$ onto $[0, (A - 1)/A]$ and $[\frac{1}{4}A, 1]$ onto $[0, 1/A]$.

Discuss briefly the attractors of F for $a = A$.

[In fact $A \approx 3.6790$. It may be helpful to sketch graphs of G and F, and some important points and their maps, for $a = A$. Also numerical experiments may suggest the nature of the attractors.]

Q3.19 *The explicit solution of the logistic difference equation in a special case.* Show that the logistic map

$$F(x) = -2x(1 - x)$$

maps the closed interval $[-\frac{1}{2}, \frac{3}{2}]$ onto itself.

Show that if $x_n = \frac{1}{2} + \cos 2\pi\theta_n$, $\theta_{n+1} \equiv 2\theta_n$ modulo 1 and $0 \leqslant \theta_n < 1$ then $x_{n+1} = F(x_n)$. Deduce that if $x_{n+1} = F(x_n)$, $-\frac{1}{2} \leqslant x_0 \leqslant \frac{3}{2}$ and θ_0 is a root of $\cos 2\pi\theta_0 = x_0 - \frac{1}{2}$ then $x_n = \frac{1}{2} + \cos(2^{n+1}\pi\theta_0)$. Discuss briefly the periodic and aperiodic solutions $\{x_n\}$.

Q3.20 *Some statistical properties of the logistic difference equation.* The *mean* $\bar{x}$ of a sequence $\{x_n\}$ is defined by $\bar{x} = \lim_{N \to \infty} \{(x_1 + x_2 + \cdots + x_N)/N\}$ and the *variance* $\mathrm{Var}(x)$ by $\mathrm{Var}(x) = \overline{x^2} - (\bar{x})^2$. Deduce that if $x_{n+1} = 4x_n(1 - x_n)$ for $n = 0, 1, \ldots$ and $\{x_n\}$ is an aperiodic sequence then $\bar{x} = \frac{1}{2}$ and $\mathrm{Var}(x) = \frac{1}{8}$.

Q3.21 *Bifurcations of the logistic map for $a < 0$.* Take $F(a, x) = ax(1 - x)$ and use a computer to investigate the bifurcations for $a < 0$ by iterating $F^n(a, \frac{1}{2})$. Find, in particular, the period doubling and chaos as a decreases from -1.

Q3.22 *The tent map.* Define the map F by

$$F(a, x) = a(\tfrac{1}{2} - |x - \tfrac{1}{2}|)$$

for $a > 0$. Sketch the curve $y = F(a, x)$ in the (x, y)-plane for a 'typical' positive value of a. Find the fixed points of F and their stability for all $a > 0$. Sketch a bifurcation diagram in the (a, x)-plane. Find a two-cycle for $a = 2$, and show that $\{\frac{2}{7}, \frac{4}{7}, \frac{6}{7}\}$ is a three-cycle.

Show that

$$F^n(a, c + \epsilon) - F^n(a, c) = \pm a^n \epsilon \qquad \text{as } \epsilon \to 0$$

for fixed $c \neq \frac{1}{2}$, a and n. Discuss briefly the relevance of this result to sensitive dependence on initial conditions.

Q3.23 *Second-generation tent map.* Show that if $G(a, x) = F^2(a, x)$, $F(a, x) = a(\frac{1}{2} - |\frac{1}{2} - x|)$, and $1 < a \leqslant 2$ then

$$G(a, x) = \left\{ \begin{array}{ll} a^2 x & \text{for } 0 \leqslant x \leqslant 1/2a \\ a(1 - ax) & \text{for } 1/2a \leqslant x \leqslant \frac{1}{2} \\ a(1 - a + ax) & \text{for } \frac{1}{2} \leqslant x \leqslant 1 - 1/2a \\ a^2(1 - x) & \text{for } 1 - 1/2a \leqslant x \leqslant 1 \end{array} \right\}.$$

Sketch the graph $y = G(a, x)$ in the (x, y)-plane. Hence or otherwise find how many solutions of period two the difference equation $x_{n+1} = F(a, x_n)$ has for each value of $a \in (1, 2]$, and state which are stable.

Q3.24 *Topological conjugacy of the tent and logistic maps.* Two maps F and G are said to be *topologically conjugate* if there exists a smooth invertible map H such that $F = H^{-1}GH$.

Show that the logistic map, defined by $F(x) = 4x(1 - x)$ for $0 \leqslant x \leqslant 1$, and the tent map, by $G(x) = 2(\frac{1}{2} - |\frac{1}{2} - x|)$, are topologically conjugate, on taking $H(x) = 2\pi^{-1} \arcsin\sqrt{x}$. Deduce that if $x_{n+1} = G(x_n)$ for $n = 0, 1, \ldots$ and $y_n = \sin^2(\frac{1}{2}\pi x_n)$ then $y_{n+1} = F(y_n)$. What is the solution x_n in this case?

Q3.25 *A class of one-dimensional difference equations.* (a) Suppose that $\theta_{n+1} \equiv 2\theta_n$ modulo 1 and $0 \leqslant \theta_n < 1$ for $n = 0, 1, \ldots$, and define

$$x_n = \operatorname{sn}^2(2K(m)\theta_n | m) \qquad \text{for } 0 \leqslant m < 1,$$

where sn is the Jacobian elliptic function with parameter m, and K is the complete elliptic integral of the first kind. Then show that

$$x_{n+1} = 4f(m, x_n), \qquad \text{where } f(m, x) = x(1 - x)(1 - mx)/(1 - mx^2)^2.$$

[You need have no knowledge of elliptic functions to solve this problem correctly; the hints will suffice. It may help to think of sn as a generalization of sin and $K(m)$ as a generalization of $\frac{1}{2}\pi$ with the extra variable m; indeed, $\operatorname{sn}(z|0) = \sin z$ and $K(0) = \frac{1}{2}\pi$. Hint: you are given that $\operatorname{sn}(z|m)$ is an odd function of z with period $4K(m)$ and

$$\operatorname{sn}^2(2z|m) = 4\operatorname{sn}^2(z|m)\{1 - \operatorname{sn}^2(z|m)\}\{1 - m\operatorname{sn}^2(z|m)\}/\{1 - m\operatorname{sn}^4(z|m)\}^2.]$$

Hence show that $f(m, x)$ has only one maximum for $0 < x < 1$, where its value is $\frac{1}{4}$, and that $f(m, 0) = 0$, $f(m, 1) = 0$, $f_x(m, 0) = 1$.

[Hint: you are given further that $\operatorname{sn}(z|m)$ is a monotonically increasing continuous function of z for $0 \leqslant z \leqslant K(m)$, that $\operatorname{sn}(0|m) = 0$ and

$\mathrm{sn}(K(m)|m) = 1$. In fact, $f(m, x)$ is a convex function of x over $0 < x < 1$ if $0 < m < 1$.]

(b) Consider the difference equations of the form $x_{n+1} = af(m, x_n)$, where $0 \leqslant x_0 \leqslant 1$, $0 < a \leqslant 4$ and $0 \leqslant m < 1$. Ascertain the number and stability of the fixed points by graphical means. Describe qualitatively the bifurcations of the system as a increases for fixed m.

Use a computer to verify your description, finding some of the attractors for negative as well as positive values of m.

[Bristol (1987).]

Q3.26 *The explicit solution of a one-dimensional difference equation.* Consider the map F defined by

$$F(x) = 2x/(1 + x^2)$$

for $-\infty < x < \infty$. Show that F maps $(-\infty, \infty)$ onto $[-1, 1]$. Find the fixed points of F and whether they are stable.

Show that if $x_{n+1} = F(x_n)$ for $n = 0, 1, \ldots$ and $-\infty < x_0 < \infty$ then there exists a real number z_1 such that $x_1 = \tanh z_1$ provided that $x_0 \neq \pm 1$. Deduce that $x_n = \tanh(2^{n-1} z_1)$. Find the attractors of F and their domains of attraction.

Q3.27 *Another one-dimensional difference equation.* Define F by

$$F(a, x) = a\pi^{-1} \sin \pi x$$

for $a \geqslant 0$. Show that there are fixed points $X = 0$ for all a and that $X = \pm U(a)/\pi$, where U is the zero of $u - a \sin u$ such that $\frac{1}{2}\pi < U < \pi$. Sketch a graph showing clearly the values of a for which the second and third fixed points exist. What other fixed points are there?

For what values of a is the origin a stable fixed point? Show that the second and third fixed points are unstable when $a > a_1$, where $a_1 = U_1/\sin U_1$ and U_1 is the zero of $\tan u + u$ such that $\frac{1}{2}\pi < U_1 < \pi$.

Describe *qualitatively* the bifurcations as a increases from zero to π. You should use graphs of F and some higher-generation maps, but may also use heuristic arguments and quote any relevant results.

Describe briefly the various sets of cluster points when $a = \pi$.

Q3.28 *Period doubling of an exponential map.* Define F by $F(a, x) = ae^x$ for $-\infty < x < \infty$, $a \leqslant 0$. Show that F has a unique fixed point $X(a)$, which is stable if $-\mathrm{e} \leqslant a \leqslant 0$ and unstable if $a < -\mathrm{e}$. Show that $X(a) \sim a$ as $a \uparrow 0$, $X(a) = -1 + (a + \mathrm{e})/2\mathrm{e} + O\{(a + \mathrm{e})^2\}$ as $a \to -\mathrm{e}$, and $X(a) \sim -\ln(-a)$ as $a \to -\infty$.

Defining the second-generation map G of F by $G(a, x) = F(a, F(a, x))$, show that $G_x = FG$, $G_{xx} = F(1 + F)G$, $G_{xxx} = F(1 + 3F + F^2)G$, where the subscript denotes partial differentiation with respect to x. Deduce that $G_x(a, X) = X^2$, $G_{xx}(a, X) = X^2(1 + X)$, $G_{xxx}(a, X) = X^2(1 + 3X + X^2)$. By graphical consideration of the intersections of the line $y = x$ and the curve

$y = G(a, x)$ in the (x, y)-plane or otherwise, show that a stable two-cycle bifurcates from X as a decreases through $-\mathrm{e}$.

Show that if $x = G(a, x)$ and $x \neq X$ then

$$(X + 1)(X - 1) + \tfrac{1}{2}X^2(X + 1)(x - X) + \tfrac{1}{6}X^2(1 + 3X + X^2)(x - X)^2 + \cdots = 0$$

as $x \to X$. Hence or otherwise deduce that the two-cycle is given by

$$X_1, X_2 = -1 \pm \{-6(a + \mathrm{e})/\mathrm{e}\}^{1/2} + O(a + \mathrm{e}) \qquad \text{as } a \uparrow -\mathrm{e}.$$

Sketch the bifurcation diagram in the (a, x)-plane, indicating the fixed points and the two-cycle.

Q3.29 *Arnol'd tongues for the sine map.* Define the fractional part as the function $f\colon \mathbb{R} \to [0, 1)$ such that

$$f(x) \equiv x \text{ modulo } 1 \qquad \text{and } 0 \leqslant f < 1.$$

(a) The rotation map $F\colon [0, 1) \to [0, 1)$ is defined by

$$F(x) = f(x + a)$$

for $a \in [0, 1)$. Find the fixed points X, if any, of F, showing for what values of a each exists.

Prove that F has two-cycles $\{X, Y\}$ if and only if $a = \frac{1}{2}$, the two-cycles then being given by $Y = X + \frac{1}{2}$ for all $X \in [0, \frac{1}{2})$.

(b) The *sine map* $G\colon [0, 1) \to [0, 1)$ is defined by

$$G(x) = f(x + a + (2\pi)^{-1} b \sin 2\pi x)$$

for fixed parameters $a \in [0, 1)$ and $b \in [0, \infty)$. Sketch the curve $y = a + (2\pi)^{-1} b \sin 2\pi x$ in the (x, y)-plane for $0 \leqslant x \leqslant 1$, $0 < a < \frac{1}{2}$ and $2\pi a < b < 2\pi(1 - a)$. Hence or otherwise show that then G has two fixed points, and find them in explicit terms of elementary functions. Show further that G has no fixed point if $b < 2\pi a$ and one if $b = 2\pi a$.

Consider next the two-cycles $\{X, Y\}$ of G, taking $0 \leqslant X < Y < 1$ without loss of generality. Show that

$$X + a + (2\pi)^{-1} b \sin 2\pi X = Y, \tag{1}$$

$$Y + a + (2\pi)^{-1} b \sin 2\pi Y = X + 1 \tag{2}$$

for sufficiently small b. It is given that the series

$$a = \tfrac{1}{2} + ba_1 + b^2 a_2 + \ldots,$$

$$X(a) = X_0 + bX_1 + b^2 X_2 + \ldots, \qquad Y(a) = Y_0 + bY_1 + b^2 Y_2 + \ldots$$

converge for sufficiently small b. Equating coefficients of b in equations (1) and (2), deduce that

$$a_1 = 0, \qquad Y_1 = X_1 + (2\pi)^{-1} \sin 2\pi X_0.$$

Equating coefficients of b^2, deduce that if

$$\tfrac{1}{2} - b^2/8\pi + O(b^3) < a < \tfrac{1}{2} + b^2/8\pi + O(b^3) \qquad \text{as } b \downarrow 0.$$

then there exists a two-cycle for all $X_0 \in [0, \tfrac{1}{2})$ for sufficiently small $b > 0$.

Summarize your results in a sketch of the regions of the (a, b)-plane where there exist fixed points of G and where there exist two-cycles.

[There are similar regions of the (a, b)-plane where p-cycles of G exist for $p = 2, 3, \ldots$, and small b. These regions join up and chaos ensues as b increases. The regions are called *Arnol'd tongues*. Cf. Arnol'd (1961, §12).]

Q3.30 *An exponential map.* Find the fixed points of $F(a, x) = xe^{a(1-x)}$ and their stability for all $a > 0$. Discover the stable periodic orbits for $a = 2.3, 2.6$ and 2.9 by numerical experiments.

Q3.31 *A cubic one-dimensional map.* Given that $F(a, x) = ax(1 - x^2)$, show that F maps the interval $[0, 1]$ into itself if $0 \leqslant a \leqslant 3\sqrt{3}/2 = 2.598$. Show that the fixed points are $X = 0$ for all a and $X = \pm\{(a - 1)/a\}^{1/2}$ for all $a < 0$ and $a > 1$. Show that $X = 0$ is stable for $-1 < a < 1$ and $X = \{(a - 1)/a\}^{1/2}$ for $1 < a < 2$. Sketch the bifurcation diagram in the (a, x)-plane.

Use a computer to find period doubling at values a_r of a for $r = 1, 2, \ldots$, and verify that

$$a_r = a_\infty - A\delta^{-r} + o(\delta^{-r}) \qquad \text{as } r \to \infty,$$

where $a_\infty = 2.3202$ and δ is Feigenbaum's universal constant.

Q3.32 *Superstable periodic solutions of a one-dimensional map.* A fixed point X of the smooth map $f\colon \mathbb{R} \to \mathbb{R}$ is called *superstable* if it is, according to linearized theory, the most stable a fixed point can be, i.e. if it is quadratically stable. Deduce that $f'(X) = 0$.

Similarly, a p-cycle $\{X_1, \ldots, X_p\}$ is superstable if it corresponds to a superstable fixed point of f^p. Show then that $f'(X_j) = 0$ for j equal to one value of $1, 2, \ldots, p$. Deduce that there is a superstable p-cycle of f if p is the least positive integer such that $f^p(x_m) = x_m$, where x_m is a simple maximum of f.

Use a computer to calculate the first few superstable 2^r-cycles of $F(a, x) = ax(1 - x)$, basing your method on a numerical scheme (the method of false position, for example) to solve $F^{2^r}(a, \tfrac{1}{2}) = \tfrac{1}{2}$ for a iteratively. Defining A_r as the value of a at which the the 2^r-cycle is superstable, calculate the first few values of $(A_r - A_{r-1})/(A_{r+1} - A_r)$.

Similarly calculate the value of a at which the three-cycle is superstable by solving $F^3(a, \tfrac{1}{2}) = \tfrac{1}{2}$.

[It may help, and is instructive, to plot the graph of $y = F^p(a, \tfrac{1}{2}) - \tfrac{1}{2}$ in the (a, y)-plane before solving $F^p(a, \tfrac{1}{2}) = \tfrac{1}{2}$ iteratively, in order to find the value of a at which the p-cycle is superstable. Note that it is easier to calculate A_r accurately than the value a_r at which there is period doubling, because when a is close to a_r the 2^r-cycle is weakly stable and so solutions converge to it slowly.]

Q3.33 *Another form of the Hénon map.* Show that if $x_{n+1} = 1 + y_n - ax_n^2$ and $y_{n+1} = bx_n$ for $n = 0, 1, \ldots$, then $u_{n+1} = a + bv_n - u_n^2$ and $v_{n+1} = u_n$, where $u_n = ax_n$ and $v_n = ay_n/b$ for $b \neq 0$.

Discuss the solutions of the difference equation in the (u, v)-plane for various values of a when $b = 0$.

Q3.34 *Two-cycles of the Hénon map.* Consider the map $\mathbf{F}: \mathbb{R}^2 \to \mathbb{R}^2$ defined by $\mathbf{F}(\mathbf{x}) = (f(x, y), g(x, y))$,

$$f(x, y) = 1 + y - ax^2, \qquad g(x, y) = bx$$

for all $\mathbf{x} = (x, y)$, where $-1 < b < 1$. Show that if $\mathbf{X} = (X, Y)$ is a fixed point of the second-generation map $\mathbf{F}^2$ then

$$\{aX^2 + (1 - b)X - 1\}\{a^2X^2 - a(1 - b)X + (1 - b)^2 - a\} = 0.$$

Deduce that the two-cycle $\{\mathbf{X}_1, \mathbf{X}_2\}$ of $\mathbf{F}$ is given by

$$\mathbf{X}_1, \mathbf{X}_2 = (\{1 - b \pm 2(a - a_1)^{1/2}\}/2a, b\{1 - b \mp 2(a - a_1)^{1/2}\}/2a)$$

respectively for all $a > a_1 = \frac{3}{4}(1 - b)^2$. Show further that the two-cycle is stable if $a_1 < a < a_2 = (1 - b)^2 + \frac{1}{4}(1 + b)^2$ and unstable if $a > a_2$.

Q3.35 *The location of an invariant set of the Hénon map.* Show that if $a = 1.4$, $b = 0.3$ then the quadrilateral with vertices $(-1.33, 0.42)$, $(1.32, 0.133)$, $(1.245, -0.14)$, $(-1.06, -0.5)$ is mapped into itself by equations (6.1).

[Hénon & Pomeau (1976).]

Q3.36 *The Lozi map.* A piecewise linear map $\mathbf{F}: \mathbb{R}^2 \to \mathbb{R}^2$ is defined by $\mathbf{F}(\mathbf{x}) = (f(\mathbf{x}), g(\mathbf{x}))$, where $\mathbf{x} = (x, y)$,

$$f(x, y) = 1 + y - a|x|, \qquad g(x, y) = bx$$

for given real parameters a, b.

Show that the fixed points of the map $\mathbf{F}$ are X_+ for all $a > -(1 - b)$ and X_- for all $a > 1 - b$, where $\mathrm{X}_+ = (1/(1 + a - b), b/(1 + a - b))$ and $\mathrm{X}_- = (-1/(a + b - 1), -b/(a + b - 1))$.

Henceforth assume that $-1 < b < 1$. Deduce that X_- is always unstable. Given that X_+ is stable for $-(1 - b) < a \leqslant a_1(b)$, find the function a_1 of b.

Show that if $\{\mathbf{X}_1, \mathbf{X}_2\}$ is a two-cycle of $\mathbf{F}$, $a^2 \neq (1 - b)^2$, $\mathbf{X}_1 = (X_1, Y_1)$, and $\mathbf{X}_2 = (X_2, Y_2)$, then $X_1 X_2 < 0$. Hence or otherwise find each two-cycle of $\mathbf{F}$ and the condition satisfied by a and b for it to exist.

Sketch the fixed points and two-cycles of $\mathbf{F}$ in the (a, x)-plane for fixed b, indicating clearly what is a fixed point and what a two-cycle.

[It *may* help to regard the map as an analogue of the Hénon map (Lozi 1978).]

Q3.37 *The Lozi map again.* A piecewise linear map $\mathbf{F}: \mathbb{R}^2 \to \mathbb{R}^2$ is defined by $\mathbf{F}(\mathbf{x}) = (f(\mathbf{x}), g(\mathbf{x}))$, where $\mathbf{x} = (x, y)$,

$$f(x, y) = 1 + y - a|x|, \qquad g(x, y) = bx$$

for given parameters $a, b > 0$.

Show that $\mathbf{F}$ maps the right half-plane to the upper half-plane.

Defining the points $\mathrm{U} = (0, b(q - 1)/q(1 + a - b))$, $\mathrm{Z} = ((1 - q)/(1 + a - b), 0)$, and $q = -\frac{1}{2}\{a + (a^2 + 4b)^{1/2}\}$, where $a > |1 - b|$, show that $\mathbf{F}(\mathrm{U}) = \mathrm{Z}$ and that the fixed point X_+ lies in the line UZ. Find the coordinates of $\mathrm{V} = \mathbf{F}(\mathrm{Z})$ and $\mathrm{W} = \mathbf{F}(\mathrm{V})$.

*Deduce that $\mathbf{F}(\Delta) \subset \Delta$, where Δ is the triangle with vertices Z, U and V, if $0 < b < 1$ and $b + 1 < a < 2 - \frac{1}{2}b$.

[Misiurewicz (1980) *proved* that there is a strange attractor in Δ. Compute it and show it graphically for, say, $a = 1.7$ and $b = 0.5$ (Lozi 1978).]

Q3.38 *A quadratic area-preserving map.* Define the rotation map $\mathbf{R}: \mathbb{R}^2 \to \mathbb{R}^2$, the shear map $\mathbf{S}: \mathbb{R}^2 \to \mathbb{R}^2$ and $\mathbf{T}: \mathbb{R}^2 \to \mathbb{R}^2$ by

$$\mathrm{R}(\mathbf{x}) = (x \cos\alpha - y \sin\alpha, x \sin\alpha + y \cos\alpha), \quad \mathrm{S}(\mathbf{x}) = (x, y - x^2), \quad \mathrm{T} = \mathrm{RS},$$

where $\mathbf{x} = (x, y)$ and α is the angle of the rotation.

Show that

$$\mathbf{T}(\mathbf{x}) = (x \cos\alpha - (y - x^2)\sin\alpha, x \sin\alpha + (y - x^2)\cos\alpha)$$

and $\mathbf{T}$ is area preserving.

Show that if $\alpha \to 2\pi - \alpha$ then $\mathbf{T}(-x, y) \to \mathbf{T}(x, y)$. Deduce that it is sufficient to consider only $0 \leqslant \alpha \leqslant \pi$.

Show that if $\alpha = 0$ then $\mathbf{T}^n(\mathbf{x}) = (x, y - nx^2)$ and if $\alpha = \pi$ then $\mathbf{T}^2$ is the identity map.

By expressing $\mathbf{T}^{-1} = \mathbf{S}^{-1}\mathbf{R}^{-1}$ or otherwise, show that the inverse map of $\mathbf{T}$ is given by $\mathbf{T}^{-1}(\mathbf{x}) = (x \cos\alpha + y \sin\alpha, -x \sin\alpha + y \cos\alpha + (x \cos\alpha + y \sin\alpha)^2)$.

Show that if $0 < \alpha < 0$ then the fixed points of $\mathbf{T}$ are $\mathbf{0}$ and $(2\tan(\frac{1}{2}\alpha), 2\tan^2(\frac{1}{2}\alpha))$. Find the stability characteristics of each point, and state for what values of α each is stable.

Show that there is no two-cycle of $\mathbf{T}$.

Using a computer, experiment with iterations of the map $\mathbf{T}$.

[Hénon (1969).]

Q3.39 *Area-preserving maps of the plane and period doubling.* Show that the map $\mathbf{F}: \mathbb{R}^2 \to \mathbb{R}^2$, where $\mathbf{F}(\mathbf{x}) = (F(x, y), G(x, y))$,

$$F(x, y) = -y + f(x), \qquad G(x, y) = x - f(F(x, y)),$$

is area preserving for all differentiable functions f.

Consider iterated maps of this form for $f(x) = ax - (1 - a)x^2$, finding all the fixed points and conditions for their linear stability. Show, in particular, that the origin is stable for $-1 \leqslant a \leqslant 1$. Suggest a plausible reason for anticipating a flip bifurcation at $a = -1$. Sketch a bifurcation diagram in the (a, x)-plane.

Verify by use of a computer that there is period doubling at $a = a_r$ for $r = 1, 2, \ldots$, where $a_r \downarrow a_\infty$ and $(a_{r-1} - a_r)/(a_r - a_{r+1}) \to \delta$ as $r \to \infty$, $a_\infty \approx -1.266$, and $\delta \approx 8.271$.

[In fact this new δ is a universal constant for period doubling of *measure-preserving* maps (Bountis 1981), analogous to Feigenbaum's constant δ.]

Q3.40 *The standard map.* Consider the difference equations $p_{n+1} = F(k, p_n, x_n)$, $x_{n+1} = G(k, p_n, x_n)$ for $n = 0, 1, \ldots$, corresponding to the map of the unit square of the (p, x)-plane into itself, where

$$F(k, p, x) \equiv p - \frac{k}{2\pi} \sin 2\pi x \text{ modulo } 1 \quad \text{and} \quad 0 \leqslant F < 1,$$

$$G(k, p, x) \equiv x + F(k, p, x) \text{ modulo } 1 \quad \text{and} \quad 0 \leqslant G < 1.$$

Show that the map is area preserving for all k.

Show that when $k = 0$ each orbit lies in the line $p = p_0$ and the solution is periodic if p_0 is rational and aperiodic if p_0 is irrational.

Use a computer to examine how these orbits change as k increases from zero for various values of x_0 and p_0.

Q3.41 *The cat map.* Consider the map $\mathbf{F}: \mathbb{T}^2 \to \mathbb{T}^2$ of the unit square modulo 1, i.e. of the torus in which $2\pi x$ and $2\pi y$ may represent the angles around two circles which thread the torus in perpendicular directions, defined by $\mathbf{F}(\mathbf{x}) = (f(x, y), g(x, y))$, where

$$f(x, y) \equiv x + y, \qquad g(x, y) \equiv x + 2y \text{ modulo } 1$$

and $0 \leqslant f(x, y), g(x, y) < 1$.

Show that the eigenvalues of the Jacobian matrix are $q_1, q_2 = \frac{1}{2}(3 \pm \sqrt{5})$ for all x, y, and that the map of $\mathbb{T}^2$ is area preserving.

Prove that $(0, 0)$ is the fixed point of $\mathbf{F}$, and that the two-cycles of $\mathbf{F}$ are $\{(\frac{1}{5}, \frac{3}{5}), (\frac{4}{5}, \frac{2}{5})\}$, $\{(\frac{2}{5}, \frac{1}{5}), (\frac{3}{5}, \frac{4}{5})\}$.

Show that if X, Y are rational then $\mathbf{X} = (X, Y)$ is a fixed point of $\mathbf{F}^p$ for some positive integer p. Show that if $\mathbf{X}$ is a fixed point of $\mathbf{F}^p$ then X, Y are rational.

Deduce that each fixed point of $\mathbf{F}^p$ is unstable for $p = 1, 2, \ldots$.

[This map is called the cat map because Arnol'd & Avez (1968, p. 6) first posed it with a sketch of how a silhouette of a cat's head is distorted, cut up, and transposed by $\mathbf{F}$ and $\mathbf{F}^2$ (see Fig. 3.14).]

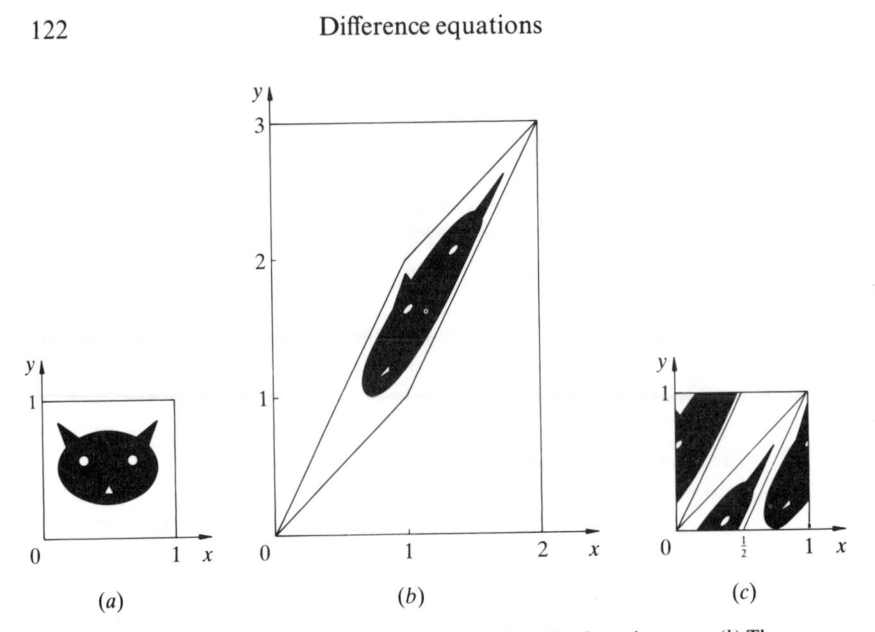

Fig. 3.14 Sketches of the cat map. (*a*) The cat's head in the unit square. (*b*) The map of the head before moduli one are taken. (*c*) The final map of the head.

Q3.42 *A map with an invariant curve.* Defining a map $\mathbf{F}: \mathbb{R}^2 \to \mathbb{R}^2$ by

$$\mathbf{F}(x, y) = (ax, a^b(y - f(x)),$$

where $f: \mathbb{R} \to \mathbb{R}$ is differentiable, and a, $b > 0$, show that $\mathbf{F}$ maps the line $x = x_0$ into the line $x = ax_0$. What are the fixed points of $\mathbf{F}$, and for what values of a, b does each exist?

Show that if $g: \mathbb{R} \to \mathbb{R}$ is a function of Weierstrass type defined by

$$g(x) = \sum_{k=0}^{\infty} a^{-bk} f(a^k x) \qquad \text{for } a > 1$$

then the curve $y = g(x)$ is mapped into itself by $\mathbf{F}$. Find the Jacobian of $\mathbf{F}$ at a typical point of the curve, and deduce that the curve is a repeller.

Taking $f(x) = \sin 2\pi x$ and $a \in \{2, 3, \ldots\}$, modify $\mathbf{F}$ so that it maps a cylinder into itself and find an invariant closed curve of the modified map.

[Falconer (1990, p. 196).]

Q3.43 *Canonical form of a quadratic map of the complex plane.* Show that if $z_{n+1} = az_n^2 + 2bz_n + c$, $w_n = az_n + b$, $d = b^2 - ac - b$ and $a \neq 0$ then $w_{n+1} = w_n^2 - d$.

Q3.44 *Stability of a fixed point of an analytic map of the complex plane.* Suppose that the map $F: \mathbb{C} \to \mathbb{C}$ is analytic at Z, where Z is a fixed point of F, i.e. $Z = F(Z)$. Then show that Z is stable if $|F'(Z)| < 1$ and unstable if $|F'(Z)| > 1$.

Find the fixed points of $F(z) = z^2 + a$ for all complex a, and for what values of a each is stable.

Show that if Z is a fixed point of F^2 then

$$(Z^2 - Z + a)(Z^2 + Z + a + 1) = 0.$$

Hence find the two-cycle of F. For what values of a is it stable?

Sketch the region of the complex a-plane in which either a fixed point or the two-cycle is stable. What does this remind you of?

Q3.45 *Analytic map of the complex plane.* Suppose that $F: \mathbb{C} \to \mathbb{C}$ is analytic and generates a map $\mathbf{G}: \mathbb{R}^2 \to \mathbb{R}^2$ by taking real and imaginary parts in the usual way. Deduce that the Jacobian determinant of $\mathbf{G}$ is $|dF/dz|^2$. What are the eigenvalues of the Jacobian matrix?

Q3.46 *The cubic map.* Defining $F: \mathbb{C} \to \mathbb{C}$ by $F(z) = z^3$ for all z, find each fixed point of F, and whether it is stable. Find the domain of attraction of each stable point.

Find each two-cycle of F and whether it is stable.

Given the difference equation $z_{n+1} = F(z_n)$ for $n = 0, 1, \ldots$, express $z_n = r_n \exp(i\theta_n)$ for $r_n \geqslant 0$ and real θ_n, and thence determine the difference equations satisfied by r_n, θ_n. Describe all the other cycles briefly.

Find at least one closed curve mapped by F onto itself.

Q3.47 *The real part of a Mandelbrot set.* Consider what *real* values of a lie in the Mandelbrot set M of $F(a, z) = z^2 + a$. Show first that $(\frac{1}{4}, \infty) \subset$ M, then that $[-\frac{3}{4}, \frac{1}{4}) \subset$ M, and $[-\frac{5}{4}, -\frac{3}{4}) \subset$ M. What happens if $a < -\frac{5}{4}$?

Q3.48 *Some simple maps of the plane.* Consider the following four maps $\mathbf{F}: \mathbb{R}^2 \to \mathbb{R}^2$ corresponding to maps $H: \mathbb{C} \to \mathbb{C}$, where $\mathbf{F} = (F, G)$, $z = x + iy$, $H(z) = F(x, y) + iG(x, y)$. Prove in each case that $\mathbf{F}$ and H are equivalent, and that $\mathbf{F}$ is area preserving.

Deduce that all compositions of the four maps $\mathbf{F}$ are area preserving.

(a) *Translations*: $\mathbf{F}(\mathbf{x}) = (x + a, y + b)$ for real a, b; $H(z) = z + a + ib$.

(b) *Reflection in the x-axis*: $\mathbf{F}(\mathbf{x}) = (x, -y)$; $H(z) = \bar{z}$.

(c) *Rotation about the origin*: $\mathbf{F}(\mathbf{x}) = (x \cos a - y \sin a, x \sin a + y \cos a)$ for real a; $H(z) = e^{ia}z$.

(d) *Shear in the x-direction*: $\mathbf{F}(\mathbf{x}) = (x + f(y), y)$ for real differentiable function f; $H(z) = z + f(\mathrm{Im}(z))$.

Q3.49 *Fixed points of a map and their stability.* Suppose that

$$x_{n+1} = ka + mx_n^r \qquad \text{if } x_n > 0, \qquad \text{for } n = 0, 1, \ldots,$$

where $m, r > 0$. Defining $p > 0$, $u_n > 0$, b by $m = p^{r-1}$, $u_n = px_n$, $b = akp$, deduce that

$$u_{n+1} = F(b, r, u_n),$$

where $F(b, r, u) = b + u^r$ for all $u > 0$. Taking $r = \frac{1}{2}, 2$ in turn, find the fixed points of F for all b and determining for what values of b each exists and each is stable.

Show, by graphical means or otherwise, that the qualitative results are typical for $r < 1$, $r > 1$ respectively.

*4

Some special topics

One gets a similar impression when making a drawing of a rising cumulus from a fixed point; the details change before the sketch can be completed. We realize thus that: big whirls have little whirls that feed on their velocity, and little whirls have lesser whirls and so on to viscosity....

L.F. Richardson (*Weather Prediction by Numerical Process*, p. 66)

1 Cantor sets

We have met some strange attractors and asserted that they were 'Cantor sets'. We shall now define Cantor sets more carefully. They are called Cantor sets because they share some important properties of *Cantor's middle-thirds set*, K say (Cantor 1883, p. 590). To define K we first define a sequence of subsets K_n of the unit interval, as shown in Fig. 4.1. Thus we define $K_0 = [0, 1]$, $K_1 = [0, \frac{1}{3}] \cup [\frac{2}{3}, 1]$, $K_2 = [0, \frac{1}{9}] \cup [\frac{2}{9}, \frac{1}{3}] \cup [\frac{2}{3}, \frac{7}{9}] \cup [\frac{8}{9}, 1]$ and so forth. Note that K_n is the union of 2^n closed subintervals of $[0, 1]$ of the form $[r/3^n, (r + 1)/3^n]$, for appropriate integers r, and so the subintervals have length 3^{-n}. The middle third of each subinterval of K_n is removed to give K_{n+1}, so that $K_{n+1} \subset K_n \subset [0, 1]$. Then K is the set of all points common to $K_0, K_1, K_2, \ldots$, i.e. $K = \bigcap_{n=0}^{\infty} K_n$.

Even this set, in spite of its being so simply constructed, appears as a repeller of a one-dimensional map (namely the tent map shown in Q4.4).

Another definition of K is the set of all numbers expressible in the ternary form $0.x_1x_2x_3\ldots = x_1/3 + x_2/3^2 + x_3/3^3 + \ldots$, where the digit $x_n = 0$ or 2 for $n = 1, 2, \ldots$. (Note that the same real number may have two ternary forms, e.g., $\frac{1}{3}$ may be expressed as either 0.1 or 0.0222... and so belongs to K.) It can be seen that each point of K_1 has the ternary form $0.0x_2x_3\ldots$ if it lies in $[0, \frac{1}{3}] = [0, 0.0222\ldots]$ or $0.2x_2x_3\ldots$ if it lies in $[\frac{2}{3}, 1] = [0.2, 0.222\ldots]$, where now we place no restriction on $x_2, x_3, \ldots$. Similarly each point of K_2 has the ternary form $0.00x_3x_4\ldots$, $0.02x_3x_4\ldots$, $0.20x_3x_4\ldots$, or $0.22x_3x_4\ldots$ according to which of the four subintervals of K_2 it lies in. Extending this idea by induction, we see that the two definitions of K are equivalent.

Cantor's middle-thirds set K has an uncountable infinity of points. To

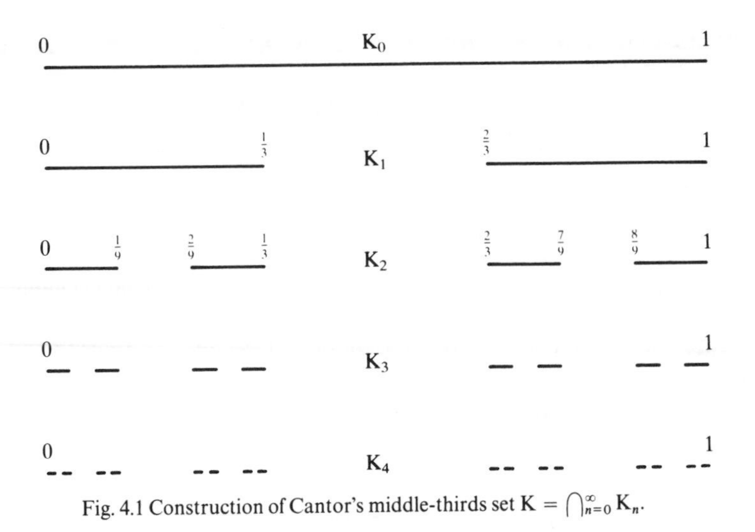

Fig. 4.1 Construction of Cantor's middle-thirds set $K = \bigcap_{n=0}^{\infty} K_n$.

prove this suppose that K is countable and use *reductio ad absurdum*. Thus suppose that all the members of K can be ordered as $x_1 = 0.x_{11}x_{12}x_{13}\ldots, x_2 = 0.x_{21}x_{22}x_{23}\ldots, \ldots$, say, in ternary form. Now define $y = 0.y_1y_2y_3\ldots$, where $y_m = 0$ if $x_{mm} = 2$ and $y_m = 2$ if $x_{mm} = 0$. Then $0 < y < 1$, $y \in K$ and $x_n \neq y \; \forall \, n$. This proves the falsity of the supposition that K is countable.

Nonetheless, K clearly contains no subinterval, however small, of [0, 1], and [0, 1] contains an infinity of subintervals which do not intersect K. Indeed the length of K is zero. To prove this, note that the length of K_0 is 1, of K_1 is $\frac{2}{3}$, and the length of K_{n+1} is two thirds of the length of K_n. Therefore the length of K_n is $(\frac{2}{3})^n$. Also $K \subset K_n$ implies that K is not longer than K_n. Therefore

$$\begin{aligned} 0 &\leqslant \text{length of K} \\ &\leqslant \text{length of } K_n \qquad \text{for } n = 1, 2, \ldots \\ &= (\tfrac{2}{3})^n \\ &\to 0 \qquad \text{as } n \to \infty. \end{aligned}$$

Therefore the length of K is zero. Therefore K contains no interval, i.e. K is *totally disconnected.*

The set K is closed, i.e. all the cluster points of K belong to K. To prove this note that each of the K_n is closed and that the intersection of a set of closed sets is closed.

Each point $x \in K$ is a cluster point of K. To prove this suppose that

$x = 0.x_1x_2\ldots$ in ternary form, and define $y_n = 0.x_1x_2\ldots x_n$. Then $y_n \in \mathrm{K}$ and $y_n \to x$ as $n \to \infty$. If $x = 0.x_1x_2\ldots x_m$ has a ternary expression which terminates then define $y_n = x + 2/3^{m+n}$ if $x \neq 1$ and $y_n = x - 2/3^n$ if $x = 1$.

Cantor defined a *perfect set* as a set which is both closed and such that each point of the set is a cluster point.

Note that two of the thirds of K_{n+1}, when magnified by a factor of three, are identical to the whole of K_n. This is an example of a *self-similar* or *scaling* property. Many interesting sets are self-similar like K.

A set which is an uncountable subset of the points of $\mathbb{R}^m$, is totally disconnected, and is perfect is called a *Cantor set*; it may or may not be self-similar.

The problems at the end of this chapter show that there is nothing very special in Cantor's choosing to remove the middle third of each interval rather than the middle rth for some r such that $0 < r < 1$. Also the ideas of self-similarity may be used to construct Cantor sets in $\mathbb{R}^m$ for $m > 1$. Again, self-similarity is not an essential property of a Cantor set, although it is a useful property in constructing a Cantor set.

2 Dimension and fractals

To each set $\mathrm{E} \subset \mathbb{R}^m$ there is assigned a *topological dimension*, d say, which is an integer such that $0 \leqslant d \leqslant m$. We are familiar with the facts, for example, that for a finite set of points $d = 0$, for a parabola $d = 1$, for the interior of a triangle $d = 2$, for the interior of a sphere $d = 3$, and for $\mathbb{R}^m$ itself $d = m$. This notion of dimension goes back to Euclid's definitions of point, line, surface and solid, and to Descartes' coordinate geometry because it is the minimum number of coordinates needed to describe all the points of E. This notion of topological dimension has been formalized in the twentieth century on the basis of the idea of Poincaré (1905, Chap. III, §3) to let a point have zero dimension, to relate the dimension of a space to the way it can be divided by boundaries, and to use mathematical induction. Thus curves can be divided by cuts which are of zero dimension (i.e. by points) and so have one dimension, surfaces can be divided by cuts of one dimension (i.e. by curves) and so have two dimensions, etc. Cantor's middle-thirds set had no place in Euclid's *Elements*, of course, but it may be said to have topological dimension $d = 0$ because it is a subset of $\mathbb{R}$ with zero length (and so has neither breadth nor length).

However, there are some useful generalizations of this definition of dimension. We shall give here an informal definition of the dimension, D say, of a set $\mathrm{E} \subset \mathbb{R}^m$. We shall see that $D = d$ for 'simple' sets and $d \leqslant D \leqslant m$,

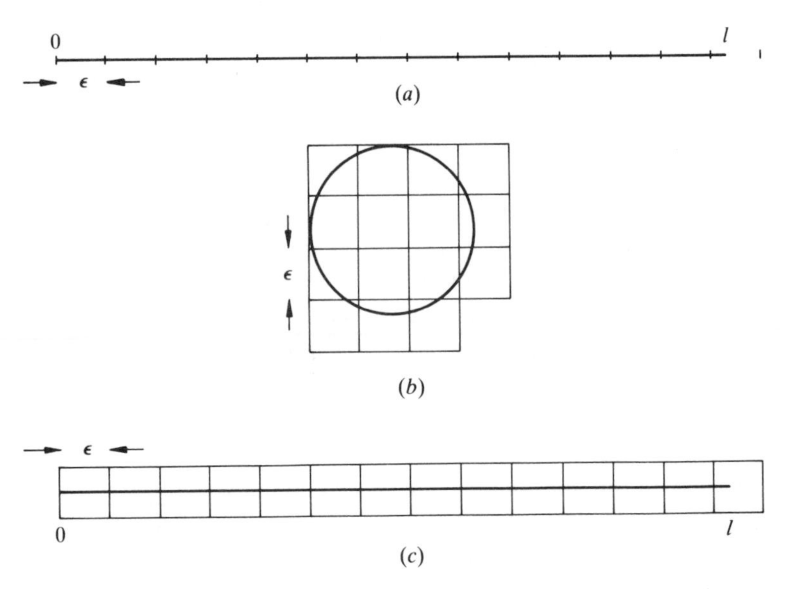

Fig. 4.2 Covering of (a) a line segment (in $\mathbb{R}$) of length l by N lines of length ε, (b) a circle of radius l and centre 0 by n squares of side ε, and (c) a line segment (in $\mathbb{R}^2$) of length l by N squares of side ε.

but D is not an integer or $D \neq d$ for some less simple sets. The ideas come from the definition of *Hausdorff dimension* (Hausdorff 1919).

We shall in fact use *box-counting dimension*, or *box dimension*, which is somewhat similar to Hausdorff dimension; the dimensions have equal values for many of the sets we shall meet. To introduce the definition of box dimension, first take an example in which E is a straight line segment of length l in $\mathbb{R}$ (see Fig. 4.2). Then the number N of equal lines of length ϵ *necessary* to cover all the points of E is given by $N(\epsilon) \sim l\epsilon^{-1}$ as $\epsilon \to 0$. Of course, we could use as many intervals of length ϵ as we wished to cover the line segment, but if overlap is avoided and wastage at the ends is minimized then the number N satisfies the limit. Next take E as the interior of a circle of radius l in $\mathbb{R}^2$. Then the number N of *equal* squares of side ϵ necessary to cover E is such that $N(\epsilon) \sim \pi l^2\epsilon^{-2}$ as $\epsilon \to 0$. These and other examples with simple sets E give a relation of the form $N(\epsilon) \sim V\epsilon^{-d}$ as $\epsilon \to 0$ for fixed V, where d is the topological dimension of E in $\mathbb{R}^m$ and E is covered by hypercubes of side ϵ. This property suggests the generalization that if a set E is covered in this way such that

$$N(\epsilon) \sim V\epsilon^{-D} \qquad \text{as } \epsilon \to 0 \tag{1}$$

for some numbers D and V then D is the dimension of E, whether D is or is not an integer. Thus we define

$$D = \lim_{\epsilon \downarrow 0} \left\{ \frac{\ln N(\epsilon)}{\ln(\epsilon^{-1})} \right\}, \tag{2}$$

if the limit exists.

Example 4.1: the box dimension of Cantor's middle-thirds set. We consider first the set K_n of 2^n closed subintervals of length 3^{-n}. So to cover K_n with lines of length $\epsilon = 3^{-n}$ we need $N(\epsilon) = 2^n$ of them. Of course K may be covered in the same way, because $K_{n+1} \subset K_n$ and $K = \bigcap_{n=0}^{\infty} K_n$. Now we seek D such that equation (1) is satisfied. We find $\epsilon \to 0$ as $n \to \infty$ and

$$\begin{aligned} N(\epsilon) &\sim 2^n \qquad \text{as } n \to \infty \\ &= e^{n \ln 2} \\ &= (e^{n \ln 3})^{\ln 2/\ln 3} \\ &= \epsilon^{-D}, \end{aligned}$$

where

$$D = \ln 2/\ln 3 = 0.63093\ldots,$$

a result due to Hausdorff (1919, p. 172).

By the definition we require ϵ to be a continuous variable which tends to zero, although here we have used $\epsilon = 3^{-n}$ as the discrete variable $n \to \infty$. If $\epsilon \neq 3^{-n}$ for any integer n, then the covering of K may be 'wasteful', but the limit is essentially the same. □

Example 4.2: the von Koch curve. An interesting curve S shaped like a snowflake was proposed by von Koch (1904). It may be defined as follows. Let S_0 be the three sides of an equilateral triangle, each side having length one. Then the length of the curve S_0 is three. Let S_1 be the curve for which the middle third of each straight part of S_1 is replaced by two equal straight lines directed outwards, so that the replacement is two sides of an equilateral triangle of which the middle third is the base. It is easiest to see the definition of S_1 by looking at Fig. 4.3. Similarly we replace the middle thirds of each straight part of S_1 to define S_2, and define S_n iteratively. Then S may be defined as the limit of S_n as $n \to \infty$, i.e. $S = \{X\colon \exists$ a point $X_n \in S_n$ for $n = 0, 1, \ldots, \& X_n \to X$ as $n \to \infty\}$. (Note that $S_{n+1} \not\subset S_n$. So the limiting set S is *not* $\bigcap_{n=0}^{\infty} S_n$.) Now here S_n consists of 3×4^n line segments, each of length 3^{-n}. Therefore S_n can be covered by 3×4^n equal squares of side $\epsilon = 3^{-n}$. This gives

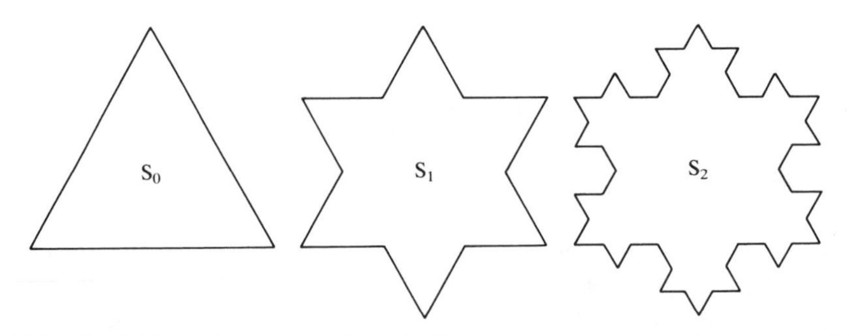

Fig. 4.3 The curves S_0, S_1 and S_2 leading to the definition of the von Koch curve S.

$$N(\epsilon) \sim 3 \times 4^n \qquad \text{as } n \to \infty$$
$$= 3\epsilon^{-D}$$

if $D = \ln 4/\ln 3 = 1.26186$. This gives the box dimension D of S plausibly. Also the length of S_n is given by

$$L(\epsilon) = \epsilon N(\epsilon) \tag{3}$$
$$= 3\epsilon^{1-D}.$$

Of course $L(\epsilon) \to \infty$ as $n \to \infty$, i.e. as $\epsilon \to 0$. Lastly note that the von Koch curve is self-similar in the sense that if a twelfth of S_{n+1} is magnified by a factor of three then it becomes identical to a third of S_n, for $n = 0, 1, \ldots$.

The von Koch curve resembles the shape of many coastlines and contours in some ways. Indeed, Richardson (1961) examined the western coastline of Britain and discovered empirically that it was approximately self-similar over many scales, the self-similarity being in a statistical sense that the coastline looks similar on any scale of magnification unless a specific feature can be recognized. He found that the length L of the coastline satisfies the relation

$$L(\epsilon) \approx V\epsilon^{1-D} \tag{4}$$

for 10 km $\lesssim \epsilon \lesssim$ 1000 km, where $D \approx 1.25$, by counting on maps, for various values of ϵ, the number $N = L/\epsilon$ of steps of length ϵ from point to point along the coast. He did this by 'walking' dividers whose points were a distance ϵ apart.

In the theory of box dimension we take the limit as $\epsilon \to 0$, but, of course, Richardson did not use measurements on scales smaller than those resolved by the maps available to him, let alone on the scales of metres, microns or infinitesimals. Nonetheless, the theoretical model of a Cantor set of non-integral dimension is a useful description of the observations made on a wide range of scales, and suggests an investigation of the geological reasons for different values of the dimension of different coastlines and contours. □

Mandelbrot (1975) defined a *fractal* as a set whose dimension is strictly greater than its topological dimension, i.e. $D > d$. However, the spirit of his approach is that, like life, a fractal is an easier thing to sense than to define rigorously. The essential property of a fractal is that its structure varies on all small length scales. This is seen, for example, in Cantor's middle-thirds set and in the attractor of the Hénon map. Similarly, definitions of strange attractor vary, but for most practical purposes we may regard attractors which are Cantor sets and fractal attractors as equivalent to one another and to a strange attractor. It is certainly helpful to describe as fractal a set whose box dimension is an integer but whose topological dimension is different: for example, the plane set of Q4.11 is fractal although its box dimension is one.

*We have simplified the concept of Hausdorff dimension in defining box dimension D. The essential difference is that for box dimension the set E is covered by hypercubes whereas for Hausdorff dimension the covering sets are quite general in shape and size. For a more formal approach to Hausdorff dimension, we define the *Hausdorff-s-measure* of $\mathrm{E} \subset \mathbb{R}^m$ as

$$\mathscr{H}^s(\mathrm{E}) = \lim_{\epsilon \to 0} \left\{ \inf \left(\sum_{i=1}^{\infty} |\mathrm{U}_i|^s \right) \right\},$$

where E is covered by sets U_i of diameter less than ϵ and the infimum is taken over all such ϵ-coverings. If there exists D such that $\mathscr{H}^s(\mathrm{E}) = 0$ for all $s > D$, and $\mathscr{H}^s(\mathrm{E}) = \infty$, for $s < D$, then D is the Hausdorff dimension of E; and we may often identify $\mathscr{H}^D(\mathrm{E})$ as being proportional to the quantity V introduced above. There are also other definitions of dimension which are not always integers. For the sets we shall meet, the values of the dimensions are mostly equal.

It is usually impossible to evaluate rigorously the dimension of a fractal, except for self-similar sets such as those in our examples. However, we can *compute* the dimension of a given set more easily. Many methods to compute D have been devised, because the calculation of D by use of the covering method is very time consuming. In these methods sometimes different definitions of dimension are used (although different definitions can occasionally give different values of the dimensions).

3 Renormalization group theory

3.1 *Introduction*

This section is an elaboration of some ideas raised in §3.4, so a review of that section would make a good foundation to the reading of this one. Note, in particular, Table 3.1 and Fig. 3.6. For new results first note that in fact not only is the nature of period doubling universal but so, in a sense to be seen soon, is the order of the p-cycles which arise at the bifurcations as the parameter increases. Thus Fig. 3.6 has the same *qualitative* appearance for a wide class of maps of which the logistic map is just one. This follows from a remarkable theorem.

Šarkovskii's theorem. If $F: \mathbb{R} \to \mathbb{R}$ is continuous, F has a k-cycle and $l \lhd k$ in the following ordering of all the positive integers, then F also has an l-cycle:

$$1 \lhd 2 \lhd 2^2 \lhd 2^3 \lhd 2^4 \lhd \cdots$$

$$\cdots$$

$$\cdots \lhd 2^3 \cdot 9 \lhd 2^3 \cdot 7 \lhd 2^3 \cdot 5 \lhd 2^3 \cdot 3$$

$$\cdots \lhd 2^2 \cdot 9 \lhd 2^2 \cdot 7 \lhd 2^2 \cdot 5 \lhd 2^2 \cdot 3$$

$$\cdots \lhd 2 \cdot 9 \lhd 2 \cdot 7 \lhd 2 \cdot 5 \lhd 2 \cdot 3$$

$$\cdots \lhd 9 \lhd 7 \lhd 5 \lhd 3.$$

This powerful theorem with so few hypotheses is due to Šarkovskii (1964); a simpler proof related by Devaney (1989) is more accessible. The theorem is valid only for one-dimensional maps. The converse of the theorem is in fact also true, i.e. if $l \lhd k$ then there exists a continuous function $F: \mathbb{R} \to \mathbb{R}$ such that F has a cycle of period l but not one of period k.

Note that first the powers of 2 are listed in ascending order, then the products of the powers of 2 (in descending order) and the odd numbers (in descending order). The theorem means, for example, that if F has a 10-cycle then it also has a 176-cycle, because $176 = 2^4 \cdot 11 \lhd 2 \cdot 5 = 10$. In particular, it implies that if F has a k-cycle where k is not a power of 2 then F has an infinity of cycles, and if F has a finite number of cycles then all their periods are powers of 2. It also has the following corollary.

Corollary. If F has a three-cycle then F has an l-cycle for all positive integers l.

This astonishing corollary has been epitomized as 'period three implies chaos' (Li & Yorke 1975). To understand the background to the epitome, recall that, although the theorem tells nothing about the stability of the l-cycles, experience of the logistic map in §3.4 suggests that almost all if not all the l-cycles will be unstable. So the cycles will play the role of the repellers in the metaphor of the pin-ball machine. Also recall that the logistic map $F(a, x) = ax(1 - x)$ has stable cycles in the 'windows' of its parameter a. For example, it can be seen in Fig. 3.5(b) that F has two stable six-cycles, first a six-cycle on its own account and secondly a six-cycle from the period doubling of the three-cycle. The six-cycle is visited by $F^n(a, x)$ in different orders in each of the two cases. Šarkovskii's theorem does not cover the multiplicity of a cycle of a given period, so it does not imply a universal order of the appearance of cycles at the bifurcations of a difference equation as a parameter increases. The theorem *suggests* period doubling of a k-cycle to $2^n \cdot k$-cycles for $k = 3, 5, \ldots$ as well as 2; in fact each of these sequences of period doublings leads to chaos with a Feigenbaum relation of the form (3.4.9) but with a different universal constant δ for each value of k. Again, it is possible that only a finite sequence of flip bifurcations occurs as a parameter increases, in which case there is no route to chaos by period doubling.

Now we move on to examine the detailed structure of period doubling. It is a good example of self-similarity. Period doubling is found to be characterized by a universal scale α for the state variable x as well as the scale δ for the parameter a. The structure of the period doubling is therefore revealed by *renormalization*, the name being used for 40 years by theoretical physicists to describe groups of scaling transformations in the theories of particle physics and of phase transitions. To explain renormalization group theory, we shall first introduce the concept of *superstability* and then the scales themselves.

Numerical calculations of the value a_r of a at which a 2^r-cycle arises from a flip bifurcation are especially difficult, because the cycle is very weakly stable when a is near to a_r and so computations over a long time are needed to calculate the eigenvalue accurately. However, calculations of the value A_r of a at which the 2^r-cycle is *most* stable are much easier. Accordingly we say that a cycle is *superstable* if it is as linearly stable as it can be, e.g. if the eigenvalue of F^{2^r} is $q = 0$ at each point of the 2^r-cycle

$\{X_1, X_2, \ldots, X_{2^r}\}$. For an example of superstability with a cycle of period one, the Newton–Raphson method to calculate a fixed point is superstable and so converges very rapidly once a close approximation to a fixed point has been found.

Example 4.3: the logistic map. It is shown in §3.4 that if $F(a, x) = ax(1 - x)$ then there is a stable fixed point, i.e. a 2^0-cycle, $X = (a - 1)/a$ for $1 < a \leqslant a_1 = 3$ with eigenvalue $q = 2 - a$. Therefore $A_0 = 2$ because $q = 0$ when $a = 2$; then $X = \frac{1}{2}$. Also there is a stable 2^1-cycle $\{X_1, X_2\}$ when $a_1 < a \leqslant a_2 = 1 + \sqrt{6}$ with $q = [\partial F^2(a, x)/\partial x]_{X_1} = F_x(a, X_1)F_x(a, X_2) = 4 + 2a - a^2$. Therefore A_1 is the zero of q such that $a_1 < A_1 < a_2$, i.e. $A_1 = 1 + \sqrt{5} = 3.236$; then $X_1 = \frac{1}{2}$, $X_2 = \frac{1}{4}(1 + \sqrt{5})$. □

Recall that, by use of the chain rule, the multiplier q determining the stability of the 2^r-cycle can be shown to have the same value $\prod_{j=1}^{2^r} F_x(a, X_j)$ at each point of the 2^r-cycle, so that $q = 0$ if and only if the derivative of F vanishes at one point of the cycle. Therefore, if F is a smooth convex function with a simple maximum at X_m then $q = 0$ if and only if $X_j = X_m$ for one value of j, i.e. if and only if X_m belongs to the 2^r-cycle, i.e.

$$F^{2^r}(A_r, X_m) = X_m.$$

Thus A_r is a value such that

$$F_x^{2^r}(A_r, X_m) = 0.$$

In fact if F is a smooth convex function then A_r is the unique value; indeed, as a increases from a_r to A_r to a_{r+1}, $F_x^{2^r}(a, X_1)$ decreases monotonically from 1 to 0 to -1.

We have established that if $a = A_r$ then X_m belongs to the 2^r-cycle. So the other points are $F^j(A_r, X_m)$ for $j = 1, 2, \ldots, 2^r - 1$. Of these points the closest to X_m is $F^{2^{r-1}}(A_r, X_m)$. To see why this is true, first note that each member of a 2^{r-1}-cycle of F is a fixed point of $F^{2^{r-1}}$ and the 2^r-cycle of F contains two 2^{r-1}-cycles of F. The 2^r-cycle of F bifurcates from the 2^{r-1}-cycle of F as a increases through a_r and the line $y = x$ cuts the curve $y = F^{2^{r-1}}(a, x)$ at the two-cycle as well as the fixed point of $F^{2^{r-1}}$ (see, e.g., Fig. 3.4). So, as a increases through a_r, the two points of the two-cycle of $F^{2^{r-1}}$ separate from the fixed point and one another. However, this leaves X_1, the point that becomes X_m when $a = A_r$, closest to $F^{2^{r-1}}(a, X_1)$ and these two points of the 2^r-cycle stay closest as a increases to A_r. To prepare to investigate the scaling of the separations of the points $\{X_1, X_2, \ldots, X_{2^r}\}$ of a 2^r-cycle for large r, define

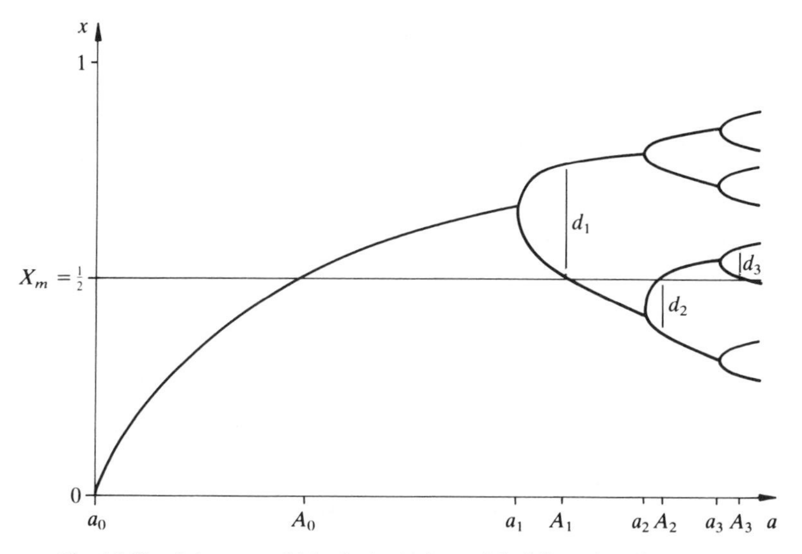

Fig. 4.4 Sketch (not to scale) in the (a, x)-plane of the bifurcation diagram of a one-dimensional map, showing the flip bifurcations and superstable 2^r-cycles.

$$d_r = F^{2^{r-1}}(A_r, X_m) - X_m \qquad \text{for } r = 1, 2, \ldots, \tag{1}$$

the distance from X_m to the nearest other member of the superstable 2^r-cycle. Then the location of the flip bifurcations and superstable cycles is summarized in Fig. 4.4.

3.2 *Feigenbaum's theory of scaling*

We are now ready to describe Feigenbaum's theory of period doubling, although in addition a knowledge of the elements of applied functional analysis will help. It is interesting that Feigenbaum's (1978) paper was rejected by the first journal to which it was submitted (Cvitanović 1984, p. 244). Feigenbaum calculated A_r and d_r numerically for several values of r and for a few functions F and concluded that

$$A_r = a_\infty - B\delta^{-r} + o(\delta^{-r}), \qquad d_r \sim D/(-\alpha)^r \qquad \text{as } r \to \infty, \tag{2}$$

where B, D are constants which depend upon the map F, but $\delta = 4.6692\ldots$ and $\alpha = 2.5029\ldots$ are 'universal' constants which do not. This shows that α is the x-scale of the route to chaos by period doubling much as δ is the a-scale. The scaling of d_r can be expressed as

$$\lim_{r\to\infty} (-\alpha)^r \{F^{2^r}(A_{r+1}, X_m) - X_m\} = -D/\alpha. \tag{3}$$

This leads to the further hypothesis that the limit

$$g_1(x - X_m) = \lim_{r\to\infty} \{g_{1r}(x - X_m)\} \tag{4}$$

exists, where $g_{1r}(x - X_m) = (-\alpha)^r\{F^{2^r}(A_{r+1}, X_m + (x - X_m)/(-\alpha)^r) - X_m\}$, for we see that (3) implies that $g_1(0) = -D/\alpha$, and calculations of $g_{1r}(x - X_m)$ for quite low values of r (rather than infinity) seem to confirm the existence of a limit g_1 independent of F. Then the scaling of $x - X_m$ shows that only the behaviour of F near to its maximum determines g_1 and so it is this behaviour which is responsible for the universality of g_1.

To make the notation a little less cumbersome it is convenient to translate the origin of x to the maximum X_m of F. So henceforth we shall simply put $X_m = 0$ without loss of generality.

Example 4.4: the logistic map. If $F(a, x) = ax(1 - x)$ then we may replace $x - X_m = x - \frac{1}{2}$ by x to get the new function $F(a, x) = a(\frac{1}{4} - x^2)$, ensuring that the maximum of F is now at $x = 0$. Then we find that

$$\begin{aligned} g_{10}(x) &= F(A_1, x) = A_1(\tfrac{1}{4} - x^2), \\ g_{11}(x) &= (-\alpha)F^2(A_2, x/(-\alpha)) \\ &= (-\alpha)A_2\{\tfrac{1}{4} - A_2^2(\tfrac{1}{4} - x^2/\alpha^2)^2\} \\ &= \alpha A_2\{\tfrac{1}{4}(\tfrac{1}{4}A_2^2 - 1) - \tfrac{1}{2}A_2^2x^2/\alpha^2 + A_2^2x^4/\alpha^4\}. \end{aligned}$$

We can similarly find $g_{1r}(x)$ for $r = 2, 3$, etc. and plot the curves $y = g_{1r}(x)$ in the (x, y)-plane to see that a limiting function g_1 seems to emerge as r increases. □

The essence of this scaling of x is captured by the operator T defined by

$$\mathrm{T}\psi(x) = -\alpha\psi(\psi(-x/\alpha)) \tag{5}$$

for all continuous functions ψ. Then

$$\begin{aligned} \mathrm{T}g_1(x) &= -\alpha g_1(g_1(-x/\alpha)) \\ &= -\alpha \lim_{r\to\infty} \{(-\alpha)^r F^{2^r}(A_{r+1}, (-\alpha)^r F^{2^r}(A_{r+1}, x/(-\alpha)^{r+1}))\}. \end{aligned}$$

Now define ϕ by $\phi(y) = (-\alpha)^r F^{2^r}(A_{r+1}, y/(-\alpha)^r)$, so $\phi^2(y) = \phi(\phi(y)) = (-\alpha)^r F^{2^{r+1}}(A_{r+1}, y/(-\alpha)^r)$. Then taking $y = x/(-\alpha)$, we deduce that

$$\begin{aligned} \mathrm{T}g_1(x) &= \lim_{r\to\infty} \{(-\alpha)^{r+1} F^{2^{r+1}}(A_{r+1}, x/(-\alpha)^{r+1})\} \\ &= \lim_{q\to\infty} \{(-\alpha)^q F^{2^q}(A_q, x/(-\alpha)^q)\}, \\ &= g_0(x), \qquad \text{say.} \end{aligned}$$

Similarly, it can be shown that

$$\mathrm{T}g_k(x) = g_{k-1}(x) \qquad \text{for } k = 2, 3, \ldots, \tag{6}$$

where g_k is defined by

$$g_k(x) = \lim_{r \to \infty} \{(-\alpha)^r F^{2^r}(A_{r+k}, x/(-\alpha)^r)\}. \tag{7}$$

Taking the limit as $k \to \infty$ in equation (6), we conclude plausibly that there exists a function

$$g(x) = \lim_{k \to \infty} g_k(x) \tag{8}$$

such that

$$\mathrm{T}g = g, \tag{9}$$

i.e. that there exists a 'fixed point' g of the nonlinear functional operator T. The famous equation (9) was discovered in a discussion between Cvitanović and Feigenbaum (1978, p. 46). We shall incidentally show later that the fixed point g is unstable.

Although we in fact know α from numerical solutions of difference equations, its proper status at this stage of the theory is a constant to be determined from equation (9). To find it first note that if $g(x)$ is a solution of equation (9) then so is $\mu g(x/\mu)$ for all $\mu \neq 0$. So we may, by convention, choose a particular value of μ such that

$$g(0) = 1. \tag{10}$$

Then, on putting $x = 0$ into equation (9) and using (5), it follows that

$$\alpha = -1/g(1). \tag{11}$$

Feigenbaum (1979) verified numerically the above scaling structure, and sought to find g as an even function by expanding $g(x)$ as a series in powers of x^2, truncating the series, and equating coefficients of successive powers of x^2 in equation (9). In this way he found that

$$g(x) = 1 - 1.52763x^2 + 0.10482x^4 - 0.02671x^6 + \cdots, \tag{12}$$

and $\alpha = -1/g(1) = 2.5029\ldots$. Thus α appears as a sort of nonlinear eigenvalue of the functional equation (9).

Example 4.5: quadratic approximation to g. To solve equation (9) for all x, where g is an even function and $g(0) = 1$, and then find $\alpha = -1/g(1)$ approximately, assume that $g(x) = 1 + bx^2$ for some constant b and neglect all higher powers of x. Then substitution into equation (9) gives

$$1 + bx^2 = -\alpha[1 + b\{1 + b(-x/\alpha)^2\}^2]$$
$$= -\alpha(1 + b + 2b^2x^2/\alpha^2 + b^3x^4/\alpha^4).$$

Equating coefficients of x^0 and x^2, and neglecting the term in x^4, we find that

$$1 = -\alpha(1 + b), \qquad b = -2b^2/\alpha.$$

Therefore $\alpha = -1/(1 + b) = -1/(1 - \frac{1}{2}\alpha)$. This gives a quadratic equation for α with solution $\alpha = 1 \pm \sqrt{3}$. But we require $\alpha > 1$. Therefore $\alpha = 1 + \sqrt{3} = 2.73\ldots$, and $b = -\frac{1}{2}\alpha = -1.37\ldots$. It is a crude approximation, but the example shows how to calculate g and α to higher approximations. □

Next we move on to find the scaling of the parameter a in the route to chaos by period doubling, evaluating δ. We shall show that

$$g_k(x) - g(x) \sim \text{constant} \times \delta^{-k}u_1(x) \qquad \text{as } k \to \infty, \tag{13}$$

where δ is the eigenvalue belonging to the first eigenfunction u_1 of the linear operator J_g defined as the Fréchet derivative of the nonlinear operator T evaluated at the 'point' g in the space of continuous functions.

Example 4.6: calculation of the Fréchet derivative of T. The *Fréchet derivative* J_ψ of the operator T 'at' ψ is defined by linearization of T about ψ, i.e. by the equation

$$\mathrm{T}(\psi + \epsilon\phi) = \mathrm{T}\psi + \epsilon \mathrm{J}_\psi\phi + o(\epsilon) \qquad \text{as } \epsilon \to 0 \tag{14}$$

for all (well-behaved) functions ϕ. To find J_ψ we expand

$$\mathrm{T}(\psi + \epsilon\phi) = -\alpha(\psi + \epsilon\phi)(\psi(-x/\alpha) + \epsilon\phi(-x/\alpha))$$
$$= -\alpha\psi(\psi(-x/\alpha) + \epsilon\phi(-x/\alpha)) - \alpha\epsilon\phi(\psi(-x/\alpha) + \epsilon\phi(-x/\alpha))$$
$$= -\alpha\psi(\psi(-x/\alpha)) - \alpha\psi'(\psi(-x/\alpha))\cdot\epsilon\phi(-x/\alpha)) + O(\epsilon^2)$$
$$- \alpha\epsilon\phi(\psi(-x/\alpha)) + O(\epsilon^2) \qquad \text{as } \epsilon \to 0,$$

on assuming that ψ, ϕ are well-behaved and taking a Taylor series,

$$= \mathrm{T}\psi + \epsilon \mathrm{J}_\psi\phi + O(\epsilon^2),$$

where J_ψ is defined by

$$\mathrm{J}_\psi\phi = -\alpha\psi'(\psi(-x/\alpha))\phi(-x/\alpha) - \alpha\phi(\psi(-x/\alpha)). \; \square \tag{15}$$

To proceed to find the a-scaling as $a \to a_\infty$, we expand

$$F(a,x) = F(a_\infty, x) + (a - a_\infty)f(x) + O\{(a - a_\infty)^2\} \quad \text{as } a \to a_\infty, \tag{16}$$

where $f(x) = F_a(a_\infty, x)$. (Of course, if $F(a,x) = af(x)$ then equation (16) is exact for all a on omission of the remainder term $O\{(a - a_\infty)^2\}$.) Therefore the Taylor expansion of the operator T acting on equation (16) gives

$$\begin{aligned}
\mathrm{T}F(a,x) &= -\alpha F(a, F(a, -x/\alpha)) \\
&= -\alpha F(a, F(a_\infty, -x/\alpha) + (a - a_\infty)f(-x/\alpha) + \cdots) \\
&= -\alpha F(a_\infty, F(a_\infty, -x/\alpha)) - \alpha(a - a_\infty)f(F(a_\infty, -x/\alpha)) + \cdots \\
&\quad - \alpha(a - a_\infty)f(-x/\alpha)F_x(a_\infty, F(a_\infty, -x/\alpha)) + \cdots \\
&= \mathrm{T}F(a_\infty, x) + (a - a_\infty)\mathrm{J}_{F(a_\infty, x)}f(x) + O\{(a - a_\infty)^2\} \quad \text{as } a \to a_\infty.
\end{aligned}$$

On iteration, this process gives

$$\begin{aligned}
\mathrm{T}^k F(a,x) &= \mathrm{T}^k F(a_\infty, x) + (a - a_\infty)\mathrm{J}_{\mathrm{T}^{k-1}F(a_\infty, x)}f(x) + O\{(a - a_\infty)^2\} \\
&= g(x) + (a - a_\infty)\mathrm{J}_g^k f(x) + O\{(a - a_\infty)^2\} + o(1)
\end{aligned} \tag{17}$$

as $a \to a_\infty$, $k \to \infty$.

To simplify equation (17) consider the eigenvalue problem

$$\mathrm{J}_g u = \lambda u$$

and suppose that it has eigenvalue λ_j belonging to the eigenfunction u_j for $j = 1, 2, \ldots$, where $\{u_j\}$ is a complete set of continuous functions over the interval on which f is positive, for example a complete set for $\mathrm{C}[-\frac{1}{2}, \frac{1}{2}]$ if $F(a,x) = a(\frac{1}{4} - x^2)$. Then

$$f(x) = \sum_{j=1}^{\infty} \xi_j u_j(x)$$

for some constants ξ_j. Therefore

$$\begin{aligned}
\mathrm{J}_g^k f(x) &= \sum_{j=1}^{\infty} \xi_j \lambda_j^k u_j(x) \\
&\sim \xi_1 \lambda_1^k u_1(x) \qquad \text{as } k \to \infty,
\end{aligned}$$

if we assume that $\xi_1 \neq 0$ and we may take $|\lambda_1| > |\lambda_j|$ for $j = 2, 3, \ldots$. Then let $\delta = \lambda_1$ and $h(x) = \xi_1 u_1(x)$. Therefore equation (17) gives

$$\mathrm{T}^k F(a,x) = g(x) + (a - a_\infty)\delta^k h(x) + o(1) + O\{(a - a_\infty)^2\}$$

$$\text{as } k \to \infty, a \to a_\infty.$$

Therefore

$$\mathrm{T}^r F(A_r, 0) - g(0) \sim (A_r - a_\infty)\delta^r h(0) \qquad \text{as } r \to \infty. \tag{18}$$

Now

$$\begin{aligned} \mathrm{T}F(A_r, 0) &= -\alpha F(A_r, F(A_r, 0)) \\ &= -\alpha F^2(A_r, 0). \end{aligned}$$

On iteration, this gives

$$\mathrm{T}^r F(A_r, 0) = (-\alpha)^r F^{2^r}(A_r, 0).$$

But X_m belongs to each superstable cycle, so $F^{2^r}(A_r, X_m) = X_m$, and, after translation of the maximum to the origin, this gives $F^{2^r}(A_r, 0) = 0$. Therefore

$$\mathrm{T}^r F(A_r, 0) = 0.$$

Therefore relation (18) gives

$$\begin{aligned} A_r - a_\infty &\sim -g(0)/\delta^r h(0) \qquad \text{as } r \to \infty \\ &= -\delta^{-r}/h(0), \end{aligned}$$

which was anticipated in the first of relations (2).

Feigenbaum (1980) also examined the Fourier spectrum of $\{F^n(x_0)\}$ for 2^r-cycles as $r \to \infty$.

All this, then, is Feigenbaum's heuristic theory of scaling of x and a in the route to chaos by period doubling. The astonishing ubiquity of Feigenbaum's sequence in period doubling of maps of $\mathbb{R}^m$ for $m > 1$, of solutions of differential equations, and of phenomena in laboratory experiments, stems from this theory for one-dimensional maps.

4 Liapounov exponents

In studying chaotic solutions (§§3.3, 3.4) we have met sensitive dependence on initial conditions and met simple examples of neighbouring orbits which separate exponentially. To be more formal we may define an infinite invariant set S of a map $F: \mathbb{R} \to \mathbb{R}$ to *have sensitive dependence on initial conditions* if there exists $\delta > 0$ such that for all $x \in \mathrm{S}$ and all neighbourhoods N (however small) of x there exists $y \in \mathrm{N}$ and $n > 0$ such that $|F^n(x) - F^n(y)| > \delta$. So neighbouring orbits, however close initially, separate from one another, although each keeps close to the invariant set.

It is, moreover, a characteristic of neighbouring chaotic orbits that their separation is an exponential function *on average*, though not necessarily an exact exponential function. It is this rapid separation which makes it impossible in practice to predict the behaviour of a chaotic solution far into the future. This is in contrast to the behaviour of an orbit near an attractor which is a fixed point or a periodic solution. These ideas can be quantified by use of what are called *Liapounov exponents.*

Consider then a continuously differentiable map $F: \mathbb{R} \to \mathbb{R}$ and suppose that there exists λ such that

$$|F^n(x_0 + \epsilon) - F^n(x_0)| \sim \epsilon \mathrm{e}^{n\lambda} \qquad \text{as } \epsilon \to 0, n \to \infty$$

provided that $\epsilon \mathrm{e}^{n\lambda} \to 0$ also, i.e.

$$\epsilon \left| \frac{\mathrm{d}F^n(x_0)}{\mathrm{d}x_0} \right| \sim \epsilon \mathrm{e}^{n\lambda} \qquad \text{as } n \to \infty,$$

to express the average exponential separation of the orbit starting at $x_0 + \epsilon$ from the orbit starting at x_0. Therefore

$$\lambda = \lim_{N\to\infty} \left\{ \frac{1}{N} \ln \left| \frac{\mathrm{d}F^N(x_0)}{\mathrm{d}x_0} \right| \right\}$$

$$= \lim_{N\to\infty} \{N^{-1} \ln |F'(x_{N-1}) F'(x_{N-2}) \ldots F'(x_0)|\},$$

where $x_n = F^n(x_0)$, on differentiating a function of a function and using induction (cf. Q3.5),

$$= \lim_{N\to\infty} \left\{ \frac{1}{N} \sum_{n=0}^{N-1} \ln |F'(x_n)| \right\}. \tag{1}$$

This shows that λ is a measure of the exponential separation of the neighbouring orbits averaged over all points of an orbit around an attractor.

We now may formally define the *Liapounov exponent* λ of an invariant set of F by the limit (1), if it exists. Sometimes e^{λ} is called a *Liapounov multiplier* or *Liapounov number*. In general λ depends on the initial point x_0 of the orbit, but it is the same for almost all x_0 in the domain of attraction of a given attractor. We see that for a stable cycle $\lambda < 0$ and neighbouring orbits converge (Q3.5), but that for a chaotic attractor $\lambda > 0$. The Liapounov exponent may be interpreted in terms of information theory as giving the rate of loss of information about the location of the initial point x_0 (Shannon & Weaver 1949) or in terms of Kolmogorov entropy as measuring the disorder of the system (Kolmogorov 1959).

In general λ can only be found by computation, but it can be evaluated analytically in some simple cases.

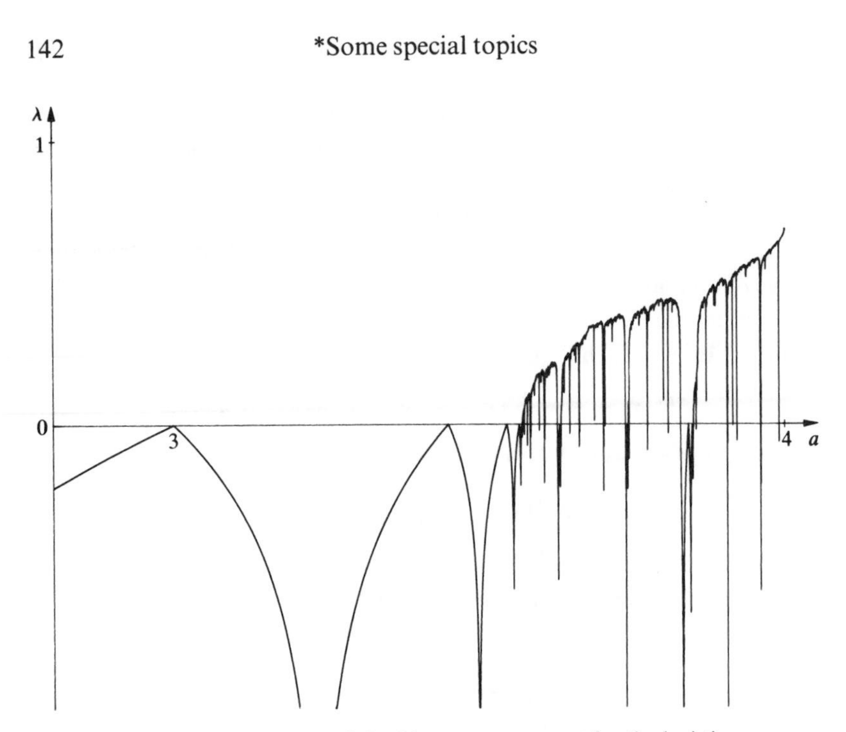

Fig. 4.5 Computed values of the Liapounov exponent for the logistic map $F(a, x) = ax(1 - x)$ for $2.9 < a < 4$.

Example 4.7: some simple one-dimensional maps. (i) First take the linear map $F(a, x) = ax$. Therefore $F^n(a, x) = a^n x$ and $|\mathrm{d}F^n(a, x)/\mathrm{d}x| = |a^n| = \mathrm{e}^{n \ln |a|}$. Therefore $\lambda = \ln |a|$. If the fixed point $X = 0$ is stable then $|a| \leqslant 1$ and $\lambda \leqslant 0$. Of course if $a > 1$ then orbits separate exponentially as they are repelled by the fixed point, but there is no attractor and no chaos.

(ii) For the Bernoulli shift in §3.3 we found that $\sigma^n(\theta_0 + \epsilon) - \sigma^n(\theta_0) = 2^n \epsilon$ as $\epsilon \to 0$ for fixed n, where $\theta_0 \in (0, 1)$ is fixed and σ is the sawtooth map. Therefore $\lambda = 2$. This is of the essence of sensitive dependence, although the ω-limit set of an aperiodic sequence is an invariant set but not an attractor because it does not attract all neighbouring points. None the less, it seems natural to define chaos so that the aperiodic solutions of the Bernoulli shift are chaotic.

(iii) For the logistic map $F(a, x) = ax(1 - x)$ computations are in general necessary to find λ. Some results are shown in Fig. 4.5. Note how the windows where a corresponds to a stable cycle have $\lambda < 0$ and the chaotic attractors have $\lambda > 0$. Also note the stable three-cycle and its period doubling as a increases above 3.8284. The imperfect resolution of the diagram smooths out some of the fine structure of the curve. □

There is no widely accepted definition of chaos for all nonlinear systems, but sensitive dependence is often used as a defining property. We may define F to be chaotic on an infinite invariant set S if F has sensitive dependence on initial conditions, if periodic points of F are dense in S, and if F is *topologically transitive*, i.e. if for all pairs of open sets A, B $\subset$ S there exists $n > 0$ such that $F^n(\mathrm{A}) \cap \mathrm{B} \neq \varnothing$. The condition of topological transitivity is added to ensure that the invariant set S is not decomposable into the union of two or more invariant sets.

Liapounov exponents may also be defined for higher-dimensional maps. Suppose that $\mathbf{F}\colon \mathbb{R}^m \to \mathbb{R}^m$ is a continuously differentiable map with $m \times m$ Jacobian matrix $\mathbf{J}(\mathbf{x})$, the element in the ith row and jth column being $\partial F_i/\partial x_j$. Let $q_1(\mathbf{F}^N(\mathbf{x})), \ldots, q_m(\mathbf{F}^N(\mathbf{x}))$ be the m eigenvalues of $\mathbf{J}(\mathbf{F}^N(\mathbf{x}))$, ordered according to decreasing magnitudes of their moduli. Then the m Liapounov exponents of an orbit $\{\mathbf{F}^N(\mathbf{x})\}$ as it approaches an attractor are defined by

$$\lambda_j = \lim_{N\to\infty} \left(\frac{1}{N} \ln|q_j(\mathbf{F}^N(\mathbf{x}))|\right) \qquad \text{for } j = 1, 2, \ldots, m, \tag{2}$$

$$= \lim_{N\to\infty} \left(\frac{1}{N} \sum_{n=0}^{N-1} \ln|q_j(\mathbf{F}^n(\mathbf{x}))|\right), \tag{3}$$

on using the chain rule with the Jacobian matrix of $\mathbf{F}^N$. Of course it is the largest of the Liapounov exponents which will in general give the exponential separation of neighbouring orbits.

Kaplan & Yorke (1979) conjectured that the Hausdorff dimension of an attractor in $\mathbb{R}^m$ is given by

$$D = k + \left(\sum_{j=1}^{k} \lambda_j\right) \Big/ |\lambda_{k+1}|, \tag{4}$$

where $\lambda_1 > \lambda_2 > \cdots > \lambda_m$ and k is the largest integer such that $\sum_{j=1}^{k} \lambda_j > 0$. It is now known that equation (4) is correct for most of the attractors likely to be met in practical problems but that the right-hand side of (4) is only an upper bound for the Hausdorff dimension. The advantage of computing D by means of formula (4) is that it is quicker to compute Liapounov exponents than to compute D by covering methods.

Further reading

§4.2 For a development of the theory of dimension and fractals, the book by Falconer (1990) is recommended. It covers many aspects with rigour and without excessive technical difficulties.

§4.3 Collect & Eckmann (1980) prove the properties of period doubling of one-dimensional maps.

§4.4 Falconer (1990, Chap. 13) treats the theory of Liapounov exponents with greater rigour.

Problems

Q4.1 *Self-similarity of Cantor's middle-thirds set.* Let K denote Cantor's middle-thirds set, and $\tau: [0,1] \to [0,1)$ the map defined by $\tau(x) \equiv 3x$ modulo 1, $0 \leqslant \tau < 1$. Then show that τ maps $K \cap [0, \frac{1}{3}]$ onto $K \cap [0,1)$, $K \cap [0, \frac{1}{9}]$ onto $K \cap [0, \frac{1}{3}]$, etc.

Q4.2 *Cantor's middle-thirds set.* (i) Prove that $\frac{1}{4}$ belongs to Cantor's middle-thirds set K. (ii) Defining $x = 2\sum_{n=1}^{\infty} 3^{-n(n+1)}$, show that $x \in K$ and that x is irrational.

Q4.3 *Self-similarity of a middle-halves set.* Define the map $\tau: \mathbb{R} \to [0,1)$ by $\tau(x) \equiv 4x$ modulo 1 and $0 \leqslant \tau < 1$. Then show that τ maps S and $S \cap [0, \frac{1}{4}]$ onto S (less the single point 1), where S is the middle-halves set of Q4.7.

Q4.4 *Invariance of Cantor's middle-thirds set under a tent map.* Consider the map $F: (1, \infty) \times \mathbb{R} \to \mathbb{R}$ defined by

$$F(a, x) = a(\tfrac{1}{2} - |x - \tfrac{1}{2}|).$$

Find the fixed points of F for $a > 1$ and their linear stability. Show that if $x < 0$ then $F(a,x) = ax$ and that if $x > 1$ then $F(a,x) < 0$. Deduce that $F^n(x) \to -\infty$ as $n \to \infty$ if $x < 0$ or $x > 1$.

Taking $a = 3$, show that if $\frac{1}{3} < x < \frac{2}{3}$ then $F(3,x) > 1$, that if either $0 \leqslant x \leqslant \frac{1}{3}$ or $\frac{2}{3} \leqslant x \leqslant 1$ then $0 \leqslant F(3,x) \leqslant 1$, and that if either $\frac{1}{9} < x < \frac{2}{9}$ or $\frac{7}{9} < x < \frac{8}{9}$ then $\frac{1}{3} < F(3,x) < \frac{2}{3}$. Hence or otherwise show plausibly that Cantor's middle-thirds set is invariant under the map F.

[Mandelbrot (1982, p. 181).]

Q4.5 *Invariance of a Cantor set under the logistic map.* Consider the map $F: (4, \infty) \times \mathbb{R} \to \mathbb{R}$ defined by $F(a,x) = ax(1-x)$. Show that if $x < 0$ then $F(a,x) < ax$ and that if $x > 1$ then $F(a,x) < 0$. Deduce that $F^n(x) \to -\infty$ as $n \to \infty$ if $x < 0$ or $x > 1$.

Show that if $\frac{1}{2} - (1 - 4/a)^{1/2} < x < \frac{1}{2} + (1 - 4/a)^{1/2}$ then $F(a,x) > 1$.

Speculate on the nature of the invariant set of F in the interval $[0,1]$.

[Fatou (1906), Mandelbrot (1982, pp. 182, 192).]

Q4.6 *A devil's staircase.* You are given that the continuous piecewise linear function F_n is defined over the interval $[0,1]$ by $F_n(0) = 0$, $F_n'(x) = 1/l_n$ if $x \in K_n$ and $F_n'(x) = 0$ if $x \notin K_n$, where K_n is the union of the 2^n closed intervals of length 3^{-n} used to define Cantor's middle-thirds set K and l_n is the length of K_n. Also F is the function defined by $F(x) = \lim_{n\to\infty} F_n(x)$ pointwise for $0 \leqslant x \leqslant 1$. Further, C_n is defined as the curve in the (x,y)-plane with equation $y = F_n(x)$ for $0 \leqslant x \leqslant 1$, i.e. as the graph of F_n, and C is defined as the graph of F.

(a) Show that $F_n(1) = 1$, $F_n(\frac{1}{2}) = \frac{1}{2}$ for $n = 0, 1, \ldots$. Sketch C_0 and C_2.

(b) Find the length of C_n and deduce that the length of C is 2.

(c) Show that $F'(x) = 0$ if $x \notin K$ and that F is continuous over $[0, 1]$. Deduce that F is not differentiable at $x \in K$.

(d) Interpret F as the probability distribution function of a random variable which is uniformly distributed over K.

Q4.7 *A middle-halves set.* Define $S_0 = [0, 1]$, $S_1 = [0, \frac{1}{4}] \cup [\frac{3}{4}, 1]$, $S_2 = [0, \frac{1}{16}] \cup [\frac{3}{16}, \frac{1}{4}] \cup [\frac{3}{4}, \frac{13}{16}] \cup [\frac{15}{16}, 1], \ldots$, so that S_{n+1} is S_n after the removal of the middle half of each subinterval, and thence define $S = \bigcap_{n=0}^{\infty} S_n$.

Show that each number $x \in S$ has the quaternary form $x = 0.x_1x_2x_3\ldots$, where the digit x_n is either 0 or 3 for $n = 1, 2, \ldots$.

Show that the box dimension of S is $\frac{1}{2}$.

[Cf. Falconer (1985, p. 15).]

Q4.8 *The box dimension of the Sierpiński gasket.* Let S_0 be the interior of an equilateral triangle with sides of length l. Regard S_0 as the union of the interiors of four equilateral triangles with sides length $\frac{1}{2}l$, and define S_1 as S_0 less the interior of the middle of the four triangles. Similarly, remove the middles of the three triangles of S_1 to construct S_2, and so forth (see Fig. 4.6). Define $S = \bigcap_{n=0}^{\infty} S_n$.

Show that the box dimension of S is $D = \ln 3/\ln 2 = 1.58496$.

[Sierpiński (1915), Eggleston (1953).]

Q4.9 *The box dimension of self-similar sets.* Define the set S as $S = \bigcap_{n=0}^{\infty} S_n$, where $S_n \subset \mathbb{R}$, S_0 is the unit interval, and S_{n+1} is similar to S_n such that S_{n+1} contains R times as many equal subintervals as S_n does, the length of each being of r times the length of those of S_n. Deduce that the box dimension is

$$D = \ln R/\ln(1/r).$$

State the values of R and r for Cantor's middle-thirds set and for the von Koch curve. Define the set $S \subset [0, 1]$ such that if $x \in S$ has decimal expression $x = 0.x_1x_2\ldots$ then each digit x_n is even, and show that $D = \log_{10} 5$.

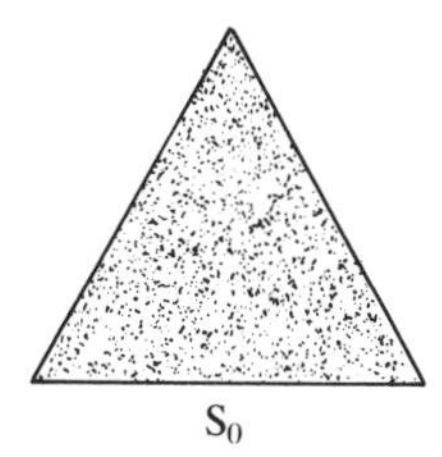

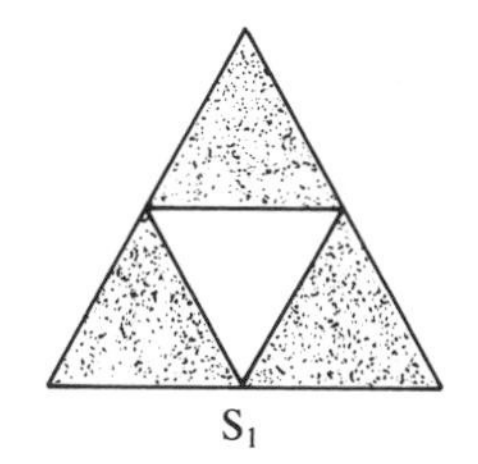

Fig. 4.6 Sketch of the sets for Q4.8.

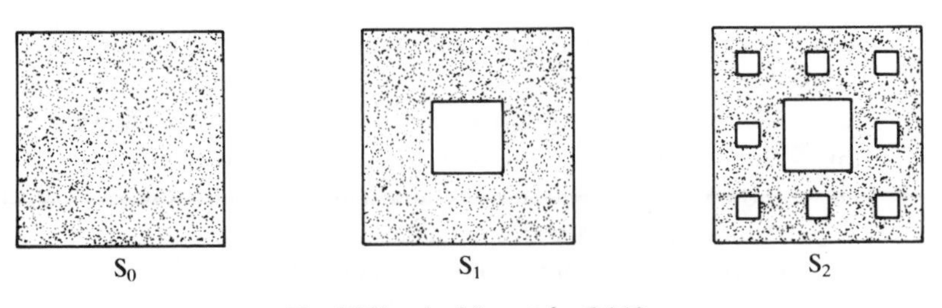

Fig. 4.7 Sketch of the sets for Q4.10.

Justify the application of the formula displayed above to the box dimension of self-similar nesting sets in $\mathbb{R}^m$, generalizing the definition of r.

[Mandelbrot (1982, p. 37).]

Q4.10 *The Sierpiński carpet.* Define self-similar sets $S_n \subset \mathbb{R}^2$ as follows, for $n = 0, 1, \ldots$. First define $S_0 = \{(x, y)\colon 0 \leqslant x, y \leqslant 1\}$, the unit square; $S_1 = \{(x, y)\colon (x, y) \in S_0 \;\&\; x \text{ or } y \notin (\frac{1}{3}, \frac{2}{3})\}$, i.e. S_1 is the union of eight of the nine subsquares of S_0 with sides of length $\frac{1}{3}$, the central subsquare being removed; S_2 as S_1 with the central subsubsquare of side $1/3^2$ removed from each subsquare of S_1, as shown in Fig. 4.7; and so forth. Hence define $S = \bigcap_{n=0}^{\infty} S_n$.

Show that the area of S is zero and that the box dimension of S is $D = \ln 8/\ln 3 = 1.8928$.

[Sierpiński (1916).]

Q4.11 *A self-similar plane set with dimension of a line segment.* Let S_0 be the square $[0, 1] \times [0, 1]$. Divide S_0 into 16 subsquares with sides of length $\frac{1}{4}$, and define S_1 as the union of the four corner subsquares. Similarly, remove the 12 non-corner subsquares of sides of length $1/4^2$ from each subsquare of S_1 to get S_2, as shown in Fig. 4.8, and define $S_3, S_4, \ldots$ similarly. Then define $S = \bigcap_{n=0}^{\infty} S_n$.

Show that $S = \{(x, y)\colon x = 0.x_1x_2\ldots \;\&\; y = 0.y_1y_2\ldots$ in quaternary form, where $x_n, y_n = 0$ or 3 for all $n\}$.

Show that the box dimension of S is $D = 1$.

Show that the orthogonal projection of S_n onto the line $y = -\frac{1}{2}x$ is the segment AB, where $A = (-\frac{2}{5}, \frac{1}{5})$ and $B = (\frac{4}{5}, -\frac{2}{5})$. Divide AB into 4^n equal

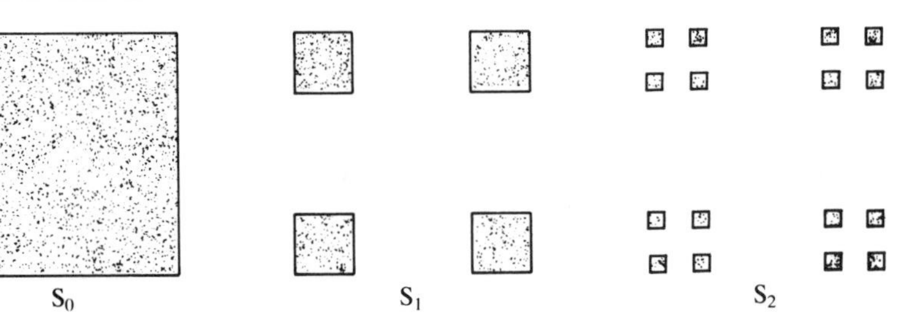

Fig. 4.8 Sketch of the sets for Q4.11.

subintervals. Deduce that to each subinterval there corresponds a subsquare of S_n and to each subsquare of S_n there corresponds a subinterval of AB.

[Cf. Falconer (1985, p. 16).]

Q4.12 *The Cartesian product of Cantor's middle-thirds set and an interval.* Show that if the plane set E is the Cartesian product of Cantor's middle-thirds set K and the unit interval, i.e. if $E = \{(x, y): x \in K, 0 \leqslant y \leqslant 1\}$, then the box dimension of E is $D = 1 + \ln 2/\ln 3$.

What is the box dimension of the Cartesian product of K and an m-dimensional hypercube with sides of unit length?

Q4.13 *The baker's transformation.* The map $\mathbf{F}: S \to S$ is defined by $\mathbf{F}(\mathbf{x}) = (\sigma(x), g(a, x, y))$, where $S = [0, 1) \times [0, 1)$ is the half-open unit square, $\mathbf{x} = (x, y)$,

$$\sigma(x) \equiv 2x \text{ modulo } 1 \qquad \text{and} \qquad 0 \leqslant \sigma < 1,$$

and

$$g(a, x, y) \equiv \left\{\begin{matrix} \tfrac{1}{2}ay & \text{for } 0 \leqslant x < \tfrac{1}{2} \\ \tfrac{1}{2}(ay + 1) & \text{for } \tfrac{1}{2} \leqslant x < 1 \end{matrix}\right\} \text{modulo } 1 \qquad \text{and} \qquad 0 \leqslant g < 1.$$

Show plausibly that if $0 < a < 1$ then the attractor of the orbit $\{\mathbf{F}^n(\mathbf{x})\}$ is topologically equivalent to the Cartesian product of the interval $0 \leqslant x < 1$ and a Cantor set in the y-direction, and the box dimension of the attractor is $D = 1 + \ln 2/(\ln 2 - \ln a)$.

[See Example 3.13.]

Q4.14 *A non-uniform Cantor set.* Let $S_0 = [0, 1]$, $S_1 = [0, \tfrac{1}{4}] \cup [\tfrac{2}{3}, 1]$, $S_2 = [0, \tfrac{1}{16}] \cup [\tfrac{1}{6}, \tfrac{1}{4}] \cup [\tfrac{2}{3}, \tfrac{3}{4}] \cup [\tfrac{8}{9}, 1], \ldots$, whereby the right-hand half of the left-hand half and the left-hand third of the right-hand half of each subinterval of S_n is removed to form S_{n+1}. The set S is defined by $S = \bigcap_{n=0}^{\infty} S_n$. Show that the box dimension of S is $D = \ln 2/\ln 3$.

Q4.15 *Proof of a very special case of Šarkovskii's theorem.* Prove that if $F: \mathbb{R} \to \mathbb{R}$ is continuous and F has a two-cycle then F has a fixed point.

[Hint: it may help to consider the sign of $F(x) - x$.]

Q4.16 *Proof of a special case of the converse of Šarkovskii's theorem.* Define the continuous function $F: [1, 5] \to [1, 5]$ such that $F(1) = 3$, $F^2(1) = 4$, $F^3(1) = 2$, $F^4(1) = 5$, $F^5(1) = 1$, and F is piecewise linear; and note that F has the five-cycle $\{1, 3, 4, 2, 5\}$. Sketch the graph of $y = F(x)$ in the (x, y)-plane for $1 \leqslant x \leqslant 5$. Show that F has a unique fixed point and evaluate it. Show that $F^3[1, 2] = [2, 5]$, $F^3[2, 3] = [3, 5]$, $F^3[4, 5] = [1, 4]$, and F^3 is monotonically decreasing on $[3, 4]$; hence or otherwise prove that F^3 has no three-cycle.

*Q4.17 *An approximate model of period doubling.* Consider period doubling for the logistic difference equation in the form $x_{n+1} = F(a, x_n)$, where

$$F(a, x) = 1 - ax^2.$$

Show that the fixed points are $X_\pm = \{-1 \pm (1+4a)^{1/2}\}/2a$ respectively, and that X_+ becomes unstable at a flip bifurcation as a increases through $\frac{3}{4}$. Show further that the second-generation map is given by $F(a, F(a,x)) = 1 - a + 2a^2x^2 - a^3x^4$.

We next assume that, in the Feigenbaum sequence, a stable 2^r-cycle becomes unstable as a increases through a_r, that $a_r \to a_\infty$, that the limit is determined by the behaviour of $F(a,x)$ near its maximum where $x = 0$, and that there are scaling properties as $r \to \infty$. Accordingly we may assume that x_n and $a - a_\infty$ are small, and proceed to develop an approximate model as follows.

Renormalize x_n by the factor α_0, where $\alpha_0 = -1/(a-1)$ and neglect x_n^4 to deduce that $x_{n+2}/\alpha_0 \simeq 1 - b_1(x_n/\alpha_0)^2$, where $b_1 = \phi(a)$ and $\phi(a) = 2a^2(a-1)$. Write this as $x_{n+2} = 1 - b_1x_n^2$, by change of notation. Repeat this scaling argument to show similarly that $x_{n+2^r} = 1 - b_rx_n^2$ approximately, where $b_r = \phi(b_{r-1})$. Deduce that if the 2^r-cycle becomes unstable when $a = a_r$, then $b_r = \frac{3}{4}$ and $\frac{3}{4} = \phi^{r-1}(a_r)$ approximately. This gives $a_r \to a_\infty$, where $a_\infty = \phi(a_\infty)$ approximately, i.e. $a_\infty \approx \frac{1}{2}(1+\sqrt{3}) = 1.37$, and $\alpha_r \to \alpha$ as $r \to \infty$, where $\alpha = -1/(a_\infty - 1) \approx -2.8$.

From the approximate relation $a_r = \phi(a_{r+1})$, deduce that

$$(a_\infty - a_{r+1})/(a_\infty - a_r) \to \delta \qquad \text{as } r \to \infty,$$

where $\delta \approx \phi'(a_\infty) \approx 4 + \sqrt{3} = 5.73$.

[Landau & Lifshitz (1987, §32).]

Q4.18 *The tent map.* Show that the Liapounov exponent of the tent map defined in Q3.22 is $\ln|a|$.

Q4.19 *The baker's transformation.* Show that the Liapounov exponents are $\lambda_1 = \ln 2 > 0$ and $\lambda_2 = -\ln(2/a) < 0$ for the baker's transformation defined in Example 3.13 and Q4.13. Verify that formula (4.4) gives the box dimension of the attractor correctly.

Q4.20 *Arnol'd's cat map.* Define $\mathbf{F}: \mathbb{T}^2 \to \mathbb{T}^2$, where $\mathbb{T}^2 = \mathbb{R}^2/\mathbb{Z}^2$ is the torus, by

$$\mathbf{F}(x,y) = (x + y \text{ modulo } 1, x + 2y \text{ modulo } 1).$$

Show that $\mathbf{F}$ is area-preserving.

Find the Liapounov exponents of $\mathbf{F}$.

[Falconer (1990, p. 196).]

5

Ordinary differential equations

Time present and time past
Are both present in time future,
And time future contained in time past.
If all time is eternally present
All time is unredeemable.
What might have been is an abstraction
Remaining a perpetual possibility
Only in a world of speculation.
What might have been and what has been
Point to one end, which is always present.

T.S. Eliot (*Burnt Norton*, I)

1 Introduction

We have introduced ordinary differential equations in Chapter 1, and considered some bifurcations of their equilibrium points. We shall next consider the points, their stability, attractors and bifurcations more systematically and in greater depth. In this chapter we make a few general remarks and definitions, and consider a few general properties, before examining special classes of equations in detail in later chapters. We shall observe many similarities but some dissimilarities between the properties of difference and differential equations.

We can represent an mth-order differential equation as a system of m first-order equations, much as we did for difference equations. (Again, the converse is in general false.) So we need consider only

$$\frac{\mathrm{d}\mathbf{x}}{\mathrm{d}t} = \mathbf{F}(\mathbf{x}, t) \tag{1}$$

for $\mathbf{F}: \mathbb{R}^m \times \mathbb{R} \to \mathbb{R}^m$. We shall assume that $\mathbf{F}$ is sufficiently well behaved for there to exist a unique solution $\mathbf{x}(t)$ of equation (1) such that $\mathbf{x}(0) = \mathbf{x}_0$ for given $\mathbf{x}_0 \in \mathbb{R}^m$; moreover we assume that $\mathbf{x}(t)$ is nonsingular, at least for t less than some finite value depending on $\mathbf{x}_0$ as well as $\mathbf{F}$. The system (1) is said to be *autonomous* if $\mathbf{F}$ does not depend on the independent variable t explicitly, and to be *non-autonomous* otherwise. We can transform a non-autonomous system (1) in $\mathbb{R}^m$ to an autonomous system in $\mathbb{R}^{m+1}$, much as we did for difference equations. To do this, identify $\mathbf{y} = [\mathbf{x}, y_{m+1}]^T \in \mathbb{R}^{m+1}$, choose $y_{m+1}(0) = 0$ and write equation (1) as the *suspended system*,

$$\frac{\mathrm{d}\mathbf{y}}{\mathrm{d}t} = \begin{bmatrix} \mathbf{F}(\mathbf{x}, t) \\ 1 \end{bmatrix}. \tag{2}$$

The $(m + 1)$th component ensures that $y_{m+1}(t) = t$ for all $t > 0$. Therefore

$$\frac{\mathrm{d}\mathbf{y}}{\mathrm{d}t} = \mathbf{G}(\mathbf{y}),$$

say, where $\mathbf{G}: \mathbb{R}^{m+1} \to \mathbb{R}^{m+1}$ is defined by $\mathbf{G}(\mathbf{y}) = [\mathbf{F}(\mathbf{y}), 1]^T$ and $\mathbf{x} = [y_1, \ldots, y_m]^T$. So we have been able to replace the explicit dependence of $\mathbf{F}$ on t by an equivalent implicit dependence of $\mathbf{G}$ on t. As a result, we may consider only autonomous systems, without loss of generality.

Example 5.1: Duffing's equation. Suppose that

$$\frac{\mathrm{d}^2 x}{\mathrm{d}t^2} + x - \tfrac{1}{3}x^3 = \Gamma \cos \omega t$$

for real parameters Γ and ω. Then define $x_1 = x$, $x_2 = \mathrm{d}x/\mathrm{d}t$, x_3 by $\mathrm{d}x_3/\mathrm{d}t = 1$ and $x_3(0) = 0$, and $\mathbf{x} = [x_1, x_2, x_3]^T \in \mathbb{R}^3$. Therefore $x_3(t) = t$ on integration, $\mathrm{d}x_1/\mathrm{d}t = \mathrm{d}x/\mathrm{d}t = x_2$, and $\mathrm{d}x_2/\mathrm{d}t = \mathrm{d}^2x/\mathrm{d}t^2 = \Gamma \cos \omega t - x + \frac{1}{3}x^3 = \Gamma \cos \omega x_3 - x_1 + \frac{1}{3}x_1^3$. Therefore $\mathrm{d}\mathbf{x}/\mathrm{d}t = \mathbf{F}(\mathbf{x})$, where $\mathbf{F}(\mathbf{x}) = [x_2, \Gamma \cos \omega x_3 - x_1 + \frac{1}{3}x_1^3, 1]^T$. □

Example 5.2: equivalence of an mth-order equation to m first-order equations. Given that $x^{(m)} = f(x, x', x'', \ldots, x^{(m-1)}, t)$, where $x' = \mathrm{d}x/\mathrm{d}t$ etc., define $\mathbf{y} = [x, x', \ldots, x^{(m-1)}]^T \in \mathbb{R}^m$. It follows that $\mathrm{d}\mathbf{y}/\mathrm{d}t = [x', x'', \ldots, x^{(m)}]^T = \mathbf{F}(\mathbf{y}, t)$, say, where we define $\mathbf{F}(\mathbf{y}, t) = [y_2, y_3, \ldots, y_m, f(\mathbf{y}, t)]^T \in \mathbb{R}^m$.

It is conventional to write computer routines to integrate a system of m ordinary differential equations rather than an mth-order equation, so the method of this example is important in computing solutions of a differential equation. □

Consider *orbits* $\{\mathbf{x}(t)\}$ by solving the equation

$$\frac{\mathrm{d}\mathbf{x}}{\mathrm{d}t} = \mathbf{F}(\mathbf{x}) \tag{3}$$

for $t > 0$, with initial condition $\mathbf{x}(0) = \mathbf{x}_0$ for given $\mathbf{x}_0 \in \mathbb{R}^m$. The orbit may be regarded as the path of a particle which starts at a given point $\mathbf{x}_0$ and moves with velocity $\mathbf{F}$ in the phase space $\mathbb{R}^m$ for $t > 0$. To describe the paths it is helpful to define the map $\boldsymbol{\phi}_s: \mathbb{R}^m \times \mathbb{R} \to \mathbb{R}^m$, called the *flow*, which maps each point $\mathbf{x}_0$ in phase space for each 'time' s to $\mathbf{x}(s)$, where $\mathbf{x}(t)$ is the solution of the above system such that $\mathbf{x}(0) = \mathbf{x}_0$. Thus the flow $\boldsymbol{\phi}_t$ is the analogue of the iterated map $\mathbf{F}^n$ for a difference equation $\mathbf{x}_{n+1} = \mathbf{F}(\mathbf{x}_n)$. Note that $\boldsymbol{\phi}_s(\boldsymbol{\phi}_t(\mathbf{x}_0)) = \boldsymbol{\phi}_s(\mathbf{x}(t)) = \mathbf{x}(s + t) = \boldsymbol{\phi}_{s+t}(\mathbf{x}_0)$, and $\boldsymbol{\phi}_t(\boldsymbol{\phi}_s(\mathbf{x}_0)) = \boldsymbol{\phi}_{s+t}(\mathbf{x}_0)$, similarly, so that $\boldsymbol{\phi}_s\boldsymbol{\phi}_t = \boldsymbol{\phi}_{s+t} = \boldsymbol{\phi}_t\boldsymbol{\phi}_s$, just as $\mathbf{F}^m\mathbf{F}^n = \mathbf{F}^{m+n} = \mathbf{F}^n\mathbf{F}^m$ for $m, n = 0, 1, 2, \ldots$. Therefore all the flows $\boldsymbol{\phi}_t$ for $t > 0$ are the elements of an Abelian semigroup (not a group because the inverse $\boldsymbol{\phi}_t^{-1} = \boldsymbol{\phi}_{-t}$ may not exist).

An *equilibrium point*, or *critical point*, $\mathbf{X}$ is a zero of $\mathbf{F}$, for if $\mathbf{x}_0 = \mathbf{X}$ and $\mathbf{F}(\mathbf{X}) = \mathbf{0}$ it follows that $\mathbf{x}(t) = \mathbf{X}$ for all $t > 0$. Note that an equilibrium point is a fixed point of the flow $\boldsymbol{\phi}_t$ for all t.

An equilibrium point $\mathbf{X}$ is said to be *stable* if $\forall\, \epsilon > 0\ \exists\ \delta(\epsilon)$ such that

$$|\mathbf{x}(t) - \mathbf{X}| < \epsilon\ \forall\, t > 0$$

$\forall\ \mathbf{x}_0$ such that $|\mathbf{x}_0 - \mathbf{X}| < \delta$. This definition of stability, analogous to the definition of stability of a fixed point of a difference equation in Chapter 3, is that the solutions $\mathbf{x}$ are uniformly continuous for $t > 0$ at $\mathbf{x} = \mathbf{X}$ with respect to variations of initial points $\mathbf{x}_0$. The point is *asymptotically stable* if it is stable as above and moreover

$$\mathbf{x}(t) \to \mathbf{X} \qquad \text{as } t \to \infty$$

$\forall\ \mathbf{x}_0$ such that $|\mathbf{x}_0 - \mathbf{X}| < \delta$. It is *globally asymptotically stable* if it is stable and $\mathbf{x}(t) \to \mathbf{X}$ for all $\mathbf{x}_0 \in \mathbb{R}^m$.

To consider the stability of a given point $\mathbf{X}$ of equilibrium we may, without loss of generality, suppose that $\mathbf{X} = \mathbf{0}$. To show this, define the perturbation

$$\mathbf{x}' = \mathbf{x} - \mathbf{X}.$$

Therefore

$$\begin{aligned} \frac{d\mathbf{x}'}{dt} &= \frac{d\mathbf{x}}{dt} \\ &= \mathbf{F}(\mathbf{x}) \\ &= \mathbf{F}(\mathbf{X} + \mathbf{x}') \\ &= \mathbf{H}(\mathbf{x}'), \end{aligned}$$

say. Then $\mathbf{H}(\mathbf{0}) = \mathbf{F}(\mathbf{X}) = \mathbf{0}$, as required.

So we shall examine the stability of the null solution of equation (3), assuming that $\mathbf{F}(\mathbf{0}) = \mathbf{0}$. Stability concerns small perturbations, so we shall linearize the system by approximating $\mathbf{F}$ for small $\mathbf{x}$. Now if $\mathbf{F}$ is continuously twice differentiable, then

$$\mathbf{F}(\mathbf{x}) = \mathbf{J}\mathbf{x} + O(\mathbf{x}^2) \qquad \text{as } \mathbf{x} \to \mathbf{0},$$

where the $m \times m$ constant matrix $\mathbf{J}$ is defined as the Jacobian matrix with element $[\partial F_i/\partial x_j]_0$ in its ith row and jth column. Therefore the linearized system of (3) at $\mathbf{X} = \mathbf{0}$ is

$$\frac{d\mathbf{x}}{dt} = \mathbf{J}\mathbf{x}. \tag{4}$$

To solve this linearized equation, we use the method of normal modes, assuming that $\mathbf{x}(t) = e^{st}\mathbf{u}$. Therefore

$$\mathbf{J}\mathbf{u} = s\mathbf{u}; \tag{5}$$

so the exponent s is the eigenvalue of $\mathbf{J}$ belonging to eigenvector $\mathbf{u}$. Let us suppose that the real matrix $\mathbf{J}$ has eigenvalue s_j belonging to eigenvector $\mathbf{u}_j$ for $j = 1, 2, \ldots, m$ and that the set $\{\mathbf{u}_1, \ldots, \mathbf{u}_m\}$ spans $\mathbb{R}^m$. (The case when $\mathbf{J}$ does not have m independent eigenvectors requires special treatment.) Then for all $\mathbf{x}_0 \in \mathbb{R}^m$ there exist $\xi_1, \xi_2, \ldots, \xi_m$ such that

$$\mathbf{x}_0 = \sum_{j=1}^{m} \xi_j \mathbf{u}_j. \tag{6}$$

It follows that the solution of equation (4) is

$$\mathbf{x}(t) = \sum_{j=1}^{m} \xi_j \exp(s_j t)\mathbf{u}_j \qquad \text{for } t \geqslant 0. \tag{7}$$

(Note that ξ_j, s_j and $\mathbf{u}_j$ are either a real triplet or one of a complex conjugate pair of triplets because $\mathbf{J}$ and $\mathbf{x}_0$ are real.) We infer asymptotic stability, with exponential decay of all modes if $\mathrm{Re}(s_j) < 0$ for $j = 1, 2, \ldots, m$ and instability of $\mathbf{X}$ if $\mathrm{Re}(s_j) > 0$ for at least one value of j. It is said that the equilibrium point $\mathbf{0}$ is *linearly stable* or *infinitesimally stable* if it is a stable point of equilibrium of the linearized system (4), whether it is a stable or an unstable point of equilibrium of the nonlinear system (3).

A *cluster point* or *limit point* $\mathbf{x}_\infty$ of an orbit $\{\mathbf{x}(t)\}$ as $t \to \infty$ is defined as a point such that $\forall\, \epsilon, \tau > 0\ \exists\, t_1$ such that

$$|\mathbf{x}(t_1) - \mathbf{x}_\infty| < \epsilon \qquad \text{and} \qquad t_1 > \tau.$$

We may qualify ϵ by 'however small' and τ by 'however large'. The definition implies the existence of a sequence $t_1 < t_2 < \cdots$ such that

$$\mathbf{x}(t_n) \to \mathbf{x}_\infty \qquad \text{and} \qquad t_n \to \infty \qquad \text{as } n \to \infty.$$

In particular, if $\mathbf{x}(t) \to \mathbf{X}$ as $t \to \infty$ then $\mathbf{X}$ is an equilibrium point which is a cluster point of the orbit. The set of cluster points is called the *ω-limit set* of the orbit. An *attractor* is an ω-limit set to which all neighbouring orbits tend as $t \to \infty$. Thus an attractor may be an asymptotically stable point of equilibrium or a *solution of period T*, i.e. a solution for which T is the least number such that $\mathbf{x}(t + T) = \mathbf{x}(t)\ \forall\ t$, to which neighbouring orbits tend. The *domain of attraction*, or *basin of attraction*, of an attractor $\mathrm{A} \subset \mathbb{R}^m$ is defined as

$$\mathrm{D(A)} = \{\mathbf{x}_0\colon \mathrm{d}\mathbf{x}/\mathrm{d}t = \mathbf{F}(\mathbf{x}), \mathbf{x}(0) = \mathbf{x}_0 \ \&\ \text{the } \omega\text{-limit set of } \{\mathbf{x}(t)\} \text{ is contained in A}\}.$$

An *α-limit set* and a *repeller* may be defined as analogues of ω-limit set and attractor respectively as $t \to -\infty$.

To consider bifurcations next, suppose that

$$\frac{\mathrm{d}\mathbf{x}}{\mathrm{d}t} = \mathbf{F}(\mathbf{a}, \mathbf{x})$$

for $\mathbf{F}\colon \mathbb{R}^l \times \mathbb{R}^m \to \mathbb{R}^m$. If the topology of the phase portrait changes with $\mathbf{a}$ in all neighbourhoods of $\mathbf{a}_0$ then we say that there is a bifurcation at $\mathbf{a}_0$, and $\mathbf{a}_0$ is a *bifurcation value* of $\mathbf{a}$. This definition, at the cost of vagueness, deals with the bifurcations of unsteady as well as steady solutions of the differential system.

2 Hamiltonian systems

Next we examine the evolution of the volume of a set of points governed by equation (1.3), and, in particular the case in which the volume is constant for all initial sets.

First trace the evolution of a single orbit for a little time. Now

$$\mathbf{x}(t + h) = \mathbf{x}(t) + h\mathbf{F}(\mathbf{x}(t)) + o(h) \qquad \text{as } h \to 0$$

if $\mathbf{F}$ is continuously differentiable. Therefore the Jacobian of the transformation $\boldsymbol{\phi}_h\colon \mathbf{x}(t) \mapsto \mathbf{x}(t + h)$ is

$$\det\left\{\frac{\partial x_i(t + h)}{\partial x_j(t)}\right\} = \det\left\{\delta_{ij} + h\left[\frac{\partial F_i}{\partial x_j}\right]_{\mathbf{x}(t)} + o(h)\right\}$$
$$= 1 + h \operatorname{div} \mathbf{F} + o(h) \qquad \text{as } h \to 0,$$

where $\operatorname{div} \mathbf{F}$, i.e. $\boldsymbol{\nabla} \cdot \mathbf{F}$, $= [\partial F_i/\partial x_i]_{\mathbf{x}(t)} = \operatorname{trace} \mathbf{J}(\mathbf{x})$, because the determinant consists of ones in the diagonal with a term of order h added to each element (whether the element is on or off the diagonal). It follows (§3.3.2) that the system is measure-preserving, i.e. that the volume of a set of the 'particles' is independent of t, if

$$\operatorname{div} \mathbf{F} = 0. \tag{1}$$

This may be recognized as the equation of continuity of an incompressible fluid composed of the particles.

An especially important subset of measure-preserving systems is the set of Hamiltonian systems, important because of their omnipresence in the theories of classical and quantum mechanics.

A *Hamiltonian system* is of the form

$$\frac{\mathrm{d}\mathbf{q}}{\mathrm{d}t} = \frac{\partial H}{\partial \mathbf{p}}, \qquad \frac{\mathrm{d}\mathbf{p}}{\mathrm{d}t} = -\frac{\partial H}{\partial \mathbf{q}} \tag{2}$$

for a *Hamiltonian function*, or simply *Hamiltonian*, $H(\mathbf{p}, \mathbf{q}, t)$. It arises (Q5.12) from a generalized form of Newton's laws of motion, where $\mathbf{p} \in \mathbb{R}^n$ is the generalized momentum and $\mathbf{q} \in \mathbb{R}^n$ the generalized coordinate of a mechanical system of degree of freedom n; then H is usually the sum of the kinetic and potential energies. We may rewrite the system in the form

$$\frac{\mathrm{d}\mathbf{x}}{\mathrm{d}t} = \mathbf{F}(\mathbf{x}, t),$$

where $\mathbf{F}: \mathbb{R}^{2n} \times \mathbb{R} \to \mathbb{R}^{2n}$, $\mathbf{x} = (\mathbf{q}, \mathbf{p})$ and $\mathbf{F} = (\partial H/\partial \mathbf{p}, -\partial H/\partial \mathbf{q})$. We see that

$$\operatorname{div} \mathbf{F} = \sum_{i=1}^{n} \frac{\partial}{\partial q_i}\left(\frac{\partial H}{\partial p_i}\right) + \sum_{i=1}^{n} \frac{\partial}{\partial p_i}\left(-\frac{\partial H}{\partial q_i}\right) = 0.$$

It follows that a Hamiltonian system preserves volume in the $2n$-dimensional phase space. This is *Liouville's theorem* of analytical dynamics.

Example 5.3: the simple harmonic oscillator. When $H(p, q) = \frac{1}{2}(p^2/m + m\omega^2 q^2)$

$$\frac{\mathrm{d}q}{\mathrm{d}t} = \frac{\partial H}{\partial p} = p/m, \qquad \frac{\mathrm{d}p}{\mathrm{d}t} = -\frac{\partial H}{\partial q} = -m\omega^2 q,$$

so

$$\frac{\mathrm{d}^2 q}{\mathrm{d}t^2} + \omega^2 q = 0. \square$$

Note that if a Hamiltonian system is stationary, i.e. if $H = H(\mathbf{p}, \mathbf{q})$, then

$$\begin{aligned}\frac{\mathrm{d}H}{\mathrm{d}t} &= \frac{\partial H}{\partial \mathbf{p}} \cdot \frac{\mathrm{d}\mathbf{p}}{\mathrm{d}t} + \frac{\partial H}{\partial \mathbf{q}} \cdot \frac{\mathrm{d}\mathbf{q}}{\mathrm{d}t} \\ &= \frac{\mathrm{d}\mathbf{q}}{\mathrm{d}t} \cdot \frac{\mathrm{d}\mathbf{p}}{\mathrm{d}t} - \frac{\mathrm{d}\mathbf{p}}{\mathrm{d}t} \cdot \frac{\mathrm{d}\mathbf{q}}{\mathrm{d}t} \\ &= 0. \end{aligned} \tag{3}$$

Therefore H is constant. If we identify H as the total energy of the mechanical system then this represents the conservation of energy. Accordingly we sometimes call systems with $\operatorname{div}\mathbf{F} = 0$ measure-preserving or *non-dissipative*, and those with $\operatorname{div}\mathbf{F} < 0$ *dissipative*. In this way measure preservation is associated with frictionless systems. We see that each orbit is confined to the hypersurface, $H(\mathbf{p}, \mathbf{q}) = \text{constant}$, in $\mathbb{R}^{2n}$ if H is independent of t; this hypersurface has topological dimension $2n - 1$ in general.

3 The geometry of orbits

In the phase plane of a two-dimensional autonomous system, the attractors are no more complicated than a point or a closed curve, giving a periodic solution, because of the special topology of a plane: if an orbit starts inside or outside a closed orbit then it remains inside or outside respectively for all time, because two orbits can cross only at an equilibrium point, i.e. at a zero of $\mathbf{F}$. We shall discuss the phase plane in detail in the next chapter. First, we shall take a glimpse at attractors in $\mathbb{R}^m$ for $m \geqslant 3$, where a closed curve does not have an inside and an outside, and see that chaotic attractors are possible. Note, however, that a non-autonomous second-order system is equivalent to a third-order system and so may have chaotic attractors; also $\mathbf{F}(\mathbf{x}, t)$ may have different values at the same point $\mathbf{x}$ of the phase plane at different times on an orbit, so the orbit may cross itself.

When $m \geqslant 3$ attractors may have rich structures. A closed orbit still represents a periodic solution. Also an orbit may, for example, wind around a torus $\mathbb{T}^2$. To illustrate this possibility we can consider a geometrical model without reference to a particular differential system. Suppose then that a point P moves on $\mathbb{T}^2 \subset \mathbb{R}^3$ with angles (θ, ϕ) shown in Fig. 5.1

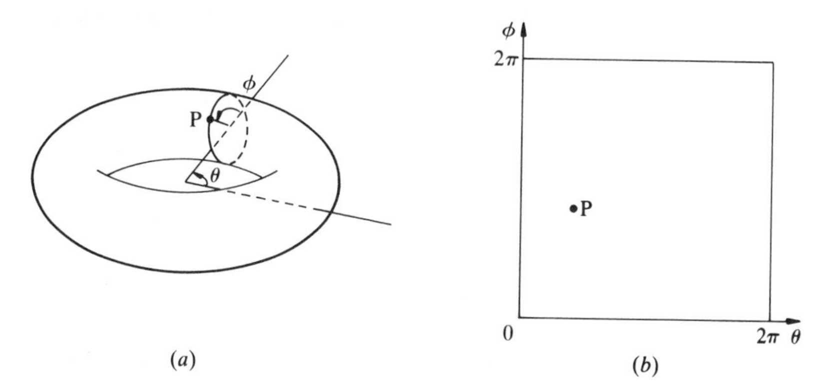

Fig. 5.1 (*a*) The torus $\mathbb{T}^2$ and its polar angles θ and ϕ. (*b*) Map of the torus onto the half-open square $[0, 2\pi) \times [0, 2\pi)$.

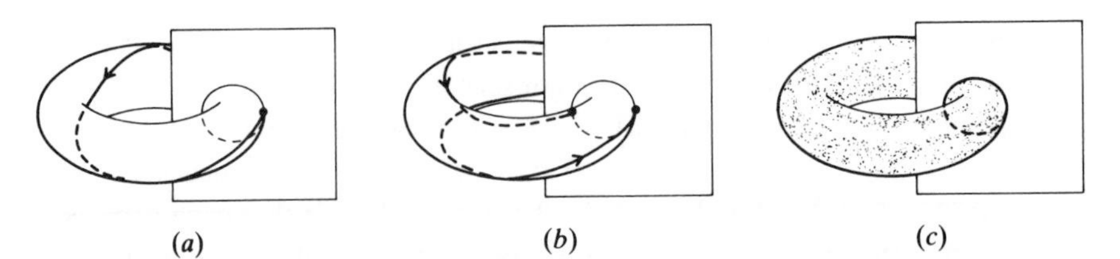

Fig. 5.2 Orbits on a torus. (*a*) Periodic solution $\mathbf{x}(t)$ and fixed point of the Poincaré map. (*b*) Periodic solution $\mathbf{x}(t)$ and two-cycle of the Poincaré map. (*c*) Quasi-periodic solution $\mathbf{x}(t)$ cutting circle densely in the section of the Poincaré map.

such that

$$\theta = 2\pi kt \qquad \text{and} \qquad \phi = 2\pi lt \qquad \text{for } t \geqslant 0,$$

where k, l are real. To picture the orbits on the torus it is helpful to use the Poincaré map on the plane section $\theta = 0$. Now P returns to the plane whenever $2\pi kt = 2n\pi$ for $n = 1, 2, \ldots$, i.e. when $t = n/k$, at which times $\phi = 2\pi ln/k$. Therefore the Poincaré map is the rotation map of points on the circle cut by the torus on the plane $\theta = 0$, $\phi_{n+1} = \phi_n + 2\pi l/k$, or $\phi_{n+1}/2\pi = \phi_n/2\pi + l/k$ (see Example 3.9). It follows that the orbit of the Poincaré map and thence the orbit of P on the torus is periodic if and only if l/k is rational; in that case $l/k = q/p$ for coprime integers p and q, and P winds p times around the torus in the θ-way and q times in the ϕ-way before it returns to its starting point. A few cases, relating the Poincaré map to orbits on the torus, are sketched in Fig. 5.2.

The case with irrational l/k gives an example $\mathbf{x}(t)$ of what is called a quasi-periodic function with two fundamental frequencies, namely $2\pi k$ and $2\pi l$. Quasi-periodicity is a natural generalization of periodicity. A *quasi-periodic function* $\mathbf{x}: \mathbb{R} \to \mathbb{R}^m$ is defined as one for which there exists a function $\mathbf{P}: \mathbb{R}^n \to \mathbb{R}^m$ and *fundamental frequencies* $f_1, f_2, \ldots, f_n$ such that

(a) $\mathbf{x}(t) = \mathbf{P}(f_1 t/2\pi, f_2 t/2\pi, \ldots, f_n t/2\pi)$ for all t, where $\mathbf{P}$ has period 1 in each of its n arguments, and
(b) $n \geqslant 2$ and no frequency is a rational multiple of another.

(Note that we need condition (b) to ensure that quasi-periodic functions differ from periodic functions; because if $n = 1$ or, for example, $f_1 = 2f_2 = \cdots = nf_n$, then $\mathbf{x}$ becomes a periodic function.)

We have not met analogues of quasi-periodic solutions in Chapter 3 because difference equations have solutions of only integral periods, and all integers are rational multiples of one another.

Example 5.4: some quasi-periodic functions. (i) $x(t) = \cos t + 3\cos\sqrt{2}t$ gives a quasi-periodic function $x\colon \mathbb{R} \to \mathbb{R}$ with fundamental frequencies 1 and $\sqrt{2}$. (ii) $\mathbf{x}(t) = [5\cos 2t \sin \pi t, 1 + \sin 2t + \sin 3\pi t]^{\mathrm{T}}$ gives a quasi-periodic function $\mathbf{x}\colon \mathbb{R} \to \mathbb{R}^2$ with fundamental frequencies 2 and π. □

Reverting to our discussion of the orbit of P on the torus, we may envisage a topologically equivalent attractor, so that neighbouring orbits approach the given attractor as $t \to \infty$. Then the Poincaré map on a plane section would give a closed curve as an attractor, and the map of the curve would be topologically equivalent to the rotation map.

More complicated solutions $\{\mathbf{x}(t)\} \subset \mathbb{R}^m$ for $m \geqslant 3$ may approach attractors of non-integral dimension. They are sometimes called *strange attractors* and found to be associated with chaotic solutions which have sensitive dependence on initial conditions. The attractors with $m = 3$, let alone $m > 3$, have not been classified yet.

In summary of this chapter so far, the concepts of steady solution, periodic solution, quasi-periodic solution, stability, a flow, an orbit, its ω-limit set, an attractor, and a domain of attraction have been described. These are the fundamental concepts of the qualitative theory of ordinary differential equations. A change in the topological character of an attractor as a parameter varies is a bifurcation, which may lead to symmetry breaking or to chaos. Detailed applications of these concepts will be developed in the remainder of the book, starting with a description of the stability of a periodic solution in the next section.

*4 The stability of a periodic solution

As a parameter increases, a periodic solution of a nonlinear differential system may become unstable and bifurcate, thereby leading to a quasi-periodic solution with two fundamental frequencies, just as a steady solution may become unstable and lead to a periodic solution at a Hopf bifurcation. The bifurcation sequence of an equilibrium point, followed by a periodic solution and then a quasi-periodic solution is represented geometrically by attractors in phase space as the sequence of fixed point, S^1 and $\mathbb{T}^2$ when the parameter increases. So the stability of a periodic solution is fundamental to this sequence of bifurcations.

We shall in this section first define the stability of a periodic solution and then sketch a theory of finding when a periodic solution is stable. The theory is a well-known method to treat the linear stability of a periodic solution, analogous to the method of normal modes to treat the linear stability of a steady solution. It is called *Floquet theory*; physicists sometimes call it *Bloch theory*.

First we shall define the stability of a periodic solution in two different ways. Suppose that the differential system

$$\frac{d\mathbf{x}}{dt} = \mathbf{F}(\mathbf{x}), \tag{1}$$

for $\mathbf{F}: \mathbb{R}^m \to \mathbb{R}^m$, has a known solution $\mathbf{X}$ of period T, so that $\mathbf{X}(t + T) = \mathbf{X}(t)$ for all t. Then $\mathbf{X}$ is represented by a closed curve in phase space. The solution is said to be *stable in the sense of Liapounov*, or simply *stable*, if we replace $\mathbf{X}$ by $\mathbf{X}(t)$ in the definition of stability of an equilibrium point in §1, i.e. if $\mathbf{x}(t) - \mathbf{X}(t)$ is small for all $t \geqslant 0$ whenever $\mathbf{x}(0) - \mathbf{X}(0)$ is sufficiently small. It follows then that the limit cycle of §1.6 is stable.

However, this definition of stability is often stricter than is desirable, because we are often more concerned with the orbit in phase space rather than where on the orbit the solution is at a given time. To understand this, consider the example of the system

$$\frac{dx}{dt} = -y(x^2 + y^2)^{1/2}, \qquad \frac{dy}{dt} = x(x^2 + y^2)^{1/2}.$$

On taking plane polar coordinates such that $x = r\cos\theta$, $y = r\sin\theta$, $r \geqslant 0$, it follows easily (§6.1) that

$$\frac{dr}{dt} = 0, \qquad \frac{d\theta}{dt} = r.$$

Therefore

$$r(t) = r_0, \qquad \theta(t) = r_0 t + \theta_0,$$

i.e.

$$x(t) = r_0\cos(r_0 t + \theta_0), \qquad y(t) = r_0\sin(r_0 t + \theta_0),$$

and the solution has period $2\pi/r_0$. This solution, $\mathbf{X}$ say, is unstable, because if there is a neighbouring solution, $\mathbf{X}_1$ say, such that $r = r_0 + \delta$, $\theta = \theta_0$ at $t = 0$ for $\delta > 0$ then

$$|\mathbf{X}_1(t) - \mathbf{X}(t)| = \{(r_0 + \delta)^2 - 2r_0(r_0 + \delta)\cos(\delta t) + r_0^2\}^{1/2}$$

does not remain small for all $t > 0$, however small $|\mathbf{X}_1(0) - \mathbf{X}(0)| = \delta$ may be. Nonetheless, the *orbit* of $\mathbf{X}_1$ in the phase plane is the circle $r = r_0 + \delta$, which does remain close to the orbit $r = r_0$ of $\mathbf{X}$ for all $t > 0$ when δ is sufficiently small. It is only the phases θ of $\mathbf{X}_1$ and $\mathbf{X}$ which do not remain close. So we might wish to deem this solution $\mathbf{X}$ stable by defining stability differently.

Accordingly, a solution $\mathbf{X}$ is said to be *orbitally stable*, or *stable in the sense of Poincaré*, if the orbits $\{\mathbf{x}(t)\}$ for $t \geqslant 0$ of all neighbouring solutions remain close to the orbit of $\mathbf{X}$ in phase space. Then stability in the sense of Liapounov implies stability in the sense of Poincaré. However, stability in the sense of Poincaré does not imply stability in the sense of Liapounov, the system of the previous paragraph being a counterexample: the periodic solution described there is unstable but orbitally stable.

The orbit $\{\mathbf{X}(t)\}$ for $t \geqslant 0$ of a given solution, periodic or otherwise, may be pictured as a curve surrounded by a tube of radius ϵ in phase space $\mathbb{R}^m$. If all orbits starting at $t = 0$ within distance δ of $\mathbf{X}(0)$ remain within the tube for all $t > 0$ then $\mathbf{X}$ is orbitally stable. If, moreover, there is only small 'shear' between the orbit of $\mathbf{X}$ and neighbouring orbits within the tube then $\mathbf{X}$ is stable in the sense of Liapounov.

The essence of Floquet theory is to treat the stability of the periodic solution $\mathbf{X}$ of system (1) as follows and as illustrated in Fig. 5.3. Consider an orbit of $\mathbf{x}(t)$ which starts at $\mathbf{x}(0)$ near $\mathbf{X}(0)$, so that the perturbation $\mathbf{x}' = \mathbf{x} - \mathbf{X}$ of the orbit is governed by the linearized system of (1). Then $\mathbf{L}$: $\mathbf{x}'(0) \mapsto \mathbf{x}'(T)$ generates a linear map of all $\mathbf{x}(0)$ near $\mathbf{X}(0)$ and the evolution of the solution $\mathbf{x}(t)$ for $t \geqslant T$ may be found by iterating the map $\mathbf{L}$. Suppose that $\mathbf{L}$ has eigenvalues $q_1, q_2, \ldots, q_m$. Therefore $\mathbf{L}^n\{\mathbf{x}'(0)\} \to \mathbf{0}$ as

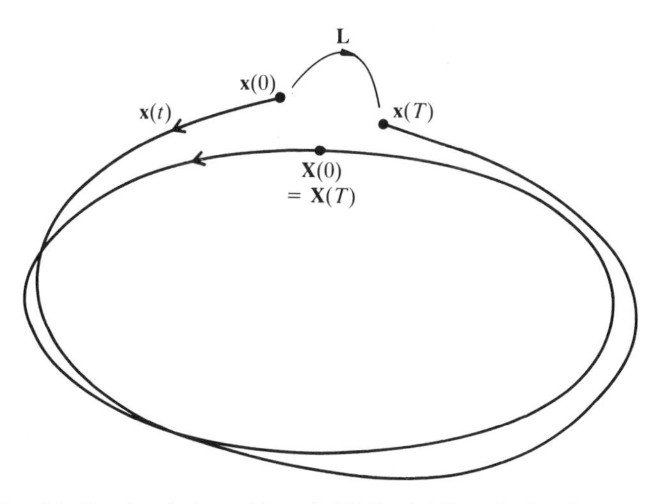

Fig. 5.3 Sketch of the orbits of $\{\mathbf{X}(t)\}$, $\{\mathbf{x}(t)\}$ and the linear map $\mathbf{L}$: $\mathbf{x}(0) - \mathbf{X}(0) \mapsto \mathbf{x}(T) - \mathbf{X}(0)$.

$n \to \infty$, i.e. $\mathbf{x}(nT)$ approaches $\mathbf{X}(nT) = \mathbf{X}(0)$, if $|q_j| < 1$ for $j = 1, 2, \ldots, m$; in this case $\{\mathbf{x}(t)\}$ is attracted by the closed orbit $\{\mathbf{X}(t)\}$ and so the periodic solution $\mathbf{X}$ is asymptotically stable (and orbitally stable). Similarly, if $|q_j| > 1$ for at least one eigenvalue q_j then $\mathbf{X}$ is unstable (but could be orbitally stable if the eigenvector $\mathbf{u}_j$ were tangential to the orbit of $\mathbf{X}$ at $t = 0$).

The method may be expressed in more detail analytically. The linearized system of (1) for small perturbations $\mathbf{x}' = \mathbf{x} - \mathbf{X}$ is

$$\frac{\mathrm{d}\mathbf{x}'}{\mathrm{d}t} = \mathbf{J}(\mathbf{X}(t))\mathbf{x}', \tag{2}$$

where the Jacobian matrix $\mathbf{J}$ has elements $[\partial F_i/\partial x_j]_\mathbf{x}$ as usual. Now the linearized system (2) has coefficients which are functions of t with period T. There is a complete analytic theory of the solution of such linear systems due to Floquet (1883). In particular (cf. Q5.17), it can be proved that for each *fundamental matrix* $\mathbf{\Phi}$ of system (2), i.e. for each $m \times m$ nonsingular matrix such that

$$\frac{\mathrm{d}\mathbf{\Phi}}{\mathrm{d}t} = \mathbf{J}(\mathbf{X}(t))\mathbf{\Phi},$$

there exists an $m \times m$ nonsingular matrix $\mathbf{P}$ with period T and a constant $m \times m$ matrix $\mathbf{R}$ such that

$$\mathbf{\Phi}(t) = \mathbf{P}(t)\mathrm{e}^{t\mathbf{R}}. \tag{3}$$

We can in general diagonalize $\mathbf{R}$ by use of similar matrices. Then there exist m linearly independent solutions of the Floquet system (2) such that

$$\phi_j(t) = p_j(t)\exp(s_j t) \qquad \text{for } j = 1, 2, \ldots, m, \tag{4}$$

where $p_j(t + T) = p_j(t)$ for all t, and s_j is an eigenvalue of $\mathbf{R}$. The *Floquet exponent* s_j is related to the *Floquet multiplier* q_j by $q_j = \exp(s_j T)$. So the condition for linear stability is that $|q_j| \leqslant 1$, i.e. $\mathrm{Re}(s_j) \leqslant 0$, for all j. To calculate $\mathbf{R}$, and hence the Floquet exponents, it is necessary to integrate m independent solutions of the system (2) over only one period, from $t = 0$ to T, say.

If a solution $\mathbf{X}$ of period T depends on a parameter such that a Floquet multiplier q_1 decreases through -1 as the parameter increases, then the perturbation develops a period double that of the basic solution $\mathbf{X}$. This is called a *subharmonic instability*. It may lead, as the parameter increases, to a bifurcation with the origin of a solution with double the period of the old one. Again, if $q_1, q_2 = \bar{q}_1$ have a modulus which increases through one as the parameter increases then there may be a bifurcation with the origin of a quasi-periodic solution whose fundamental frequencies are $2\pi/T$ and $\text{Im}(\ln q_1)/T$ initially; such a bifurcation is similar to a Hopf bifurcation in the sense that the solution acquires an extra frequency (and phase) at the bifurcation.

Further reading

§5.1 We have assumed that the right-hand side of the differential system (1) is well enough behaved for the solutions of the initial-value problem to exist and to be unique. Many good books on the analytic theory of ordinary differential equations, e.g. Coddington & Levinson (1955, Chap. 1) give the relevant conditions and theorems. Also some conditions for which stability or instability for the linearized system determine the stability of an equilibrium point are stated and proved by, e.g., Coddington & Levinson (1955, Chap. 13).

It is difficult to exaggerate the contributions of Poincaré to the theory of ordinary differential equations described in Chapters 5–8; much of his work was first published in a series of papers in the 1880s. Some is reported in his famous book on the three-body problem, *Les Méthodes Nouvelles de la Mécanique Céleste* (Poincaré 1892, 1893, 1899). In addition to finding results which bear his name, he, for example, framed the ideas and coined the names of limit cycle, node, saddle point, focus, and he anticipated chaos.

§5.4 For further reading on Floquet theory, Coddington & Levinson (1955, pp. 78–81, 218–20) and Jordan & Smith (1987, pp. 245–60) are recommended.

Problems

Q5.1 *A Poincaré map of an equivalent autonomous system.* Show that the non-autonomous first-order equation

$$\frac{dz}{dt} = -z + \epsilon \cos t$$

is equivalent to the autonomous second-order system,

$$\frac{dz}{dt} = -z + \epsilon \cos \theta, \qquad \frac{d\theta}{dt} = 1, \qquad \text{with } \theta(t_0) = t_0.$$

Represent the orbit $\{\mathbf{x}(t)\} \subset S^1 \times \mathbb{R}$ with cylindrical polar coordinates (r, θ, z) as points $(1, \theta(t), z(t))$ on the surface of a cylinder with unit radius, so that $(1, \theta, z)$ and $(1, \theta + 2n\pi, z)$ coincide for $n = \pm 1, \pm 2, \ldots$. Then construct the Poincaré map $P^{\theta_0}: \Sigma \to \Sigma$, where $\Sigma = \{(r, \theta, z): r = 1, \theta = \theta_0\}$ is the line of intersection of the orbits with the half-plane $\theta = \theta_0$, deducing that

$$P^{\theta_0}(1, \theta_0, z) = (1, \theta_0, z e^{-2\pi} + \tfrac{1}{2}\epsilon(1 - e^{-2\pi})(\cos\theta_0 + \sin\theta_0)).$$

Find the fixed point Z of P^{θ_0} and show that $Z \to 0$ as $\epsilon \to 0$ for all θ_0. For what values of θ_0, ϵ is Z stable?

Q5.2 *Gradient systems.* Show that if $V: \mathbb{R}^m \to \mathbb{R}$ is continuously differentiable,

$$\frac{d\mathbf{x}}{dt} = -\operatorname{grad} V(\mathbf{x}),$$

and $\mathbf{x}_\infty$ belongs to the ω-limit set of an orbit, then $\mathbf{x}_\infty$ is an equilibrium point.

Q5.3 *An unstable point of equilibrium which is linearly stable.* Show that if

$$\frac{dx}{dt} = 2x^2 y, \qquad \frac{dy}{dt} = -2xy^2,$$

then the origin is an equilibrium point. Show also that xy is constant on each orbit. Deduce that the origin is unstable.

Show that the origin is a stable point of equilibrium for the linearized system, namely

$$\frac{dx}{dt} = 0, \qquad \frac{dy}{dt} = 0.$$

Q5.4 *Euler's equations of motion of a rigid body about a fixed point.* It is given that a rigid body freely rotating with angular velocity $\boldsymbol{\omega} = (\omega_1, \omega_2, \omega_3)$ about a fixed point (i.e. a point fixed to an inertial frame) is governed by the equations

$$A\frac{d\omega_1}{dt} = (B - C)\omega_2\omega_3, \qquad B\frac{d\omega_2}{dt} = (C - A)\omega_3\omega_1,$$

$$C\frac{d\omega_3}{dt} = (A - B)\omega_1\omega_2,$$

where A, B, C are the principal moments of inertia of the body about its fixed point. Show that

$$T = \tfrac{1}{2}(A\omega_1^2 + B\omega_2^2 + C\omega_3^2), \qquad h^2 = A^2\omega_1^2 + B^2\omega_2^2 + C^2\omega_3^2$$

are constants of the motion.

Show that the steady solution $\omega_1 = n$, $\omega_2 = \omega_3 = 0$ is stable, with small oscillations of period $2\pi n^{-1}\{BC/(A - B)(A - C)\}^{1/2}$, if either $A > B > C$ or $A < B < C$, and that it is unstable if $B > A > C$. Discuss the relevance of this result to the motion of a tennis racket which is spun and thrown into the air.

Q5.5 *The Rössler system.* Consider the system

$$\frac{dx}{dt} = -y - z, \qquad \frac{dy}{dt} = x + ay, \qquad \frac{dz}{dt} = b + z(x - c),$$

Find the equilibrium points, and conditions for their existence. Sketch the bifurcation diagram in the (c, x)-plane for fixed $a, b > 0$, naming the bifurcation points.

[Rössler (1976).]

Q5.6 *The pitchfork bifurcation of the Lorenz system.* Show that $\mathbf{X} = \mathbf{0}$ is an equilibrium point of the system

$$\frac{d\mathbf{x}}{dt} = [\sigma(y - x), rx - y - zx, -bz + xy]^{\mathrm{T}}$$

for $\mathbf{x} = [x, y, z]^{\mathrm{T}}$, and positive parameters r, b, σ. Linearize the system about the null solution to find the eigenvalues s and eigenvectors of the normal modes proportional to e^{st}.

Show that the eigenvalue s_1 of one mode satisfies $s_1 \sim \sigma(r - 1)/(\sigma + 1)$ and that a corresponding eigenvector $\mathbf{u}_1 \to [1, 1, 0]^{\mathrm{T}}$ as $r \to 1$.

(a) Show that the system for steady solutions may be written without approximation as

$$x - y = 0,$$

$$x - y = zx - (r - 1)x,$$

$$bz = xy.$$

Also define $\delta = \frac{1}{2}[1, 1, 0]\mathbf{x}$, so that

$$x + y = 2\delta.$$

Now, assuming that there exist steady solutions of the form

$$\mathbf{x} = \delta\mathbf{x}_1 + \delta^2\mathbf{x}_2 + \ldots, \qquad r = 1 + \delta r_1 + \delta^2 r_2 + \cdots \qquad \text{as } \delta \to 0$$

for fixed b and σ, substitute these expansions into the four equations, and show that $\mathbf{x}_1 = [1, 1, 0]^{\mathrm{T}}$ and $r_1 = 0$. Find r_2 and $\mathbf{x}_2$.

(b) Next define $\epsilon = (r - 1)^{1/2}$, $A = [1, 1, 0]\mathbf{x}/2\epsilon$ and $u = \epsilon^2 t$ for $r > 1$, and assume that there exist unsteady solutions of the form

$$\mathbf{x}(t) = \epsilon\mathbf{x}_1(u) + \epsilon^2\mathbf{x}_2(u) + \ldots, \qquad A' = F_0(A) + \epsilon F_1(A) + \cdots \qquad \text{as } \epsilon \downarrow 0,$$

where a prime denotes differentiation with respect to u. Show first that the system may be expressed as

$$x - y = -\epsilon^2 x'/\sigma,$$

$$x - y = zx - \epsilon^2 x + \epsilon^2 y',$$

$$bz = xy - \epsilon^2 z',$$

without approximation. Then show that

$$\mathbf{x}_1 = A[1, 1, 0]^{\mathrm{T}}, \qquad F_0(A) = \frac{\sigma}{\sigma + 1} A(1 - A^2/b).$$

Also find $\mathbf{x}_2$.

[See §8.1.]

Q5.7 *Nonlinear oscillations of a conservative system.* Show that the equation

$$\frac{\mathrm{d}^2 x}{\mathrm{d}t^2} + V'(x) = 0$$

of §1.7 can be represented by the Hamiltonian $H(p, q) = \frac{1}{2}p^2 + V(q)$.

Q5.8 *Mathieu's equation.* Show that the equation,

$$\frac{\mathrm{d}^2 x}{\mathrm{d}t^2} + (a - 2q \cos 2t)x = 0,$$

where a, q are constants, is given by the Hamiltonian $H(y, x) = \frac{1}{2}(ax^2 + y^2) - qx^2 \cos 2t$.

Q5.9 *The Hénon–Heiles system.* Find $\mathrm{d}p_1/\mathrm{d}t$, $\mathrm{d}p_2/\mathrm{d}t$, $\mathrm{d}q_1/\mathrm{d}t$, $\mathrm{d}q_2/\mathrm{d}t$ for the system with Hamiltonian $H(p_1, p_2, q_1, q_2) = \frac{1}{2}(p_1^2 + p_2^2 + q_1^2 + q_2^2) + q_1^2 q_2 - \frac{1}{3}q_2^3$.

[Hénon & Heiles (1964).]

Q5.10 *The Toda lattice.* Find the equations of the system with Hamiltonian

$$H(p_i, q_i) = \frac{1}{2} \sum_{i=1}^{n} p_i^2 + \sum_{i=1}^{n-1} \exp(q_i - q_{i+1}).$$

[This is a model of a monatomic crystal as a lattice of particles connected by nonlinear springs (Toda 1967).]

Q5.11 *Finite-difference schemes of approximation of a differential equation.* Suppose that

$$\frac{\mathrm{d}\mathbf{x}}{\mathrm{d}t} = \mathbf{F}(\mathbf{x}),$$

where $\mathbf{F}: \mathbb{R}^2 \to \mathbb{R}^2$. This differential equation may be approximated by the difference equation

$$\mathbf{x}_{n+1} = \mathbf{x}_n + h\mathbf{f}(h, \mathbf{x}_n)$$

for small $h > 0$ and a function $\mathbf{f}: \mathbb{R} \times \mathbb{R}^2 \to \mathbb{R}^2$ such that $\mathbf{f}(h, \mathbf{x}) = \mathbf{F}(\mathbf{x}) + o(h)$ as $h \to 0$. Various functions $\mathbf{f}$ are chosen by numerical analysts, for example, $\mathbf{f}(h, \mathbf{x}_n) = \mathbf{F}(\mathbf{x}_n)$ (*explicit Euler scheme*), $\mathbf{f}(h, \mathbf{x}_n) = \mathbf{F}(\mathbf{x}_{n+1})$ (*implicit Euler scheme*), $\mathbf{f}(h, \mathbf{x}_n) = \frac{1}{2}\{\mathbf{F}(\mathbf{x}_n) + \mathbf{F}(\mathbf{x}_{n+1})\}$ (*trapezoidal scheme*), and $\mathbf{f} = \mathbf{F}(\mathbf{x}_n + \frac{1}{2}h\mathbf{f})$ (*implicit mid-point scheme*). It can be shown by, say, expanding $\mathbf{f}$ in powers of h, that area preservation of the differential equation does not imply area preservation of the difference equation unless $\mathbf{f}$ is chosen carefully.

Taking $\mathbf{F}(\mathbf{x}) = [y, -x]^{\mathrm{T}}$, show that if

$$\frac{dx}{dt} = y, \qquad \frac{dy}{dt} = -x$$

then the system is area preserving and $x^2 + y^2$ is constant on each orbit.

(a) Show that by the explicit Euler scheme

$$x_{n+1} = x_n + hy_n, \qquad y_{n+1} = y_n - hx_n,$$

and $x_{n+1}^2 + y_{n+1}^2 = (1 + h^2)(x_n^2 + y_n^2)$.

(b) Show that by the implicit Euler scheme

$$x_{n+1} = x_n + hy_{n+1}, \qquad y_{n+1} = y_n - hx_{n+1},$$

and $x_{n+1}^2 + y_{n+1}^2 = (x_n^2 + y_n^2)/(1 + h^2)$.

(c) Show that by the trapezoidal scheme

$$x_{n+1} = x_n + \tfrac{1}{2}h(y_n + y_{n+1}), \qquad y_{n+1} = y_n - \tfrac{1}{2}h(x_n + x_{n+1}),$$

the system is area preserving, and $x_{n+1}^2 + y_{n+1}^2 = x_n^2 + y_n^2$.

(d) Show that by the implicit mid-point scheme

$$x_{n+1} = x_n + h(y_n - \tfrac{1}{2}hx_n)/(1 + \tfrac{1}{4}h^2), \quad y_{n+1} = y_n - h(x_n + \tfrac{1}{2}hy_n)/(1 + \tfrac{1}{4}h^2),$$

the system is area preserving, and $x_{n+1}^2 + y_{n+1}^2 = x_n^2 + y_n^2$.

Q5.12 *Deduction of Hamilton's equations from Lagrange's equations.* You are given that a system of particles with one degree of freedom is governed by the variational principle,

$$\delta \int_{t_1}^{t_2} L(q, \dot{q}, t)\,dt = 0,$$

where q is the *generalized coordinate*, $\dot{q} = dq/dt$ and L is the *Lagrangian*. Deduce that

$$\frac{d}{dt}\frac{\partial L}{\partial \dot{q}} - \frac{\partial L}{\partial q} = 0.$$

[The solution $q(t)$ from $t = t_1$ to t_2 gives an extremum of the integral for given function L and values $q(t_1)$ and $q_2(t_2)$. Also $L = T - V$, where T is the kinetic energy and V is the potential energy of the system.]

Define the *Hamiltonian* H by $H(p, q, t) = p\dot{q} - L(q, \dot{q}, t)$ and the new variable p, called the *generalized momentum*, by $p = \partial L/\partial \dot{q}$, where $\dot{q}$ is considered to be some function of p and q. Thence show that the differential

$$\frac{\partial H}{\partial p}dp + \frac{\partial H}{\partial q}dq + \frac{\partial H}{\partial t}dt = dH$$

$$= \left(p - \frac{\partial L}{\partial \dot{q}}\right)d\dot{q} + \dot{q}\,dp - \frac{\partial L}{\partial q}dp - \frac{\partial L}{\partial t}dt$$

for all dp, dq, dt. Deduce that

$$\frac{dq}{dt} = \frac{\partial H}{\partial p}, \qquad \frac{dp}{dt} = -\frac{\partial H}{\partial q}, \qquad \frac{\partial H}{\partial t} = -\frac{\partial L}{\partial t}.$$

Q5.13 *ABC flows.* Consider the flow of a fluid with velocity distribution $\mathbf{u}: \mathbb{R}^3 \to \mathbb{R}^3$ such that $\mathbf{u}(\mathbf{x}) = (B\cos y + C\sin z, C\cos z + A\sin x, A\cos x + B\sin y)$ and $\mathbf{x} = (x, y, z)$ for given real constants A, B, and C. Show that $\operatorname{curl}\mathbf{u} = \mathbf{u}$. [*It follows that the flow is a *Beltrami flow*, a Beltrami flow being defined as one for which $\operatorname{curl}\mathbf{u}$ is parallel to $\mathbf{u}$.] Show also that

$$\tan X \tan Y \tan Z = -1, \qquad A^2 = B^2\sin^2 Y + C^2\cos^2 Z$$

at a *stagnation point* $\mathbf{X} = (X, Y, Z)$, i.e. a point where $\mathbf{u}(\mathbf{X}) = \mathbf{0}$. Taking A, B, $C \geqslant 0$, deduce that there is no stagnation point in the flow unless there exists an acute-angled triangle with sides of lengths A, B and C. Show that if (X, Y, Z) is a stagnation point then so is $(X + \pi, Y + \pi, Z + \pi)$.

Denote the half-open cube $[0, 2\pi) \times [0, 2\pi) \times [0, 2\pi)$ by $\mathbb{T}^3 = \mathbb{R}^3/(2\pi\mathbb{Z})^3$. Then consider the flow in $\mathbb{T}^3$ given by the orbits for $t > 0$, where

$$\frac{d\mathbf{x}}{dt} = \mathbf{u}(\mathbf{x}),$$

$\mathbf{x}(0) = (x_0, y_0, z_0)$, and the components of $\mathbf{x}$ are reduced periodically modulo 2π so that $0 \leqslant x_0,\ y_0,\ z_0 < 2\pi$. Show that the flow is volume preserving (i.e. incompressible) and that a stagnation point is an equilibrium point.

Show that if $A = B = C \neq 0$ then there are eight equilibrium points, namely $(\frac{1}{4}\pi, \frac{5}{4}\pi, \frac{3}{4}\pi)$, $(\frac{5}{4}\pi, \frac{1}{4}\pi, \frac{7}{4}\pi)$, their cyclic permutations, $(\frac{3}{4}\pi, \frac{3}{4}\pi, \frac{3}{4}\pi)$ and $(\frac{7}{4}\pi, \frac{7}{4}\pi, \frac{7}{4}\pi)$. By examination of the linearized system of differential equations, show that $(\frac{1}{4}\pi, \frac{5}{4}\pi, \frac{3}{4}\pi)$ is unstable.

Deduce that if $B = C = 0$ and $A \neq 0$ then

$$\mathbf{x}(t) \equiv (x_0, y_0 + At\sin x_0, z_0 + At\cos x_0) \quad \text{modulo } 2\pi.$$

Use this solution to determine the Poincaré map of the half-open face of the cube with $z = 0$ and $0 \leqslant x, y < 2\pi$, i.e. show that if $\mathbf{x}(0) = (x_n, y_n, 0)$ then the orbit next cuts the face in the point $(x_{n+1}, y_{n+1}, 0)$, where

$$x_{n+1} = x_n, \qquad y_{n+1} \equiv y_n + 2\pi\tan x_n \quad \text{modulo } 2\pi.$$

and $0 \leqslant y_{n+1} < 2\pi$, provided that $A\cos x_n > 0$. Hence describe briefly the iterates of the Poincaré map according to the values of the initial point (x_0, y_0, z_0).

[Hénon (1966), Dombre *et al.* (1986).]

*Q5.14 *Invariant set of a measure-preserving map and recurrence.* Consider the system $\frac{d\mathbf{x}}{dt} = \mathbf{F}(\mathbf{x})$, where $\mathbf{F}: \mathbb{R}^m \to \mathbb{R}^m$ is continuously differentiable and $\operatorname{div}\mathbf{F} = 0$ for all $\mathbf{x}$. Suppose further that there exists a bounded invariant set

I of the flows $\boldsymbol{\phi}_t$, and $V_0 \subset I$ is a set of points with positive measure, i.e. $\mu(V_0) > 0$, and define $V(t) = \boldsymbol{\phi}_t V_0$ for all $t > 0$.

Show first, by *reductio ad absurdum* or otherwise, that at least two of the sets $V(0), V(1), V(2), \ldots$ have a point in common. Deduce that there exists $t_1 > 0$ such that $V(0)$ and $V(t_1)$ have a common point. Hence or otherwise show that there exists an infinite sequence $t_1 < t_2 < \ldots$ such that $t_n \to \infty$ as $n \to \infty$ and $V(0)$ and $V(t_n)$ have a common point for $n = 1, 2, \ldots$. Deduce that if $\mathbf{x}(0) \subset I$ then $\mathbf{x}(t)$ lies arbitrarily close to $\mathbf{x}(0)$ for a sufficiently large value of t.

*Q5.15 *Stability in the senses of Liapounov and Poincaré.* (a) Show that if

$$\frac{d^2x}{dt^2} + x = 0,$$

then the periodic solution $X(t) = a\cos t$ is stable in the senses of both Poincaré and Liapounov.

(b) Show that if

$$\frac{d^2x}{dt^2} + \sin x = 0,$$

then a periodic solution of amplitude a is stable in the sense of Poincaré but unstable in the sense of Liapounov. [Cf. Example 1.2.]

*Q5.16 *Stability of a periodic solution.* Show that the equation

$$\frac{d^2x}{dt^2} + b\left\{\left(\frac{dx}{dt}\right)^2 + x^2 - a\right\}\frac{dx}{dt} + x = 0$$

can be expressed as the system

$$\frac{dr}{dt} = b(a - r^2)r\cos^2\theta, \qquad \frac{d\theta}{dt} = -1 + \tfrac{1}{2}b(a - r^2)\sin 2\theta,$$

where $x = r\cos\theta$, $dx/dt = r\sin\theta$, $r \geqslant 0$. Deduce that for $a > 0$ there is a solution of period 2π given by $r = \sqrt{a}$, $\theta = t_0 - t$.

Linearizing the system about the solution $X(t) = \sqrt{a}\cos t$, and defining $r' = r - \sqrt{a}$, show that

$$r'(t) = r'(0)\exp\{-ab(t + \tfrac{1}{2}\sin 2t)\}.$$

Deduce that the limit cycle X is stable in the sense of Liapounov if, moreover, $b > 0$.

Defining the Poincaré map $P_a\colon \Sigma \to \Sigma$ of the system by $P_a\{r(0)\} = r(2\pi)$ for all $r(0) \geqslant 0$, where Σ is the half-line $\theta = 0$ in the phase plane, show that $\sqrt{a}$ is an asymptotically stable fixed point of P_a for all $a, b > 0$.

*Q5.17 *Floquet theory.* Consider the system

$$\frac{d\mathbf{x}(t)}{dt} = \mathbf{A}(t)\mathbf{x}(t),$$

where $\mathbf{x}$ is an $m \times 1$ vector and $\mathbf{A}$ is a real $m \times m$ matrix of period T.

Defining $\mathbf{\Phi}$ as the fundamental matrix of the system such that

$$\frac{\mathrm{d}\mathbf{\Phi}(t)}{\mathrm{d}t} = \mathbf{A}(t)\mathbf{\Phi}(t), \qquad \mathbf{\Phi}(0) = \mathbf{I},$$

$\mathbf{I}$ as the $m \times m$ unit matrix, and $\mathbf{\Psi}(t) = \mathbf{\Phi}(t + T)$, show that

$$\frac{\mathrm{d}\mathbf{\Psi}(t)}{\mathrm{d}t} = \mathbf{A}(t)\mathbf{\Psi}(t),$$

and hence that $\mathbf{\Phi}(t + T) = \mathbf{\Phi}(t)\mathbf{C}$, for some nonsingular matrix $\mathbf{C}$. Deduce that $\mathbf{C} = \mathbf{\Phi}(T)$. Assuming that the nonsingularity of $\mathbf{C}$ implies that there exists $\mathbf{R}$ such that $\mathbf{C} = \mathrm{e}^{T\mathbf{R}}$, define $\mathbf{P}(t) = \mathbf{\Phi}(t)\mathrm{e}^{-t\mathbf{R}}$ and prove that $\mathbf{P}$ has period T. Deduce that $\mathbf{\Phi}(t) = \mathbf{P}(t)\mathrm{e}^{t\mathbf{R}}$.

[Note that integration over a period T determines $\mathbf{P}$, and then $\mathbf{P}$ determines the fundamental matrix and hence all solutions for $-\infty < t < \infty$.]

The *Floquet multipliers* are defined as the eigenvalues $q_1, q_2, \ldots, q_m$ of $\mathbf{C}$ and the *Floquet exponents* as the eigenvalues $s_1, s_2, \ldots, s_m$ of $\mathbf{R}$. Show that $q_j = \exp(s_j T)$ for $j = 1, 2, \ldots, m$ if the exponents are ordered appropriately. Prove the *Jacobi–Liouville formula* that

$$\det \mathbf{\Phi}(T) = q_1 q_2 \ldots q_m = \exp\left\{\int_0^T \operatorname{trace} \mathbf{A}(t)\, \mathrm{d}t\right\},$$

and deduce that no Floquet multiplier is zero.

[Cf. Coddington & Levinson (1955, Chap. 3, §5).]

*Q5.18 *Meissner's equation.* Find the Floquet multipliers of the equation,

$$\frac{\mathrm{d}^2 x}{\mathrm{d}t^2} + a^2 f(t)x = 0,$$

where $f(t) = 1$ if $0 \leqslant t < \frac{1}{2}$ and $f(t) = -1$ if $\frac{1}{2} \leqslant t < 1$, and f has period 1, for $a > 0$.

*Q5.19 *Hill's and Mathieu's equations.* (a) Suppose that x satisfies Hill's equation, namely

$$\frac{\mathrm{d}^2 x}{\mathrm{d}t^2} + P(t)x = 0,$$

where P is a real function of period π. Use the results of Q5.17 to show that q_1, q_2 are either both real or complex conjugates and that $q_1 q_2 = 1$. Deduce that there exist independent solutions x_1, x_2 such that $x_j(t + \pi) = q_j x_j(t)$ for $j = 1, 2$, and that if the null solution of the equation is on the margin of stability then x_j has period π (synchronous) or 2π (subharmonic) for $j = 1$ or 2.

[Cf. Coddington & Levinson (1955, Chap. 8, §4).]

(b) Seek marginally stable subharmonic solutions of Mathieu's equation,

$$\frac{d^2x}{dt^2} + (a - 2\epsilon \cos 2t)x = 0,$$

for small ϵ by expanding

$$a(\epsilon) = 1 + \epsilon a_1 + \epsilon^2 a_2 + \dots, \qquad x(t, \epsilon) = x_0(t) + \epsilon x_1(t) + \dots$$

as $\epsilon \to 0$. First take $x_0(t) = \cos t$ and show that $a_1 = 1$, $x_1(t) = -\frac{1}{8}\cos 3t$. Then take $x_0(t) = \sin t$ and find a_1, x_1.

*Q5.20 *Mathieu's equation.* Show that each solution x of the equation,

$$\frac{d^2x}{dt^2} + (a - 2\epsilon \cos 2t)x = 0,$$

is either even or odd itself, or may be expressed as the sum of an even solution and an odd solution.

Seeking a solution x of period π to determine the margin of stability of the null solution, assume that x is even and

$$x(t) = \sum_{n=0}^{\infty} A_{2n} \cos 2nt$$

for some constants $A_0, A_2, \dots$. Deduce that

$$aA_0 - \epsilon A_2 = 0,$$

$$(a - 4)A_2 - \epsilon(2A_0 + A_4) = 0,$$

$$(a - 4n^2)A_{2n} - \epsilon(A_{2n-2} + A_{2n+2}) = 0 \qquad \text{for } n \geqslant 2.$$

Find those points on the margin of stability $a = a_{2n}(\epsilon)$ in the (ϵ, a)-plane on which there is an even solution of period π by assuming that $a_{2n} = 4n^2 + \epsilon a_{n1} + \epsilon^2 a_{n2} + \dots$, $x_n(t) = \cos 2nt + \epsilon x_{n1}(t) + \epsilon^2 x_{n2}(t) + \dots$ as $\epsilon \to 0$, for $n = 0, 1, \dots$. Deduce that

$$a_0(\epsilon) = -\tfrac{1}{2}\epsilon^2 + O(\epsilon^3),$$

$$a_2(\epsilon) = 4 + \tfrac{5}{12}\epsilon^2 + O(\epsilon^3),$$

$$a_{2n}(\epsilon) = 4n^2 + \epsilon^2/2(4n^2 - 1) + O(\epsilon^3) \qquad \text{for } n \geqslant 2 \text{ as } \epsilon \to 0.$$

6

Second-order autonomous differential systems

Some instances of typical relaxation oscillations are: the aeolian harp, a pneumatic hammer, the scratching noise of a knife on a plate, the waving of a flag in the wind, the humming noise sometimes made by a water-tap, the squeaking of a door, ... the tetrode multivibrator, the periodic sparks obtained from a Wimshurst machine, ... the intermittent discharge of a condenser through a neon tube, the periodic re-occurrence of epidemics and of economical crises, the periodic density of an even number of species of animals living together, and the one species serving as food for the other, the sleeping of flowers, the periodic re-occurrence of showers behind a depression, the shivering from cold, menstruation, and, finally, the beating of the heart.

B. van der Pol & J. van der Mark (*Phil. Mag.* 1928)

1 Introduction

An important class of systems of ordinary differential equations is that of second-order autonomous systems, i.e. systems of the form

$$\frac{\mathrm{d}x}{\mathrm{d}t} = F(x, y), \qquad \frac{\mathrm{d}y}{\mathrm{d}t} = G(x, y). \tag{1}$$

They are important because they are fundamental to much of the behaviour of higher-order systems and because they may be used to illustrate much of the behaviour simply. They are important also because they have many applications; indeed, many mechanical, electronic, chemical and biological phenomena have been successfully modelled by systems of the form (1). Poincaré thoroughly investigated their properties, by both analytical and geometrical methods. In particular, he devised the use of the phase plane, examining orbits in the (x, y)-plane, i.e. the curves with equation

$$\frac{\mathrm{d}y}{\mathrm{d}x} = \frac{G(x, y)}{F(x, y)}. \tag{2}$$

He also looked at the local behaviour of orbits near equilibrium points, classifying all types of these points (Poincaré 1881, 1882).

An important subclass of systems of the form (1) is the class of those for which $F(x, y) = y$ and therefore

$$\frac{\mathrm{d}^2 x}{\mathrm{d}t^2} = G\left(x, \frac{\mathrm{d}x}{\mathrm{d}t}\right). \tag{3}$$

Conversely, any second-order differential equation (3) can be put in the form (1) with $\mathrm{d}x/\mathrm{d}t = y$ and $\mathrm{d}y/\mathrm{d}t = G(x, y)$. It is customary for computer software to give numerical algorithms to integrate differential equations in the canonical form (1), not (3). However, many well-known nonlinear oscillations are modelled by equations of the form (3), for example

$$\frac{\mathrm{d}^2 x}{\mathrm{d}t^2} + f(x)\frac{\mathrm{d}x}{\mathrm{d}t} + g(x) = 0 \qquad \text{(Liénard's equation)}, \tag{4}$$

$$\frac{\mathrm{d}^2 x}{\mathrm{d}t^2} + \epsilon\left\{\frac{1}{3}\left(\frac{\mathrm{d}x}{\mathrm{d}t}\right)^2 - 1\right\}\frac{\mathrm{d}x}{\mathrm{d}t} + x = 0 \qquad \text{(Rayleigh's equation)}, \tag{5}$$

$$\frac{\mathrm{d}^2 x}{\mathrm{d}t^2} + \epsilon(x^2 - 1)\frac{\mathrm{d}x}{\mathrm{d}t} + x = 0 \qquad \text{(van der Pol's equation)}. \tag{6}$$

We shall meet these and other examples later.

It will be occasionally easier to use plane polar coordinates, such that $x = r\cos\theta$, $y = r\sin\theta$ for $r \geqslant 0$, rather than Cartesian coordinates. Then it can be shown by elementary calculus that

$$x\frac{\mathrm{d}x}{\mathrm{d}t} + y\frac{\mathrm{d}y}{\mathrm{d}t} = r\frac{\mathrm{d}r}{\mathrm{d}t}, \qquad x\frac{\mathrm{d}y}{\mathrm{d}t} - y\frac{\mathrm{d}x}{\mathrm{d}t} = r^2\frac{\mathrm{d}\theta}{\mathrm{d}t}, \tag{7}$$

so that equations (1) become

$$\frac{\mathrm{d}r}{\mathrm{d}t} = \cos\theta\{F(r\cos\theta, r\sin\theta)\} + \sin\theta\{G(r\cos\theta, r\sin\theta)\}, \tag{8}$$

$$\frac{\mathrm{d}\theta}{\mathrm{d}t} = r^{-1}\cos\theta\{G(r\cos\theta, r\sin\theta)\} - r^{-1}\sin\theta\{F(r\cos\theta, r\sin\theta)\}. \tag{9}$$

We shall sometimes use the equations in polar form.

The qualitative character of all the solutions of system (1) can be epitomized by a sketch of the solution curves in the (x, y)-plane, i.e. by a *phase portrait*. It is desirable that the sketch includes one each of all the topologically different kinds of orbit. In principle this requires the integration of equation (2). In practice sketching phase portraits is partly an art based on experience. The practical rules are to seek any simple symmetry the system has, to find all the equilibrium points, to find the local behaviour of the orbits near each equilibrium point (by use of the solutions of the linearized system), to identify any simple solution curve (e.g., a coordinate axis), and to make such other deductions as are apparent (e.g., about orbits at infinity). Also equation (2) gives the slopes of the curves everywhere. There should come a stage of this process (for textbook and class examples, if not for a complicated example from real life, which requires numerical integration) when these details can be synthesized in your mind so that a global picture of the solution curves is understood, perhaps in a flash of realization. It must be added that the advent of accessible and easy numerical programs with simple graphics on cheap computers has already made computation a valuable complement to these traditional methods.

The rest of the chapter covers aspects of this problem, substantiating some of the details which make up a phase portrait; the examples provide a little experience of sketching phase portraits. The local behaviour of the orbits near an equilibrium point is found in §2 by explicit solution of the linearized equations. The Liapounov direct method of *proving* stability or instability of an equilibrium point is given in §3. In §5 limit cycles, which correspond to closed curves in the phase plane, are explained and the Poincaré-Bendixson theorem is given to prove their existence. Some asymptotic methods to give quantitative results are introduced in §4 and §6. It will become increasingly apparent, in sketching phase portraits, that the equilibrium points and the limit cycles are very important in determining the global pattern of the orbits.

2 Linear systems

This section is a review of the linear theory, some of which has already been used in Chapter 1.

Suppose that the system (1.1) has a point $\mathbf{X}$ of equilibrium, i.e. that $\mathbf{F}(\mathbf{X}) = \mathbf{0}$, where $\mathbf{X} = [X, Y]^{\mathrm{T}}$ and $\mathbf{F} = [F, G]^{\mathrm{T}}$. Without loss of generality we may translate the equilibrium point to the origin so that $\mathbf{X} = \mathbf{0}$. Then $F(0,0) = G(0,0) = 0$. To consider orbits of (1.1) near $\mathbf{0}$, assume that $\mathbf{F}$ is continuously twice differentiable, so

$$F(x, y) = ax + by + O(x^2 + y^2),$$

$$G(x, y) = cx + dy + O(x^2 + y^2) \qquad \text{as } x, y \to 0,$$

where $a = [\partial F/\partial x]_0$, $b = [\partial F/\partial y]_0$, $c = [\partial G/\partial x]_0$, and $d = [\partial G/\partial y]_0$. Then the behaviour of the system (1.1) may in general be approximated locally by that of the linearized system

$$\frac{d\mathbf{x}}{dt} = \mathbf{J}\mathbf{x}, \tag{1}$$

where the column vector $\mathbf{x} = [x, y]^{\mathrm{T}}$ and the 2×2 matrix $\mathbf{J} = \begin{bmatrix} a & b \\ c & d \end{bmatrix}$. We shall examine this approximation more critically later. For the present we shall classify the types of equilibrium points of the system (1) according to the properties of $\mathbf{J}$.

The linear system (1) with constant coefficients can be solved in general by using the method of normal modes, i.e. by seeking solutions $\mathbf{x}(t) = e^{st}\mathbf{u}$ for a constant vector $\mathbf{u}$. Then

$$\mathbf{J}\mathbf{u} = s\mathbf{u},$$

i.e. s is the eigenvalue of $\mathbf{J}$ belonging to eigenvector $\mathbf{u}$. Therefore

$$0 = \begin{vmatrix} a - s & b \\ c & d - s \end{vmatrix}$$

$$= s^2 - ps + q, \tag{2}$$

where $p = \operatorname{trace} \mathbf{J} = a + d$ and $q = \det \mathbf{J} = ad - bc$. Therefore $s = s_1$ or s_2, where

$$s_1, s_2 = \tfrac{1}{2}\{p \pm (p^2 - 4q)^{1/2}\}. \tag{3}$$

Therefore, if $p^2 \neq 4q$, the general solution of (1) is

$$\mathbf{x}(t) = C_1 \exp(s_1 t)\begin{bmatrix} u_1 \\ v_1 \end{bmatrix} + C_2 \exp(s_2 t)\begin{bmatrix} u_2 \\ v_2 \end{bmatrix} \tag{4}$$

for arbitrary constants C_1 and C_2, where the eigenvalue s_j belongs to eigenvector $\mathbf{u}_j = [u_j, v_j]^{\mathrm{T}}$ for $j = 1, 2$, i.e. where

$$au_j + bv_j = s_j u_j, \qquad cu_j + dv_j = s_j v_j.$$

This explicit solution, and the solution for the case $p^2 = 4q$ of a double eigenvalue, enable us to classify all the equilibrium points according to the topology of the phase portraits of neighbouring orbits. The results are as follows.

(i) *Node*: $p^2 > 4q$ and $q > 0$. Here s_1, s_2 are real, distinct and of the same sign. We take $s_1 > s_2$ without loss of generality. Therefore

$$\mathbf{x}(t) \sim C_1 \exp(s_1 t)[u_1, v_1]^{\mathrm{T}} \qquad \text{as } t \to \infty$$

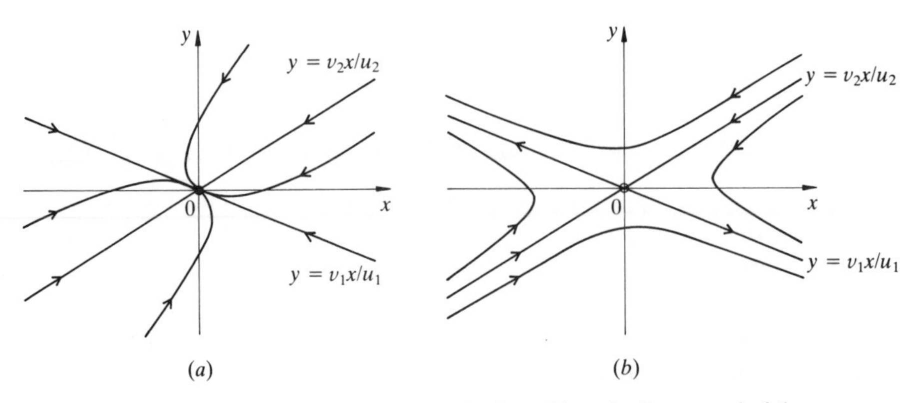

Fig. 6.1 (*a*) Sketch of the phase portrait of a stable node. On reversal of the arrows of time, we get the phase portrait of an unstable node. (*b*) Sketch of the phase portrait of a saddle point.

if $C_1 \neq 0$, and so $y/x \to v_1/u_1$ as $t \to \infty$. This describes what is called a *node*, stable if $s_1 < 0$, and so if $p < 0$, and unstable if $p > 0$. See Fig. 6.1.

(ii) *Saddle point*: $p^2 > 4q$ and $q < 0$. Here s_1, s_2 are real and of opposite signs, so we may take $s_1 > 0 > s_2$ without loss of generality. The orbits near $\mathbf{0}$ resemble hyperbolae.

(iii) *Focus* or *spiral point*: $p^2 < 4q$ and $p \neq 0$. Here s_1 and s_2 are a complex conjugate pair with non-zero real part, so we may take $s_1 = \bar{s}_2 = \frac{1}{2}\{p + \mathrm{i}(4q - p^2)^{1/2}\}$. Then

$$\mathbf{x}(t) = C\mathrm{e}^{pt/2}[\cos(\beta t + \gamma), K\cos(\beta t + \gamma + \delta)]^{\mathrm{T}},$$

where C, γ are arbitrary real constants, $\beta = (q - \frac{1}{4}p^2)^{1/2}$, and K, δ are real constants determined by a, b, c and d. Therefore the origin is a stable point of equilibrium if $p < 0$ and unstable if $p > 0$. See Fig. 6.2.

These are the generic, i.e. typical, cases, for which no two eigenvalues are equal and the real part of no eigenvalue is zero. If the real parts of all eigenvalues are non-zero then the equilibrium point is said to be *hyperbolic*. A hyperbolic point of equilibrium can be shown to be structurally stable in the sense that a small well-behaved perturbation such as weak non-linearity will not change the topology of the orbits near the point, so that a stable hyperbolic point will not thereby become unstable nor an unstable hyperbolic point become stable. There are a few special cases that remain to be described.

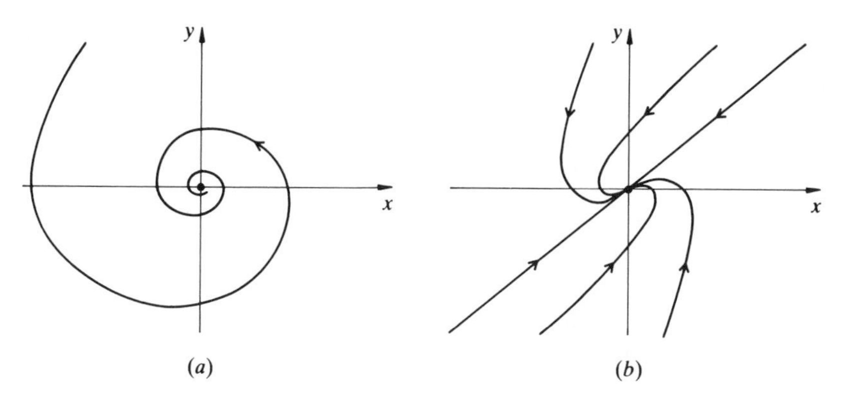

Fig. 6.2 (*a*) Sketch of the phase portrait of a stable focus. For an unstable focus the arrow is reversed. (*b*) Sketch of the phase portrait of a stable improper node. For an unstable improper node, the arrows are reversed.

(iv) *Improper node, inflected node*, or *degenerate node*: $p^2 = 4q$ and $q > 0$. Here $s_2 = s_1$ is real. The solution can in general be shown to give a limiting form of a node because the lines $y = v_1 x/u_1$ and $y = v_2 x/u_2$ coincide. The node is stable if $p < 0$ and unstable if $p > 0$. See Fig. 6.2.

(v) *Centre* or *vortex*: $p^2 < 4q$ and $p = 0$, i.e. $p = 0$ and $q > 0$. Here s_1, $s_2 = \pm iq$ are purely imaginary. Then

$$\mathbf{x}(t) = C[\cos(qt + \gamma), K\cos(qt + \gamma + \delta)]^{\mathrm{T}},$$

where C, γ are arbitrary real constants and K, δ are real constants determined by a, b, c and d. The orbits are ellipses with centre at the origin. See Fig. 6.3.

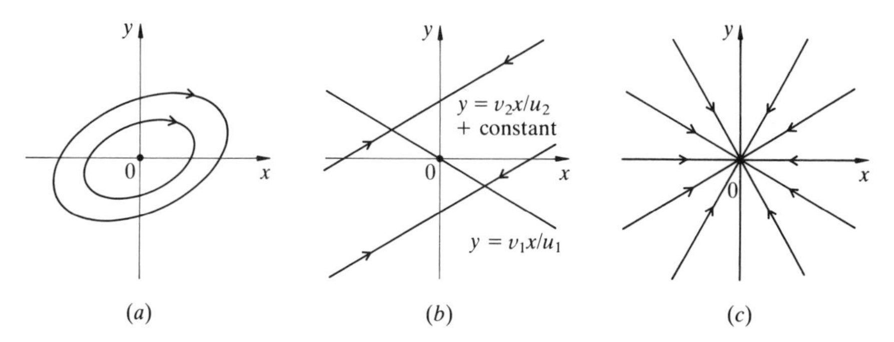

Fig. 6.3 (*a*) Sketch of the phase portrait of a centre. (*b*) Sketch of the phase portrait of a degenerate node for $p < 0$ in case (vi). The arrows are reversed if $p > 0$. (*c*) Sketch of the phase portrait of a stable sink with $a < 0$. The arrows are reversed for a source with $a < 0$.

(vi) *Degenerate node* or *improper node*: $p^2 > 4q$ and $q = 0$, i.e. $p > 0$ and $q = 0$. This is a limiting case of (i) above, for which we may put $s_1 = 0$ and $s_2 = p$. Therefore

$$\mathbf{x}(t) = C_1[u_1 v_1]^{\mathrm{T}} + C_2 e^{pt}[u_2, v_2]^{\mathrm{T}},$$

and the orbits are straight lines parallel to $y = v_2 x/u_2$, approaching the line $y = u_1 x/v_1$ as $t \to \infty$ if $p < 0$ and leaving it if $p > 0$. See Fig. 6.3.

(vii) *Source, sink, proper node* or *star point*: $p^2 = 4q$ and $q > 0$, and $a = d \neq 0$. Here $dx/dt = ax$ and $dy/dt = ay$. Therefore $x(t) = x_0 e^{at}$ and $y(t) = y_0 e^{at}$. This gives radial orbits, the origin being stable (a sink) if $a < 0$ and unstable (a source) if $a > 0$. See Fig. 6.3.

All the results are summarized in Fig. 6.4. The generic cases of an equilibrium point are those represented by points not on the parabola $p^2 = 4q$, the p-axis or the q-axis. Different authorities choose slightly different sets of names for the types of equilibrium points.

A node, a saddle point and a focus are defined for a nonlinear system

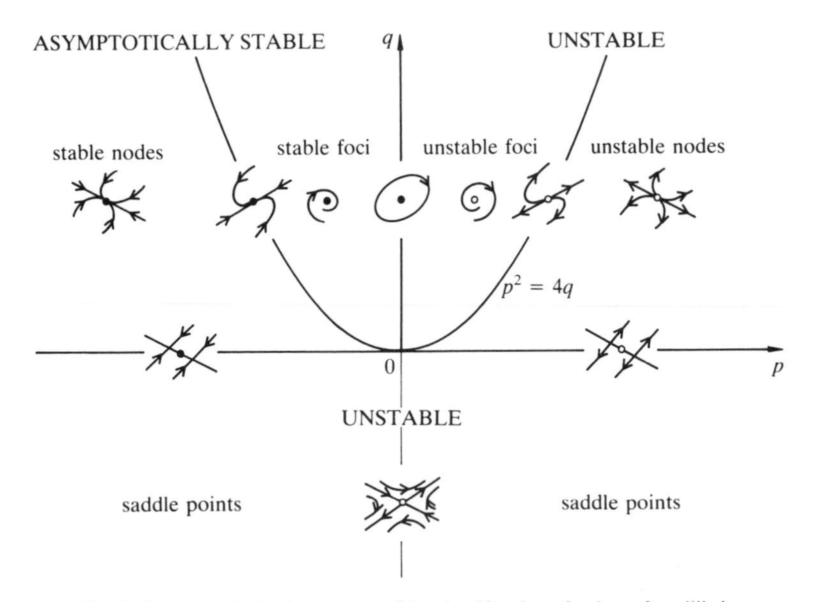

Fig. 6.4 Summary in the (p, q)-plane of the classification of points of equilibrium at $x = 0$, $y = 0$ for the linear system $dx/dt = ax + by$, $dy/dt = cx + dy$, where $p = a + d$ and $q = ad - bc$. A disc • denotes that the origin is stable, and a circle ∘ that it is unstable.

similarly, according to the topological character of the orbits near the equilibrium point; thus a node is an equilibrium point which all the nearby orbits in the plane approach (or leave) and all except possibly two approach (or leave) in the same direction, a saddle point is an equilibrium point which is approached by two orbits as $t \to \infty$ and two as $t \to -\infty$, a focus is an equilibrium point which each nearby orbit in the plane approaches (or leaves) along a spiral, and a centre is an equilibrium point surrounded by closed orbits. A stable node and a stable focus are attractors, and an unstable node and an unstable focus and repellers. A saddle point is neither an attractor nor a repeller because, although almost all orbits in its neighbourhood leave the neighbourhood, two orbits approach the equilibrium point. A centre, although stable, is not asymptotically stable and therefore is not an attractor because orbits in its neighbourhood do not approach the centre.

The nonlinear system

$$\frac{\mathrm{d}\mathbf{x}}{\mathrm{d}t} = \mathbf{J}\mathbf{x} + \boldsymbol{\xi}(\mathbf{x})$$

and the linearized system have in general the same topological character of orbits near $\mathbf{0}$ if $\boldsymbol{\xi}$ is well-behaved and $\boldsymbol{\xi} = o(|\mathbf{x}|)$ as $\mathbf{x} \to \mathbf{0}$ (cf. §3 and Coddington & Levinson (1955, Chap. 13)). So the linear theory gives the orbits near an equilibrium point. It is for this reason that the linear theory is so important in the nonlinear theory. The exceptions to the general rule arise when the equilibrium point is not hyperbolic, i.e. when $p^2 = 4q, p = 0$ or $q = 0$, because weak nonlinearity may change one kind of equilibrium point into another when the linear theory is on the margin between two kinds of point. For example, weak nonlinear dissipation may change a centre into a stable focus. We have met and will meet similar ideas in other examples and problems.

Example 6.1: a focus. If

$$\frac{\mathrm{d}x}{\mathrm{d}t} = ax - cy, \qquad \frac{\mathrm{d}y}{\mathrm{d}t} = cx + ay$$

then $p = a + d = 2a$ and $q = ad - bc = a^2 + c^2$. Therefore $p^2 - 4q = -4c^2 < 0$. Therefore $\mathbf{0}$ is a focus, stable if $a < 0$ and unstable if $a > 0$. In fact, each orbit is an equiangular spiral with $r(t) = r_0 \mathrm{e}^{at}$ and $\theta(t) = \theta_0 + ct$, where r, θ are plane polar coordinates such that $x = r\cos\theta$ and $y = r\sin\theta$. (When $c = 1$, this becomes the linearized system of §1.6.) □

3 The direct method of Liapounov

We shall next describe the *direct method* of Liapounov (1892), a powerful, but not omnipotent, method of proving the stability or instability of an equilibrium point of a nonlinear system of differential equations. It is sometimes called the *second method of Liapounov* (we shall not describe the first method, which is almost forgotten). First an example will motivate the method.

Example 6.2: the energy method. Consider the linear system,

$$\frac{dx}{dt} = ax - cy, \qquad \frac{dy}{dt} = cx + ay \qquad \text{for } a < 0, \tag{1}$$

of Example 6.1 from a different point of view, defining $H(\mathbf{x}) = \frac{1}{2}(x^2 + y^2)$. Therefore

$$\begin{aligned}\frac{dH}{dt} &= x\frac{dx}{dt} + y\frac{dy}{dt}\\ &= (ax^2 - cxy) + (cxy + ay^2)\\ &= 2aH\\ &\leqslant 0,\end{aligned} \tag{2}$$

with equality if and only if $x = y = 0$. Therefore

$$\begin{aligned}H(\mathbf{x}(t)) &= H(\mathbf{x}(0))e^{2at}\\ &\to 0 \qquad \text{as } t \to \infty\end{aligned}$$

for all $\mathbf{x}(0)$. Therefore

$$\mathbf{x}(t) \to \mathbf{0} \qquad \text{as } t \to \infty,$$

i.e. the origin is a globally asymptotically stable point of equilibrium of system (1).

Next consider a nonlinear perturbation of the system (1), i.e.

$$\frac{dx}{dt} = ax - cy + \xi(x, y), \qquad \frac{dy}{dt} = cx + ay + \eta(x, y) \tag{3}$$

for $a < 0$, where ξ, η are continuously differentiable functions such that

$$\{\xi(x, y)\}^2 + \{\eta(x, y)\}^2 = O\{(x^2 + y^2)^2\} \qquad \text{as } \mathbf{x} \to \mathbf{0}.$$

Now we find that

$$\frac{\mathrm{d}H}{\mathrm{d}t} = 2aH + x\xi + y\eta$$

$$< aH$$

for sufficiently small $\mathbf{x} \neq \mathbf{0}$ in order that $aH + x\xi + y\eta < 0$. (Remember that $a < 0$.) Therefore $H < H(0)\mathrm{e}^{at}$, and so $H(\mathbf{x}(t)) \to 0$ as $t \to \infty$ for all small $\mathbf{x}(0)$, and the origin is asymptotically stable. □

Now let us come to the direct method itself. We shall apply it to the system

$$\frac{\mathrm{d}\mathbf{x}}{\mathrm{d}t} = \mathbf{F}(\mathbf{x}), \tag{4}$$

where $\mathbf{F}: \mathbb{R}^m \to \mathbb{R}^m$ and $\mathbf{F}(\mathbf{0}) = \mathbf{0}$. The method is essentially the same for all m, so we shall give it for general m, but emphasize the case $m = 2$. We shall show that if a scalar function H with special properties may be found then the origin is a stable point of equilibrium. Finding the function is a matter of experience, and of trial and error, but the function is a generalization of the energy of a mechanical system, and the argument we use is similar to the one that if energy is dissipated and there is no available potential energy then an equilibrium point is stable.

First we need a definition. A function $f: \mathbb{R}^m \to \mathbb{R}$ is *positive definite* (or *semidefinite*) if $f(\mathbf{x}) > 0$ (or $\geqslant 0$ respectively) for all $\mathbf{x} \neq \mathbf{0}$ and $f(\mathbf{0}) = 0$. Now we may state the theorem (or, rather, three theorems).

Liapounov's theorem. If $\mathbf{0}$ is an equilibrium point of system (4) and there exists a function $H: \mathbb{R}^m \to \mathbb{R}$ such that

(i) H and its partial derivatives are continuous,
(ii) H is positive definite, and
(iii) (a) $-\mathbf{F} \cdot \operatorname{grad} H$ is positive semidefinite,
(b) $-\mathbf{F} \cdot \operatorname{grad} H$ is positive definite, or
(c) $\mathbf{F} \cdot \operatorname{grad} H$ is positive definite,

then the equilibrium point is (a) stable, (b) asymptotically stable, or (c) unstable respectively.

We shall not prove the theorem, which requires some care in case (a); but the essential idea is that $\mathrm{d}H/\mathrm{d}t = \operatorname{grad} H \cdot \mathrm{d}\mathbf{x}/\mathrm{d}t = \mathbf{F} \cdot \operatorname{grad} H$ and so that H increases or decreases monotonically with t if $\mathbf{F} \cdot \operatorname{grad} H$ is positive or negative. Because H is positive definite, $\mathbf{x}$ decreases or increases according to whether H increases or decreases everywhere near $\mathbf{0}$. The function H is called a *Liapounov function.* The only difficulty of the method lies in finding H. Even if a Liapounov function exists, it may be very hard to find.

Example 6.3: a Liapounov function. Consider the system

$$\frac{dx}{dt} = -x - 2y^2, \qquad \frac{dy}{dt} = xy - y^3.$$

By inspection, the origin is an equilibrium point. For the Liapounov function *try* $H(x, y) = \frac{1}{2}(x^2 + ay^2)$ for a value of a to be determined. Then $a > 0$ in order that H is positive definite. Therefore

$$\begin{aligned}\frac{dH}{dt} &= \mathbf{F} \cdot \operatorname{grad} H \\ &= (-x - 2y^2)x + (xy - y^3)ay \\ &= -(x^2 + 2y^4),\end{aligned}$$

on choosing $a = 2$,

$$\leqslant 0,$$

with equality if and only if $x = y = 0$. Therefore the origin is an asymptotically stable point of equilibrium. □

Example 6.4: a Hamiltonian system. Consider the system

$$\frac{d\mathbf{p}}{dt} = -\frac{\partial H}{\partial \mathbf{q}}, \qquad \frac{d\mathbf{q}}{dt} = \frac{\partial H}{\partial \mathbf{p}},$$

where $\mathbf{p}, \mathbf{q} \in \mathbb{R}^n$, the Hamiltonian function is of the form

$$H(\mathbf{p}, \mathbf{q}) = T(\mathbf{p}, \mathbf{q}) + V(\mathbf{q}),$$

$V(\mathbf{0}) = 0$, and T is a positive definite quadratic form in $\mathbf{p}$. (This represents a mechanical system with kinetic energy T and potential energy V.) Suppose further that $\mathbf{q} = \mathbf{0}$ is a simple minimum of V. Then $\mathbf{p} = \mathbf{q} = \mathbf{0}$ gives an equilibrium point. Therefore H is positive definite near $\mathbf{0}$ and $dH/dt = 0$, i.e. dH/dt is positive semidefinite. Therefore the Hamiltonian H serves as a Liapounov function and the origin is a stable point of equilibrium.

Lagrange (1788, Part II, Section 5.15) is famous for (among other things) first proving that if V has a minimum then equilibrium is stable; Liapounov (1892, §25) proved that otherwise the equilibrium is unstable. □

4 The Lindstedt–Poincaré method

There are many asymptotic methods to find approximately the behaviour of weakly nonlinear oscillations. We shall introduce one in this section, only describing it by an example. A canonical equation for oscillations of a conservative system is

$$\frac{d^2x}{dt^2} + x = \epsilon f(x), \tag{1}$$

where f is a continuously differentiable nonlinear function such that $f(0) = f'(0) = 0$, and ϵ is a parameter. (Note that any equation of the form $d^2y/d\tau^2 = g(y)$, with $g(Y) = 0$ and $g'(Y) < 0$, may be reduced to the form (1). To see this, please remember from §1.7 that a general second-order equation representing a conservative system may be expressed in the form $d^2y/d\tau^2 = g(y)$ for a well-behaved function g, and that there may be oscillations about a stable point Y of equilibrium if $g(Y) = 0$ and $g'(Y) < 0$. On our defining $x = y - Y$, $t = \{-g'(Y)\}^{1/2}\tau$ and $\epsilon f(x) = x + g(x + Y)/\{-g'(Y)\}$, this leads to equation (1) with a function f such that $f(x) = O(x^2)$ as $x \to 0$.) When ϵ is small there is weak nonlinearity. Recall that when $\epsilon = 0$ there is simple harmonic motion with period 2π independently of the amplitude of the oscillation. We anticipate, from experience of §1.7, that when ϵ is small there are periodic solutions whose periods depend on ϵ and the amplitude.

On first thoughts, one might seek to solve this problem by expanding the solution

$$x(t, \epsilon) = x_0(t) + \epsilon x_1(t) + \epsilon^2 x_2(t) + \cdots \qquad \text{as } \epsilon \to 0, \tag{2}$$

substituting this regular expansion into equation (1), equating coefficients of $\epsilon^0, \epsilon^1, \epsilon^2, \ldots$, and solving the resulting equations for $x_0, x_1, x_2, \ldots$, in turn. This method, however, may give a nonuniformly valid approximation, as $\epsilon \to 0$, over an *infinite* interval of time. To understand this nonuniformity, a simple linear example will suffice.

Example 6.5: nonuniformity of a regular perturbation of simple harmonic motion. Suppose that

$$\frac{d^2x}{dt^2} + (1 + \epsilon)^2 x = 0, \tag{3}$$

and

$$x = a, \qquad \frac{dx}{dt} = 0 \qquad \text{at } t = 0. \tag{4}$$

Then substituting expansion (2) into equation (3) and initial conditions (4), and equating coefficients of ϵ^0 gives

$$\frac{d^2 x_0}{dt^2} + x_0 = 0$$

and

$$x_0 = a, \qquad \frac{dx_0}{dt} = 0 \qquad \text{at } t = 0.$$

Therefore

$$x_0(t) = a \cos t.$$

Next, equating coefficients of ϵ, we find that

$$\frac{d^2 x_1}{dt^2} + x_1 = -2x_0 = -2a \cos t$$

and

$$x_1 = 0, \qquad \frac{dx_1}{dt} = 0 \qquad \text{at } t = 0.$$

Therefore

$$x_1(t) = -at \sin t.$$

This gives

$$x(t) = a\{\cos t - \epsilon t \sin t + O(\epsilon^2)\} \qquad \text{as } \epsilon \to 0. \tag{5}$$

We now see that, however small ϵ is, $\epsilon x_1(t)$ is as large as $x_0(t)$ after long enough a time (when $\epsilon t \approx 1$). So the solution (5) may converge uniformly to the exact solution $x(t, \epsilon)$ as $\epsilon \to 0$ if t belongs to a given finite interval, but not if t belongs to an infinite interval.

In this example of simple harmonic motion we can see in detail what is happening, because we may at once write down explicitly the exact solution,

$$\begin{aligned} x(t, \epsilon) &= a \cos\{(1 + \epsilon)t\} \\ &= a \cos \epsilon t \cos t - a \sin \epsilon t \sin t. \end{aligned}$$

The standard power series of $\cos \epsilon t$ and $\sin \epsilon t$ give all terms in the expansion (2), the first two of which can be recognized in approximation (5). □

Poincaré (1893, §123) recognized that this nonuniformity could be resolved by expanding the frequency of the periodic solution as well as the solution itself. Although he acknowledged Lindstedt's (1883) use of this idea, the method is often named after Poincaré alone. The idea serves to reduce the problem to one over a finite interval of time, namely the period of the oscillation, and so to avoid the nonuniformity of the limits as $\epsilon \to 0$ and $t \to \infty$. We shall describe the method by an example.

Example 6.6: a nonlinear spring. The equation

$$\frac{\mathrm{d}^2 x}{\mathrm{d}t^2} + x - \epsilon x^3 = 0 \qquad (6)$$

governs a soft spring if $\epsilon > 0$, Hooke's law if $\epsilon = 0$, and a hard spring if $\epsilon < 0$. It has energy integral

$$\frac{1}{2}\left(\frac{\mathrm{d}x}{\mathrm{d}t}\right)^2 + \frac{1}{2}x^2 - \frac{1}{4}\epsilon x = E.$$

The equilibrium points are $X = 0$ for all ϵ, and $X = \pm 1/\sqrt{\epsilon}$ for all $\epsilon > 0$. So there is a pitchfork bifurcation at $x = 0$, $1/\epsilon = 0$ in the $(1/\epsilon, x)$-plane. The point $x = \mathrm{d}x/\mathrm{d}t = 0$ can easily be seen to be a centre, and $x = \pm 1/\sqrt{\epsilon}$, $\mathrm{d}x/\mathrm{d}t = 0$ a saddle point in the phase plane. The phase portraits are sketched in Fig. 6.5. The separatrices have equation $\mathrm{d}x/\mathrm{d}t =$

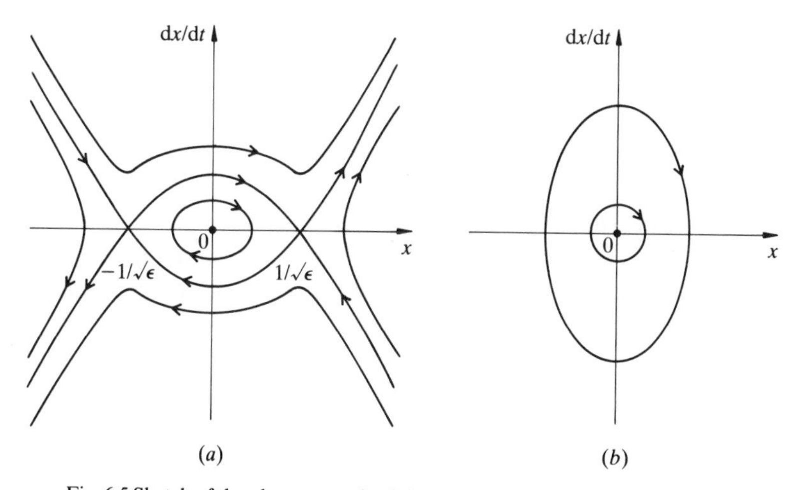

Fig. 6.5 Sketch of the phase portrait of the nonlinear spring (6) in the $(x, \mathrm{d}x/\mathrm{d}t)$-plane. (*a*) The soft spring $\varepsilon > 0$. (*b*) The hard spring $\varepsilon < 0$.

$\pm(\frac{1}{2}\epsilon)^{1/2}(x^2 - 1/\epsilon)$. Motions of small amplitude correspond to closed curves for all ϵ, and so are periodic. The period, given by equation (1.7.8), depends upon both ϵ and the amplitude (which determines the energy E) in general. However, if $\epsilon = 0$ there is simple harmonic motion of period 2π for all amplitudes.

This sets the scene. Now we shall describe the Lindstedt–Poincaré method to find the periodic solutions for small ϵ. Define ω as the unknown frequency of a solution, so the period is $2\pi/\omega$, and define $\tau = \omega t$. Then we may express equation (6) as

$$\omega^2 \frac{d^2x}{d\tau^2} + x - \epsilon x^3 = 0 \tag{7}$$

without approximation. The chosen periodic solution will also satisfy the condition of periodicity,

$$x(\tau + 2\pi, \epsilon) = x(\tau, \epsilon) \qquad \text{for all } \tau. \tag{8}$$

We apply the method by expanding the frequency

$$\omega = \omega_0 + \epsilon\omega_1 + \epsilon^2\omega_2 + \cdots \tag{9}$$

as well as the solution

$$x(\tau, \epsilon) = x_0(\tau) + \epsilon x_1(\tau) + \epsilon^2 x_2(\tau) + \cdots \qquad \text{as } \epsilon \to 0. \tag{10}$$

Further, we may define

$$a = x(0, \epsilon) \tag{11}$$

where, by translation of time, we take

$$\frac{dx}{d\tau} = 0 \qquad \text{at } \tau = 0 \tag{12}$$

without loss of generality. Then a is usually the amplitude of an oscillation.

Now it is straightforward to substitute both expansions (9) and (10) into equation (7), periodicity condition (8), and initial conditions (11) and (12). Equating coefficients of ϵ^0, we find

$$\omega_0^2 \frac{d^2x_0}{d\tau^2} + x_0 = 0,$$

$$x_0(\tau + 2\pi) = x_0(\tau) \qquad \text{for all } \tau, \quad x_0 = a, \quad \frac{dx_0}{d\tau} = 0 \text{ at } \tau = 0.$$

The differential equation has general solution

$$x_0(\tau) = a_0 \cos(\tau/\omega_0) + b_0 \sin(\tau/\omega_0)$$

for some constants a_0, b_0. Then the periodicity and initial conditions give

$$\omega_0 = 1, \qquad x_0(\tau) = a \cos \tau.$$

Equating coefficients of ϵ, we find that

$$\frac{d^2 x_1}{d\tau^2} + x_1 = -2\omega_1 \frac{d^2 x_0}{d\tau^2} + x_0^3$$

$$= (2\omega_1 a + \tfrac{3}{4}a^3)\cos \tau + \tfrac{1}{4}a^3 \cos 3\tau, \tag{13}$$

$$x_1(\tau + 2\pi) = x_1(\tau) \qquad \text{for all } \tau, \qquad x_1 = \frac{dx_1}{d\tau} = 0 \quad \text{at } \tau = 0.$$

If the term in $\cos \tau$ were not to vanish, then there would be a secular term $\frac{1}{2}(2\omega_1 a + \frac{3}{4}a^3)\tau \sin \tau$ in the particular integral of x_1, and so ϵx_1 would be unbounded and x_1 could not satisfy the periodicity condition. It follows that the coefficient of $\cos \tau$ on the right-hand side of the equation for x_1 does vanish, and therefore

$$\omega_1 = -\tfrac{3}{8}a^2.$$

Therefore

$$x_1(\tau) = a_1 \cos \tau + b_1 \sin \tau - \tfrac{1}{32} a^3 \cos 3\tau$$

for some constants a_1 and b_1. The initial conditions for x_1 determine them, finally giving

$$x_1(\tau) = \tfrac{1}{32}a^3(\cos \tau - \cos 3\tau).$$

This gives the uniformly valid approximation,

$$x(t, \epsilon) = a \cos \omega t + \tfrac{1}{32}\epsilon a^3(\cos \omega t - \cos 3\omega t) + O(\epsilon^2 a^5)$$

$$\omega = 1 - \tfrac{3}{8}\epsilon a^2 + O(\epsilon^2 a^4) \qquad \text{as } \epsilon \to 0.$$

We may find higher approximations by proceeding to find $\omega_2, x_2, \omega_3, \ldots$ in a similar way, but the leading corrections to simple harmonic motion are already found. (See also Q1.17.)

That the coefficient of $\cos \tau$ on the right-hand side of equation (13) vanishes can be deduced alternatively by use of *Lagrange's identity*, namely

$$\int_0^{2\pi} (u\mathrm{L}v - v\mathrm{L}u)\, d\tau = \left[u\frac{dv}{d\tau} - v\frac{du}{d\tau} \right]_0^{2\pi} \tag{14}$$

for all continuously twice differentiable functions u, v, where the self-adjoint linear operator L is defined by $\mathrm{L} = \mathrm{d}^2/\mathrm{d}\tau^2 + 1$. The identity can be deduced easily by integrating by parts. It becomes, on putting $u = \cos \tau$, $v = x_1$ and using equation (13),

$$\int_0^{2\pi} \cos \tau \{(2\omega_1 a + \tfrac{3}{4}a^3)\cos \tau + \tfrac{1}{4}a^3 \cos 3\tau\}\, \mathrm{d}\tau = \left[\cos \tau \frac{\mathrm{d}x_1}{\mathrm{d}\tau} + x_1 \sin \tau\right]_0^{2\pi},$$

i.e.

$$\tfrac{1}{2}(2\omega_1 a + \tfrac{3}{4}a^3) = 0,$$

on using the periodicity condition for x_1. This gives the equation for ω_1 as above. □

It should be noted that the Lindstedt–Poincaré method is applicable to periodic solutions of many more differential systems than those of the form (1). Indeed, it is applicable to partial as well as ordinary differential systems of many kinds and orders, when a linear oscillation is well enough known to perturb. In particular the method is suitable for weakly nonlinear solutions of equations of the form

$$\frac{\mathrm{d}^2x}{\mathrm{d}t^2} + x = \epsilon f(x, \mathrm{d}x/\mathrm{d}t)$$

in the limit as $\epsilon \to 0$.

5 Limit cycles

We have seen that the system of equations

$$\frac{\mathrm{d}x}{\mathrm{d}t} = F(x, y), \qquad \frac{\mathrm{d}y}{\mathrm{d}t} = G(x, y) \tag{1}$$

may have not only equilibrium points but also periodic solutions which model nonlinear oscillations; and that a periodic solution is represented by a closed curve in the phase plane. Further, damped or negatively damped oscillations may tend to a periodic solution; this limiting periodic oscillation is called a *limit cycle*. It is an attractor. An example was given in §1.6.

Note that an orbit of a non-singular autonomous second-order system (1) cannot cross itself except at an equilibrium point, where F and G both vanish. Also, because the system is well-behaved, all the orbits near each point except an equilibrium point are all in the same direction $\mathbf{F} = (F, G)$, and the equilibrium points are only of the types we have described; so no orbit can have a very complicated pattern near a point. By geometrical intuition it follows that each orbit in the phase plane goes to either (a) infinity, (b) an equilibrium point (which, like a saddle point, need not be stable), (c) a point of itself (when it is a periodic solution) or (d) a limit cycle. This is the essence of the proof of *Poincaré–Bendixson theorem*, which we shall only state. It is a theorem effectively showing that two-dimensional autonomous systems do not have chaotic solutions.

The Poincaré–Bendixson theorem. If D is a closed bounded region of the (x, y)-plane, and a solution of the non-singular system (1) is such that $\mathbf{x}(t) \in \mathrm{D}$ for all $t \geqslant 0$, then the orbit either is a closed path, approaches a closed path as $t \to \infty$, or approaches an equilibrium point.

The theorem shows that the classification of attractors of well-behaved autonomous two-dimensional systems is complete, there being no other attractors than equilibrium points and limit cycles. The remainder of the theory concerns the number, nature and location of the equilibrium points and limit cycles, and thence the phase portrait. Indeed, Hilbert's sixteenth problem concerns the number and properties of limit cycles for system (1) when F, G are polynomials. Some deep problems remain open to this day, even when F, G are quadratic. However, many useful properties of limit cycles are elementary, and we shall introduce some in this and the next section, mostly by use of examples.

Example 6.7: proof of the existence of a limit cycle. Suppose that

$$\frac{\mathrm{d}x}{\mathrm{d}t} = x - y - x(x^2 + 2y^2), \qquad \frac{\mathrm{d}y}{\mathrm{d}t} = x + y - y(x^2 + y^2).$$

It is helpful to use plane polar coordinates r, θ. Then equations (1.7) give

$$\begin{aligned} r^2\frac{\mathrm{d}\theta}{\mathrm{d}t} &= x\{x + y - y(x^2 + y^2)\} - y\{x - y - x(x^2 + 2y^2)\} \\ &= x^2 + y^2 + xy^3 \\ &= r^2 + \tfrac{1}{2}r^4 \sin^2\theta \sin 2\theta, \end{aligned}$$

$$\begin{aligned} r\frac{\mathrm{d}r}{\mathrm{d}t} &= x\{x - y - x(x^2 + 2y^2)\} + y\{x + y - y(x^2 + y^2)\} \\ &= x^2 + y^2 - (x^2 + y^2)^2 - x^2y^2 \\ &= r^2 + r^4(1 + \tfrac{1}{4}\sin 2\theta). \end{aligned}$$

Therefore $\mathrm{d}r/\mathrm{d}t > 0$ for all θ if $r < r_1$ and $\mathrm{d}r/\mathrm{d}t < 0$ for all θ if $r > r_2$, where $r_1 = 2/\sqrt{5}$ and $r_2 = 1$. Therefore if $\mathbf{x}(0) \in \mathrm{D}$, where D is the annulus defined by $D = \{\mathbf{x}: r_1 \leqslant r \leqslant r_2\}$, then $\mathbf{x}(t) \in D$ for all $t > 0$. Further $d\theta/dt \neq 0$ in D, so there is no equilibrium point in D. Then the Poincaré–Bendixson theorem gives (at least) one limit cycle in D. □

**Example 6.8: weakly nonlinear analysis of a Hopf bifurcation.* The onset of a limit cycle at a Hopf bifurcation can be analysed asymptotically by use of the Lindstedt–Poincaré method. To illustrate this we take the system

$$\frac{dx}{dt} = -y + ax + xy^2, \qquad \frac{dy}{dt} = x + ay - x^2. \tag{2}$$

It can be seen by inspection that the origin is an equilibrium point, and its linearized equations are the same as those of §1.6. There we found the onset of instability as a increased through zero, with eigenvalues $s = \pm i$ at $a = 0$. So here also we anticipate the existence of a limit cycle of period $2\pi/\omega$ such that $\omega \to 1$ and the amplitude of the cycle vanishes as $a \to 0$ from either above (supercritical bifurcation) or below (subcritical bifurcation).

To find the limit cycle asymptotically as $a \to 0$, define $\tau = \omega t$, although ω is not known yet, and deduce without approximation that

$$\omega x' = -y + ax + xy^2, \qquad \omega y' = x + ay - x^2, \tag{3}$$

where a prime denotes differentiation with respect to τ. The periodicity condition is that

$$\mathbf{x}(\tau + 2\pi, a) = \mathbf{x}(\tau, a) \qquad \text{for all } \tau, \tag{4}$$

where $\mathbf{x} = (x, y)$. It is convenient to translate the origin of time, if necessary, so that $x(\tau)$ attains a maximum at $\tau = 0$; this gives, without loss of generality, a 'maximum' condition that

$$x'(0, a) = 0. \tag{5}$$

Next assume that

$$\mathbf{x}(\tau, a) = a^{1/2}\mathbf{x}_{1/2}(\tau) + a\mathbf{x}_1(\tau) + a^{3/2}\mathbf{x}_{3/2}(\tau) + \cdots,$$
$$\omega(a) = \omega_0 + a^{1/2}\omega_{1/2} + a\omega_1 + \cdots \qquad \text{as } a \to 0.$$

The appropriateness of this ansatz to represent the true behaviour of the limit cycle for small values of a is essential for its success; experience of the Hopf bifurcation of §1.6 might lead to the choice of the ansatz, but the self-consistency evident after the calculation is more convincing. Now equate coefficients of $a^{1/2}, a, a^{3/2}, \ldots$ by turn in equations (3), (4), (5).

The coefficients of $a^{1/2}$ give

$$\omega_0 x'_{1/2} + y_{1/2} = 0, \qquad \omega_0 y'_{1/2} - x_{1/2} = 0,$$

$$\mathbf{x}_{1/2}(\tau + 2\pi) = \mathbf{x}_{1/2}(\tau) \qquad \text{for all } \tau, \quad x'_{1/2}(0) = 0.$$

On eliminating $y_{1/2}$, it follows that

$$\mathrm{L}x_{1/2} = 0,$$

where the linear differential operator $\mathrm{L} = \omega_0^2 \mathrm{d}^2/\mathrm{d}\tau^2 + 1$. Therefore

$$x_{1/2}(\tau) = A_{1/2}\cos(\tau/\omega_0 + \delta_{1/2})$$

for some constants $A_{1/2}, \delta_{1/2}$ of integration. The periodicity condition now gives

$$\omega_0 = 1,$$

and the maximum condition gives $\delta_{1/2} = 0$. Therefore $\mathrm{L} = \mathrm{d}^2/\mathrm{d}\tau^2 + 1$,

$$x_{1/2}(\tau) = A_{1/2}\cos\tau,$$

and

$$y_{1/2}(\tau) = -x'_{1/2}(\tau) = A_{1/2}\sin\tau,$$

where $A_{1/2}$ has to be determined later.

Next, coefficients of a give

$$x'_1 + y_1 = -\omega_{1/2}x'_{1/2}, \qquad y'_1 - x_1 = \omega_{1/2}y'_{1/2} - x^2_{1/2},$$

$$\mathbf{x}_1(\tau + 2\pi) = \mathbf{x}_1(\tau) \qquad \text{for all } \tau, \quad x'_1(0) = 0.$$

Therefore

$$\begin{aligned} \mathrm{L}x_1 &= \omega_{1/2}(y'_{1/2} - x''_{1/2}) + x^2_{1/2} \\ &= 2\omega_{1/2}A_{1/2}\cos\tau + \tfrac{1}{2}A^2_{1/2}(1 + \cos 2\tau). \end{aligned}$$

Therefore

$$x_1(\tau) = A_1\cos(\tau + \delta_1) + \omega_{1/2}A_{1/2}\tau\sin\tau + A^2_{1/2}(\tfrac{1}{2} - \tfrac{1}{6}\cos 2\tau)$$

for some constants A_1, δ_1 of integration. The periodicity condition implies that the secular term vanishes and therefore

$$\omega_{1/2} = 0;$$

and then the maximum condition implies that $\delta_1 = 0$. Therefore

$$x_1(\tau) = A_1\cos\tau + A^2_{1/2}(\tfrac{1}{2} - \tfrac{1}{6}\cos 2\tau),$$

$$y_1(\tau) = -x'_1(\tau) = A_1\sin\tau - \tfrac{1}{3}A^2_{1/2}\sin 2\tau.$$

Next, coefficients of $a^{3/2}$ give

$$x'_{3/2} + y_{3/2} = -\omega_1 x'_{1/2} + x_{1/2} + x_{1/2}y^2_{1/2},$$

$$y'_{3/2} - x_{3/2} = -\omega_1 y'_{1/2} + y_{1/2} + 2x_{1/2}x_1,$$

$$\mathbf{x}_{3/2}(\tau + 2\pi) = \mathbf{x}_{3/2}(\tau) \qquad \text{for all } \tau, \quad x'_{3/2}(0) = 0.$$

Therefore

$$\begin{aligned} \mathrm{L}x_{3/2} &= \omega_1(y'_{1/2} + x''_{1/2}) + x'_{1/2} - y_{1/2} + x'_{1/2}y^2_{1/2} + 2x_{1/2}y_{1/2}y'_{1/2} + 2x_{1/2}x_1 \\ &= 2A_{1/2}(\omega_1 \cos\tau - \sin\tau) + A_{1/2}A_1(1 + \cos 2\tau) \\ &\quad + A^3_{1/2}(-\tfrac{1}{4}\sin\tau + \tfrac{3}{4}\sin 3\tau + \tfrac{5}{6}\cos\tau - \tfrac{1}{6}\cos 3\tau). \end{aligned}$$

In order that $x_{3/2}$ has period 2π there must be no secular term in its particular integral, and therefore the coefficients of $\sin\tau$ and $\cos\tau$ on the right-hand side of the above equation must vanish, i.e.

$$-2A_{1/2} - \tfrac{1}{4}A^3_{1/2} = 0, \qquad 2A_{1/2}\omega_1 + \tfrac{5}{6}A^3_{1/2} = 0.$$

Therefore

$$A^2_{1/2} = -8, \qquad \omega_1 = \tfrac{10}{3}.$$

The fact that $A^2_{1/2} < 0$ implies that there is a real periodic solution only when $a < 0$, so that

$$\mathbf{x}(\tau, a) \sim (-8a)^{1/2}\begin{bmatrix}\cos\tau \\ \sin\tau\end{bmatrix},$$

$$\omega(a) = 1 + \tfrac{10}{3}a + o(a) \qquad \text{as } a \uparrow 0,$$

i.e.

$$x(t, a) \sim (-8a)^{1/2}\cos\omega t, \qquad y(t, a) \sim (-8a)^{1/2}\sin\omega t \qquad \text{as } a \uparrow 0.$$

These results give the leading asymptotic properties of the limit cycle near the subcritical Hopf bifurcation at $a = 0$. Further properties can be found by solving for $\mathbf{x}_{3/2}, \omega_{3/2}, \mathbf{x}_2, \ldots$, in turn. □

6 Van der Pol's equation

This section is devoted to a famous equation of nonlinear oscillations, which will serve as an example of a limit cycle and as an example for explaining some asymptotic methods. In modelling an electrical circuit with a thermionic valve, i.e. a vacuum tube, van der Pol (1926) derived an equation of the form

$$\frac{d^2x}{dt^2} + \epsilon(x^2 - 1)\frac{dx}{dt} + x = 0. \tag{1}$$

It may be rewritten as the system

$$\frac{dx}{dt} = v, \qquad \frac{dv}{dt} = -x - \epsilon(x^2 - 1)v; \tag{2}$$

or, on using equations (1.8), (1.9), as

$$\frac{dr}{dt} = -\epsilon(r^2\cos^2\theta - 1)r\sin^2\theta, \qquad \frac{d\theta}{dt} = -1 - \epsilon(r^2\cos^2\theta - 1)\cos\theta\sin\theta. \tag{3}$$

Note first that equation (1) is of invariant form under the transformations (a) $t \to -t$ and $\epsilon \to -\epsilon$, and (b) $x \to -x$. It follows that the *set* of all orbits is similarly invariant, although an individual orbit need not be. Also the reverse of sign of ϵ will reverse the direction of time, so we may consider $\epsilon \geqslant 0$ without loss of generality (remembering that damping for $\epsilon < 0$ corresponds to negative damping for $\epsilon > 0$).

Rayleigh (1883, 1894, Vol. I, §68a) modelled some nonlinear vibrations by an equation of the form

$$\frac{d^2u}{dt^2} + \epsilon\left\{\frac{1}{3}\left(\frac{du}{dt}\right)^3 - \frac{du}{dt}\right\} + u = 0. \tag{4}$$

On differentiation with respect to t, this becomes

$$\frac{d^3u}{dt^3} + \epsilon\left\{\left(\frac{du}{dt}\right)^2 - 1\right\}\frac{d^2u}{dt^2} + \frac{du}{dt} = 0.$$

We see, on identifying $x = du/dt$, that equations (4) and (1) are equivalent.

If $\epsilon = 0$, van der Pol's equation (1) gives simple harmonic motion of period 2π, with general solution

$$x(t) = a\cos(t - t_0) \tag{5}$$

for arbitrary real amplitude a and phase t_0.

The only equilibrium point of equation (1), i.e. system (2), is at the origin of the (x, v)-plane. It is an unstable focus for $\epsilon > 0$ (but a stable focus for $\epsilon < 0$ and a centre for $\epsilon = 0$).

System (3) gives

$$\frac{dr}{dt} = -\epsilon(x^2 - 1)r\sin^2\theta.$$

Therefore, when $\epsilon > 0$, the damping is positive where $x^2 > 1$ but negative where $x^2 < 1$. So large amplitudes are damped 'on the whole', although small ones are amplified. We anticipate that there is a stable limit cycle. Careful use of the Poincaré–Bendixson theorem leads to a proof that there exists a limit cycle, and it is unique. Indeed, the result is a special case of a more general theorem for Liénard's equation, which we state without proof.

Theorem. The equation

$$\frac{d^2x}{dt^2} + f(x)\frac{dx}{dt} + g(x) = 0$$

has a periodic solution which is unique (up to translations of t), and this solution is asymptotically orbitally stable, if

(a) f, g are continous,
(b) f is an even function,
(c) $F(x) < 0$ for $0 < x < a$, $F(x) > 0$ and is increasing for $x > a$ (so $F(x) = 0$ only at $x = 0, \pm a$), where F is defined by $F(x) = \int_0^x f(u)\,du$,
(d) g is an odd function and $xg(x) > 0$ for all $x \neq 0$.

Next we shall proceed to find this limit cycle for van der Pol's equation by asymptotic methods, first for small ϵ and then for large ϵ.

We shall find the limit cycle as $\epsilon \to 0$ by the *method of averaging*. This powerful asymptotic method in the theory of weakly nonlinear oscillations is due to Kryloff & Bogoliuboff (1943), who modified a method of van der Pol (1922). We shall first introduce the method of averaging briefly by a digression on a linear equation which we can solve explicitly for all ϵ.

Example 6.9: damped simple harmonic motion. Consider the equation

$$\frac{d^2x}{dt^2} + 2\epsilon\frac{dx}{dt} + x = 0.$$

Let $v = dx/dt$ and take polar coordinates in the phase plane, i.e. the (x, v)-plane. Then, without approximation, we deduce from equations (1.8), (1.9) that

$$\frac{dr}{dt} = -2\epsilon r \sin^2\theta, \qquad \frac{d\theta}{dt} = -1 - \epsilon \sin 2\theta.$$

It is now evident that r is nearly constant and $d\theta/dt$ nearly -1 when ϵ is small, so the motion is nearly simple harmonic with a circular orbit with centre O in the phase plane. So if the angle is θ at time t, then the time is approximately $t + 2\pi$ at angle $\theta - 2\pi$. However, $\mathrm{d}r/\mathrm{d}t$ oscillates with θ in each circuit around O and these oscillations have a non-zero mean, so that their *cumulative* effect may change r by an amount which is not small after a long enough time (of order ϵ^{-1}). We can approximate these cumulative effects for small ϵ by *averaging* the right-hand side of the equation for $\mathrm{d}r/\mathrm{d}t$ around a typical circuit. Accordingly define

$$a(t) = \oint r(t+s)\,\mathrm{d}s \Big/ \oint \mathrm{d}s,$$

where the dummy variable s of time increment is such that the angle goes from θ to $\theta - 2\pi$ in the interval of integration,

$$= \int_{2\pi}^{0} r(t+s)\frac{\mathrm{d}\theta}{\mathrm{d}\theta/\mathrm{d}s} \Big/ \int_{2\pi}^{0} \frac{\mathrm{d}\theta}{\mathrm{d}\theta/\mathrm{d}s}$$

$$= \frac{1+O(\epsilon)}{2\pi}\int_0^{2\pi} r(t+s)\,\mathrm{d}\theta \qquad \text{as } \epsilon \to 0,$$

because $\mathrm{d}\theta/\mathrm{d}s + 1 = O(\epsilon)$. Therefore

$$\frac{\mathrm{d}a}{\mathrm{d}t} = \frac{1+O(\epsilon)}{2\pi}\int_0^{2\pi} \frac{\mathrm{d}r(t+s)}{\mathrm{d}s}\,\mathrm{d}\theta$$

$$= \frac{1+O(\epsilon)}{2\pi}\int_0^{2\pi} (-2\epsilon r \sin^2\theta)\,\mathrm{d}\theta$$

$$= -\epsilon a + O(\epsilon^2) \qquad \text{as } \epsilon \to 0,$$

because $r = a + O(\epsilon)$. Similarly, we define

$$\omega(t) = \oint \frac{\mathrm{d}\theta(t+s)}{\mathrm{d}s}\,\mathrm{d}s \Big/ \oint \mathrm{d}s$$

$$= \frac{1}{2\pi}\int_0^{2\pi} \frac{\mathrm{d}\theta(t+s)}{\mathrm{d}s}\,\mathrm{d}\theta + O(\varepsilon^2) \qquad \text{as } \varepsilon \to 0.$$

Therefore, on averaging,

$$\omega + 1 = \frac{1}{2\pi}\int_0^{2\pi} (-\epsilon r \sin 2\theta)\,\mathrm{d}\theta + O(\epsilon^2)$$

$$= O(\epsilon^2) \qquad \text{as } \epsilon \to 0;$$

it so happens that the first-order correction to ω is zero. Thus $\mathrm{d}a/\mathrm{d}t = -\epsilon a$, $\omega + 1 = 0$ to leading order; and so $a(t) \sim a_0 \mathrm{e}^{-\epsilon t}$, $\theta(t) \to t_0 - t$ as

$\epsilon \to 0$. Therefore

$$x(t, \epsilon) \sim a_0 e^{-\epsilon t} \cos(t - t_0) \qquad \text{as } \epsilon \to 0.$$

Compare this with the exact solution. For $-1 < \epsilon < 1$, it is simply the solution of a focus in §2:

$$x(t, \epsilon) = a_0 e^{-\epsilon t} \cos\{(1 - \epsilon^2)^{1/2}(t - t_0)\},$$

$$v(t, \epsilon) = -a_0 e^{-\epsilon t}[(1 - \epsilon^2)^{1/2} \sin\{(1 - \epsilon^2)^{1/2}(t - t_0)\} + \epsilon \cos\{(1 - \epsilon^2)^{1/2}(t - t_0)\}],$$

for which

$$\theta(t, \epsilon) = -\arctan[(1 - \epsilon^2)^{1/2} \tan\{(1 - \epsilon^2)^{1/2}(t - t_0)\} + \epsilon]. \square$$

The method of averaging is suitable to find asymptotically the weakly nonlinear oscillations not only for equations of the form

$$\frac{d^2x}{dt} + \omega^2 x = \epsilon f(x, dx/dt, \epsilon),$$

but also for ordinary differential equations of higher order, and to find asymptotically a wide variety of weakly nonlinear waves for partial differential equations.

Let us return to van der Pol's equation. We again anticipate that when $0 < \epsilon \ll 1$ the limit cycle is close to some solution (5) of the simple harmonic motion found for $\epsilon = 0$; then $x = r \cos \theta$ and $v = r \sin \theta$, where r and $d\theta/dt$ are approximately constant. Equations (3) suggest that $dr/dt = O(\epsilon)$ and $d\theta/dt = -1 + O(\epsilon)$ as $\epsilon \to 0$, in confirmation of this. However, if we wait a long time, of the order ϵ^{-1}, the small effects of the terms of order ϵ may accumulate and change r and θ by quantities of order one. To find these quantities, we average the contributions to dr/dt and $d\theta/dt$ over the approximate period 2π of the simple harmonic motion. This gives the leading approximation to the slow changes of dr/dt and $d\theta/dt$ and thence to the changes of r and θ over long times of order ϵ^{-1}.

Accordingly define the averages of r and $d\theta/dt$ as in the example above and deduce from equations (3) that

$$\frac{da}{dt} = -\frac{\epsilon}{2\pi} \int_0^{2\pi} r\{\tfrac{1}{4}r^2 \sin^2 2\theta - \tfrac{1}{2}(1 - \cos 2\theta)\}\, d\theta + O(\epsilon^2)$$

$$= \tfrac{1}{2}\epsilon a(1 - \tfrac{1}{4}a^2) + O(\epsilon^2), \tag{6}$$

because r is within order ϵ of a for all t; and similarly

$$\omega + 1 = -\frac{\epsilon}{2\pi}\int_0^{2\pi}(r^2\cos^2\theta - 1)\cos\theta\sin\theta\,d\theta + O(\epsilon^2)$$

$$= O(\epsilon^2) \qquad \text{as } \epsilon \to 0. \tag{7}$$

Equations (6) and (7) give

$$\frac{da}{dt} = \tfrac{1}{2}\epsilon a(1 - \tfrac{1}{4}a^2), \qquad \omega = -1 \tag{8}$$

approximately for small ϵ. Equation (8) for da/dt is familiar from §1.5; its solution at once gives

$$a(t) \to 2 \qquad \text{as } t \to \infty$$

for all $a(0) > 0$. This gives the asymptotic form of the limit cycle.

$$x(t) = 2\cos(t - t_0) + O(\epsilon) \qquad \text{as } \epsilon \downarrow 0. \tag{9}$$

The orbits in the phase plane are approximately circular as they spiral clockwise about the origin but spiral inwards or outwards slowly towards the circle (9).

To find the limit cycle when $\epsilon \gg 1$ it is slightly easier to consider Rayleigh's equation (4) rather than van der Pol's equation. We write the system as

$$\frac{du}{dt} = x, \qquad \frac{dx}{dt} = -u - \epsilon(\tfrac{1}{3}x^3 - x).$$

Further, define $w = u/\epsilon$ and $F(x) = \frac{1}{3}x^3 - x$. Therefore

$$\frac{dx}{dw} = \frac{dx/dt}{dw/dt} = \frac{\epsilon dx/dt}{du/dt}$$

$$= -\epsilon^2\frac{w + F(x)}{x}.$$

This gives the orbits in the (w, x)-plane for all ϵ.

We now see that $dx/dw \to \infty$ as $\epsilon \to \infty$ *except* near the curve $w = -F(x)$. Of course, $dx/dw = 0$ on the curve for all ϵ. Therefore the direction of each orbit in the (w, x)-plane is nearly parallel to the x-axis except near the curve, the slope being large and negative where $\{w + F(x)\}/x > 0$ but large and positive where $\{w + F(x)\}/x < 0$. So the slope of an orbit changes sign at the w-axis and near the curve $w = -F(x)$.

It follows that if an orbit starts at $t = 0$ off the curve then it goes 'vertically' towards the curve, moves near the curve clockwise around the origin

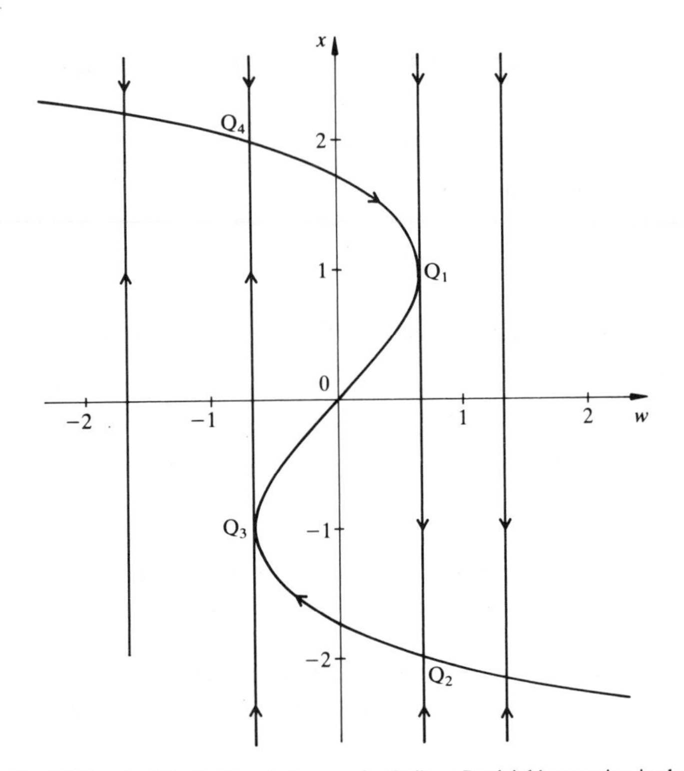

Fig. 6.6 Sketch of the limit cycle for van der Pol's or Rayleigh's equation in the (w, x)-plane for very large ε.

(clockwise because w increases where $x = \epsilon\,\mathrm{d}w/\mathrm{d}t > 0$ when $\epsilon > 0$) until the tangent to the curve is vertical. There it must leave the neighbourhood of the curve, moving vertically until it nears the curve again, and so forth, as shown in Fig. 6.6. This gives a closed orbit $Q_1Q_2Q_3Q_4Q_1$, which is the limit cycle in the (w, x)-plane as $\epsilon \to \infty$.

A more refined approximation is needed to find the details of the solution near the curve $w = -F(x)$. However, we have already found the leading-order approximation to the limit cycle, and may use it to estimate the period. On integrating around the cycle, the period may be expressed as

$$T(\epsilon) = \oint \mathrm{d}t$$

$$= \epsilon \oint \frac{\mathrm{d}w}{x}$$

because $x = \mathrm{d}u/\mathrm{d}t = \epsilon \mathrm{d}w/\mathrm{d}t$,

$$\sim 2\epsilon \int_{Q_4}^{Q_1} \frac{\mathrm{d}w}{x} \qquad \text{as } \epsilon \to \infty,$$

because '$\mathrm{d}w$' is negligible along Q_1Q_2 and Q_3Q_4 and symmetry implies that there are equal contributions to the integral along Q_2Q_3 and Q_4Q_1,

$$= 2\epsilon \int_{Q_4}^{Q_1} -\frac{F'(x)}{x}\mathrm{d}x$$

$$= 2\epsilon \int_2^1 \frac{1-x^2}{x}\mathrm{d}x$$

$$= 2\epsilon[\ln x - \tfrac{1}{2}x^2]_2^1$$

$$= (3 - 2\ln 2)\epsilon$$

$$\approx 1.614\epsilon.$$

(Note that as usual the integral $\oint \frac{\mathrm{d}w}{x}$ is not singular at $x = 0$. This is because $x = 0$ if and only if $\mathrm{d}w/\mathrm{d}t = 0$, and in the exact theory $w - w_0 \propto x^2$ as $x \to 0$ on an orbit (see §1.7).)

The solution changes rapidly along the vertical tangents Q_1Q_2 and Q_3Q_4 and slowly along the segments Q_4Q_1 and Q_2Q_3 of the curve; this has led to the name *relaxation oscillation* for this kind of stable limit cycle for large ϵ, in which the solution slowly, as if languidly, builds up and then rapidly changes.

Numerical calculations show that the limit cycle of van der Pol's equation in the (x, v)-plane and the (t, x)-plane for $\epsilon = 10$, a *fairly* large value, are as in Fig. 6.7. Note that $v = \mathrm{d}x/\mathrm{d}t = \mathrm{d}^2u/\mathrm{d}t^2$, $w = u/\epsilon$ and the point Q_j' is passed at the same time t_j as the point Q_j, for $j = 1, 2, 3, 4$; and observe the correspondence between Figs. 6.6 and 6.7.

Further reading

§6.3 A proof of Liapounov's direct method is given in many good books on ordinary differential equations, e.g., Jordan & Smith (1987, Chap. 10).

§6.4 For further reading on weakly nonlinear oscillations and asymptotic methods, the book by Kevorkian & Cole (1981, Chaps. 2, 3) is recommended.

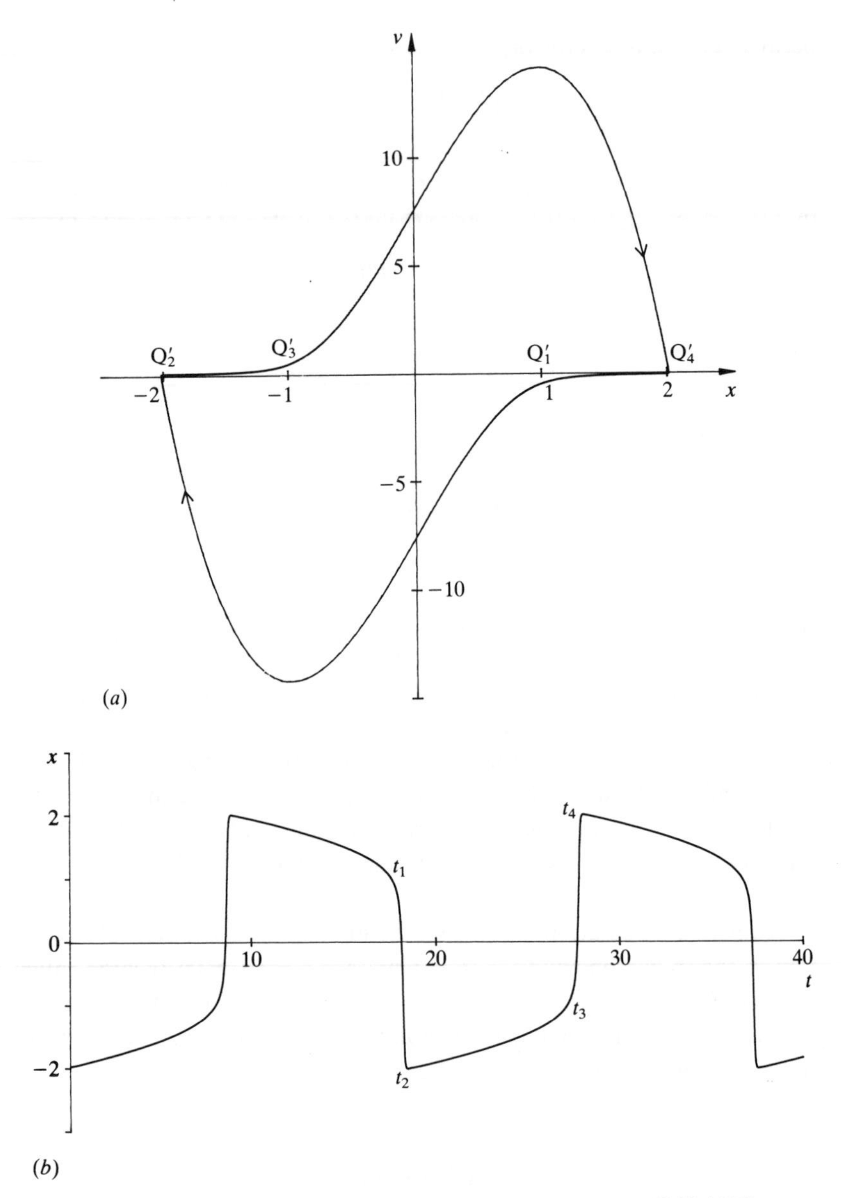

Fig. 6.7 The limit cycle of van der Pol's equation for $\varepsilon = 10$, $T = 19.08$. (*a*) The (x, v)-plane. (*b*) The (t, x)-plane.

§**6.5** Coddington & Levinson (1955, Chap. 16) prove the Poincaré–Bendixson theorem.

§**6.6** Jordan & Smith (1987) give more details of van der Pol's equation.

Problems

Q6.1 *Liénard's construction.* Show that the equation

$$\frac{d^2x}{dt^2} + f(x)\frac{dx}{dt} + g(x) = 0$$

is equivalent to the system

$$\frac{dx}{dt} = y - F(x), \qquad \frac{dy}{dt} = -g(x),$$

where $F(x) = \int_0^x f(u)\,du$. Then the (x, y)-plane is called the *Liénard plane.*

Suppose next that $g(x) = x$ for all x. Then show that the direction of the orbit at a point P in the Liénard plane may be constructed as follows. Draw the line through P parallel to the y-axis, and let Q be the point where the line cuts the curve with equation $y = F(x)$. Draw the line through Q parallel to the x-axis, and let R be the point where this line cuts the y-axis. Then the orbit through P is in the direction perpendicular to $\overrightarrow{RP}$.

[Liénard (1928).]

Q6.2 *Coulomb damping, i.e. motion of a particle under action of a spring with dry friction.* Suppose that

$$m\frac{d^2x}{dt^2} = -\lambda x - F\,\text{sgn}(dx/dt),$$

where $m, \lambda, F > 0$ and the function sgn is defined by $\text{sgn}(v) = 1$ if $v > 0$, $\text{sgn}(0) = 0$, and $\text{sgn}(v) = -1$ if $v < 0$. Sketch the phase portrait in the $(x, dx/dt)$-plane.

Q6.3 *The quintessence of a saddle-node bifurcation.* Consider the system

$$\frac{dx}{dt} = a - x^2, \qquad \frac{dy}{dt} = -y.$$

Show that there is a stable node and a saddle point if $a > 0$, but no equilibrium point if $a < 0$. Sketch the bifurcation diagram in the (a, x)-plane, and the phase portraits in the (x, y)-plane for $a > 0$, $a = 0$ and $a < 0$.

Q6.4 *Epidemics.* A simple model of an epidemic of a disease is the system,

$$\frac{dx}{dt} = -cxy, \qquad \frac{dy}{dt} = cxy - by,$$

where $x(t)$ represents the number of individuals in a population who are liable to infection, $y(t)$ is the number who are infectious at time t, $b > 0$ is the rate of recovery (or death) from the disease, and $c > 0$ is a rate of infection.

Show that if $x(0) < b/c$ then the number y of infectious individuals will decrease monotonically to zero, but if $x(0) > b/c$ then the number will increase monotonically until the number x of susceptible individuals decreases to b/c.

[Kermack & McKendrick (1927).]

Q6.5 *The Lotka–Volterra equations.* Given that the growth of a population of x individuals of a species of prey and y individuals of a species predator is governed by the equations

$$\frac{dx}{dt} = x(a - cy), \qquad \frac{dy}{dt} = -y(b - cx),$$

for constants a, b, $c > 0$, show that $\mathbf{X} = (0, 0)$ is a saddle point and $\mathbf{X} = (b/c, a/c)$ is a centre. Show further that small oscillations about the centre have period $2\pi/(ab)^{1/2}$. Prove that $dy/dx = -y(b - cx)/x(a - cy)$, and integrate this equation. Hence or otherwise sketch the phase portrait in the first quadrant of the (x, y)-plane.

[Lotka (1920) used these equations to model the chemical reactions $D + X \underset{a}{\rightarrow} 2X$, $E + Y \underset{b}{\rightarrow} E + F$, $X + Y \underset{c}{\rightarrow} 2Y$ in a well-stirred reaction vessel, where x denotes the concentration of molecule X and y of Y, where D, E are abundant molecules, and where a, b, c are the reaction coefficients. Volterra (1926) used these prey-predator equations to model the population of fish in the Adriatic Sea.]

Q6.6 *Fishing.* Suppose that a population of superpredators, e.g. fishermen, preys with equal intensity f on both species such that a is replaced by $a - f$ and b by $b + f$ in the above model (Q6.5). Then deduce that in stable equilibrium the predator species y is decreased but the prey species x is increased by the superpredators.

Q6.7 *A damped simple pendulum.* Find the equilibrium points of the equation

$$l\frac{d^2\theta}{dt^2} + 2k\frac{d\theta}{dt} + g\sin\theta = 0$$

and classify them, i.e. name their type, for all g, k, $l > 0$. Sketch the phase portrait in the $(\theta, d\theta/dt)$-plane for each topologically different case.

Q6.8 *A nonlinear discontinuous restoring force.* Find the equilibrium points of the equation

$$\frac{d^2x}{dt^2} + 2k\frac{dx}{dt} + \{\operatorname{sgn}(dx/dt)\}x = 0$$

for all $k \geqslant 0$, where the function sgn is defined by $\operatorname{sgn} x = 1$ if $x > 0$, $\operatorname{sgn} 0 = 0$, $\operatorname{sgn} x = -1$ if $x < 0$. Sketch the phase portrait in the $(x, dx/dt)$-plane for $k = 0, 0 < k < 1, k = 1, k > 1$.

Q6.9 *The Brusselator*. Biochemical oscillations are modelled by the system

$$\frac{dx}{dt} = a - x - bx + x^2 y, \qquad \frac{dy}{dt} = bx - x^2 y.$$

Find the equilibrium points and their stability for all $a, b > 0$.

Q6.10 *Phase portraits and bifurcation.* Treat each of the following real nonlinear systems in turn for all real a, (a) noting the symmetries, if any, of the system, (b) finding all the equilibrium points and for what values of a they exist, (c) ascertaining the stability or instability of each point for each value of a, (d) classifying the local behaviour of the orbits near each equilibrium point for each value of a, (e) noting any special properties of the phase portrait, (f) sketching the qualitative features in a phase portrait for each 'typical' value of a, and (g) sketching the bifurcation diagram in a suitable plane, e.g. the (a, x)-plane.

(i) $$\frac{dx}{dt} = ax - x^3 + xy^2, \qquad \frac{dy}{dt} = -y - y^3 - x^2 y.$$

(ii) $$\frac{dx}{dt} = (a + x)y, \qquad \frac{dy}{dt} = -ax + x^2 + y^2.$$

(iii) $$\frac{dx}{dt} = (a + x)y, \qquad \frac{dy}{dt} = a(y - x) + x^2 + y^2.$$

(iv) $$\frac{dx}{dt} = a(x - y) - x^2 + y^2, \qquad \frac{dy}{dt} = (a + x)y.$$

(v) $$\frac{dx}{dt} = 3a(x - y) - x^2 + y^2, \qquad \frac{dy}{dt} = x(a - y).$$

(vi) $$\frac{dx}{dt} = x(a - x - y), \qquad \frac{dy}{dt} = 2a(y - x) + y(x - y).$$

(vii) $$\frac{dx}{dt} = a(3x - 5y) - x^2 + y^2, \qquad \frac{dy}{dt} = x(2a - y).$$

[Bristol (1983, 1990), Iooss & Joseph (1990, Chap. V).]

Q6.11 *Weakly nonlinear theory of a transcritical bifurcation.* Consider the system

$$\frac{dx}{dt} = ay - y^2, \qquad \frac{dy}{dt} = x - 2y + \tfrac{1}{2}x^2,$$

and define $\mathbf{x} = [x, y]^{\mathrm{T}}$.

(a) Show that $\mathbf{x} = \mathbf{0}$ is a solution for all a. Show that it is unstable for $a > 0$, an unstable normal mode behaving like $[2, 1]^{\mathrm{T}} \exp(\frac{1}{2}at)$ as $a \to 0$. Sketch the phase portrait of the solutions near $\mathbf{x} = \mathbf{0}$ for the cases $a < -1$, $-1 < a < 0$, $a > 0$, naming the type of singularity at the origin of the phase plane in each of the three cases.

(b) Find each of the other steady solutions, stating the values of a for which it exists. Find the values of a for which it is unstable.

(c) Sketch the bifurcation diagram in the (a, x)-plane for the steady solutions of the system, and name the type of each of the bifurcations of the solutions.

(d) To investigate nonlinear solutions near one bifurcation, show that the system becomes

$$0 = a\frac{\mathrm{d}x}{\mathrm{d}u} - ay + y^2,$$

$$x - 2y = a\frac{\mathrm{d}y}{\mathrm{d}u} - \tfrac{1}{2}x^2,$$

without approximation, where $u = at$. Defining A such that $[2, 1]\mathbf{x} = aA$, i.e.

$$2x + y = aA,$$

and assuming that a solution may be expanded as

$$\mathbf{x}(a, t) = a\mathbf{x}_1(u) + a^2\mathbf{x}_2(u) + \cdots \qquad \text{as } a \to 0,$$

where

$$\frac{\mathrm{d}A}{\mathrm{d}u} = F_0(A) + aF_1(A) + \cdots,$$

show that $\mathbf{x}_1 = \frac{1}{5}A[2, 1]^{\mathrm{T}}$ and $F_0(A) = \frac{1}{2}A - \frac{1}{10}A^2$. Find $\mathbf{x}_2$ in terms of A.

Q6.12 *A sphere of combustible material in equilibrium.* It is given that the equilibrium of a sphere of a uniform combustible solid is governed by the dimensionless system,

$$\frac{\mathrm{d}^2\phi}{\mathrm{d}r^2} + \frac{2}{r}\frac{\mathrm{d}\phi}{\mathrm{d}r} + \delta \mathrm{e}^{\phi} = 0, \tag{1}$$

$$\frac{\mathrm{d}\phi}{\mathrm{d}r} = 0 \quad \text{at } r = 0, \qquad \phi = 0 \quad \text{at } r = 1. \tag{2}$$

This is a bifurcation problem to determine all dimensionless temperature distributions ϕ for all positive values of the *Frank-Kamenetskii parameter* δ.

Defining $x = \delta r^2 \mathrm{e}^{\phi}$, $y = r\mathrm{d}\phi/\mathrm{d}r + 2$, $t = \ln r$, deduce that an equivalent system is

$$\frac{\mathrm{d}x}{\mathrm{d}t} = xy, \qquad \frac{\mathrm{d}y}{\mathrm{d}t} = 2 - x - y, \tag{3}$$

where

$$(x, y) \to (0, 2) \quad \text{as } t \to -\infty, \qquad x = \delta \quad \text{at } t = 0. \tag{4}$$

Show that equation (3) has a stable focus at $(2, 0)$ in the (x, y)-plane, which solution for all t corresponds to a solution of (1) for all δ such that $\phi = 0$ at $r = 1$ only if $\delta = 2$.

Sketch the phase portrait of the solutions of equation (3) in the right-hand half of the (x, y)-plane, given that there exists an orbit H connecting $(0, 2)$ to $(2, 0)$.

Show that if a solution $(x_h(t), y_h(t))$ of equation (3) traverses the orbit H and corresponds to a solution of boundary conditions (4) then $x_h(0) = \delta$. *Hence or otherwise sketch the bifurcation diagram in the $(\delta, \phi(0))$-plane.

[Budd (1989).]

Q6.13 *A condition for stability.* Show that if

$$\frac{d\mathbf{x}}{dt} = \mathbf{F}(\mathbf{x}),$$

where $\mathbf{x} \in \mathbb{R}^m$, $\mathbf{F}\colon \mathbb{R}^m \to \mathbb{R}^m$, $\mathbf{F}(\mathbf{0}) = \mathbf{0}$, and

$$\mathbf{u}^{\mathrm{T}}\mathbf{F}(\mathbf{u}) \leqslant 0$$

for all $\mathbf{u} \in \mathbb{R}^m$, then $\mathbf{0}$ is a stable point of equilibrium.

Deduce that the solution $\mathbf{0}$ of the linear system

$$\frac{d\mathbf{x}}{dt} = \mathbf{A}\mathbf{x}$$

is asymptotically stable if all the eigenvalues of $\mathbf{A} + \mathbf{A}^{\mathrm{T}}$ are negative.

*Q6.14 *Liapounov function of a difference equation.* Show that if $\mathbf{F}\colon \mathbb{R}^m \to \mathbb{R}^m$ is continuous, $\mathbf{F}(\mathbf{0}) = \mathbf{0}$, and there exists a continuous positive definite function $H\colon \mathbb{R}^m \to \mathbb{R}$ such that $H(\mathbf{F}(\mathbf{x})) \leqslant H(\mathbf{x})$ for all $\mathbf{x}$ in a neighbourhood of $\mathbf{0}$, then $\mathbf{0}$ is a stable fixed point of $\mathbf{F}$.

Q6.15 *Asymptotic stability of an equilibrium point of Liénard's equation.* Suppose that

$$\frac{d^2x}{dt^2} + f(x)\frac{dx}{dt} + g(x) = 0,$$

where f, g are continuous, $g(0) = 0$, and $f(x) > 0$, $xg(x) > 0$ in a neighbourhood of the origin excluding $x = 0$ itself. Show that the system is equivalent to

$$\frac{dx}{dt} = y - F(x), \qquad \frac{dy}{dt} = -g(x),$$

where $F(x) = \int_0^x f(u)\,du$, and that there is an equilibrium point at the origin of the (x, y)-plane.

Taking $H(x, y) = \frac{1}{2}y^2 + G(x)$ as a Liapounov function, where $G(x) = \int_0^x g(u)\,du$, show that the origin is asymptotically stable.

Q6.16 *A linear system.* (a) Find explicitly the general solution of the system,

$$\frac{dx}{dt} = -3x, \qquad \frac{dy}{dt} = -y + 2z, \qquad \frac{dz}{dt} = -2y - z.$$

Deduce that the origin is asymptotically stable.

(b) Prove that the origin is asymptotically stable by Liapounov's direct method.

Q6.17 *A fundamental matrix of a linear system.* Suppose that

$$\frac{d\mathbf{x}}{dt} = \mathbf{A}\mathbf{x}, \tag{1}$$

where $\mathbf{x}$ is a 2×1 column vector and $\mathbf{A}(t)$ is a 2×2 real matrix. Define $\boldsymbol{\phi}_1(t)$ as the solution of the system (1) such that $\boldsymbol{\phi}_1(0) = [1,0]^T$ and $\boldsymbol{\phi}_2(t)$ as the solution such that $\boldsymbol{\phi}_2(0) = [0,1]^T$. Then a fundamental matrix $\boldsymbol{\Phi}$ of (1) is defined as the 2×2 matrix with columns $\boldsymbol{\phi}_1, \boldsymbol{\phi}_2$, i.e. $\boldsymbol{\Phi}(t) = [\boldsymbol{\phi}_1(t), \boldsymbol{\phi}_2(t)]$.

Show that $d\boldsymbol{\Phi}/dt = \mathbf{A}\boldsymbol{\Phi}$ and $\boldsymbol{\Phi}(0) = \mathbf{I}$, the unit matrix, and hence that $\mathbf{x}(t) = \boldsymbol{\Phi}(t)\mathbf{x}(0)$.

Deduce that if $\mathbf{A}$ is constant then $\boldsymbol{\Phi}(t) = \mathbf{I} + t\mathbf{A}/1! + t^2\mathbf{A}^2/2! + \cdots$, $= e^{t\mathbf{A}}$, say.

Q6.18 *A Liapounov function for an asymptotically stable linear system.* In Q6.17, suppose further that $\mathbf{A}$ is constant and has eigenvalue $s_k < 0$ belonging to the eigenvector $\mathbf{u}_k$ for $k = 1, 2$, and that the two eigenvectors are independent. Deduce that there exists a constant real invertible 2×2 matrix $\mathbf{P}$ such that $\mathbf{P}^{-1}\mathbf{A}\mathbf{P} = \begin{bmatrix} s_1 & 0 \\ 0 & s_2 \end{bmatrix}$.

Defining $\mathbf{X} = [X_1, X_2]^T$ such that $\mathbf{x} = \mathbf{P}\mathbf{X}$, deduce that the system (1) of Q6.17 is equivalent to

$$\frac{dX_1}{dt} = s_1 X_1, \qquad \frac{dX_2}{dt} = s_2 X_2. \tag{2}$$

Defining

$$H(\mathrm{x}) = X_1^2 + X_2^2, \tag{3}$$

show that H is positive definite and that $-dH/dt$ is positive definite if $\mathbf{x}$ satisfies system (1). Hence prove the asymptotic stability of the null solution of system (1) (although this result follows directly in the present case because the solution of the linear system (1) can be found explicitly).

Q6.19 *The asymptotic stability of the null solution for both a nonlinear system and its linearized system.* Suppose yet further that

$$\frac{d\mathbf{x}}{dt} = \mathbf{A}\mathbf{x} + \boldsymbol{\xi}(\mathbf{x}) \tag{4}$$

for the same real constant matrix of Q6.18, where $|\boldsymbol{\xi}(x)| = o(|\mathbf{x}|)$ as $\mathbf{x} \to \mathbf{0}$. Using the same Liapounov function H of equation (3) of Q6.18, prove that the null solution of system (4) is asymptotically stable if the null solution of its linearized system is stable.

Q6.20 *Instability of an equilibrium point of a nonlinear system.* Suppose that

$$\frac{dx}{dt} = ax + by + \xi(x, y), \qquad \frac{dy}{dt} = cx + dy + \eta(x, y), \tag{5}$$

and $p, q > 0$, where $\xi(x, y), \eta(x, y) = o(|\mathbf{x}|)$ as $\mathbf{x} \to \mathbf{0}$, $p = a + d$, $q = ad - bc$.

Show that the origin is an unstable point of equilibrium of the linearized system.

Define K by $K(x, y) = (ax + by)^2 + (cx + dy)^2$, and show that

$$K(r\cos\theta, r\sin\theta) \geqslant mr^2 \qquad \text{for all } \theta,$$

where

$$m = \tfrac{1}{2}[a^2 + b^2 + c^2 + d^2 - \{(a^2 - b^2 + c^2 - d^2)^2 + 4(ab + cd)^2\}^{1/2}] > 0.$$

Defining H by

$$H(x, y) = (ax + by)^2 + (cx + dy)^2 + q(x^2 + y^2),$$

show that

$$H(x, y) \geqslant (m + q)(x^2 + y^2) \qquad \text{for all } x, y,$$

$$\operatorname{grad} H \cdot d\mathbf{x}/dt > mp(x^2 + y^2) \qquad \text{for sufficiently small } \mathbf{x} \neq \mathbf{0}.$$

Hence prove that the origin is an unstable point of equilibrium of the nonlinear system (5).

Q6.21 *Liapounov function for a damped hard spring.* Consider the equation

$$\frac{d^2x}{dt^2} + 2k\frac{dx}{dt} + x + lx^3 = 0$$

for $k, l > 0$. Defining the Liapounov function $H(x, dx/dt) = \frac{1}{2}(dx/dt)^2 + \frac{1}{2}x^2 + \frac{1}{4}lx^4$, show that $x(t) \to 0$ as $t \to \infty$ for all initial conditions.

Q6.22 *Asymptotic stability of an equilibrium point.* Show that the null solution of the system

$$\frac{dx}{dt} = 2xy^2 - x^3, \qquad \frac{dy}{dt} = \tfrac{2}{5}x^2y - y^3$$

is asymptotically stable.

[Hint: try a Liapounov function of the form $H(x, y) = \frac{1}{2}(x^2 + ay^2)$.]

Q6.23 *Instability of an equilibrium point.* Show that the null solution is the only equilibrium point of the system

$$\frac{dx}{dt} = xy^2 + x^2y + x^3, \qquad \frac{dy}{dt} = y^3 - x^3,$$

and that it is unstable.

Q6.24 *An example of stability with an unstable linearized system.* Show that the origin is a stable point of equilibrium of the system

$$\frac{dx}{dt} = y - x^3, \qquad \frac{dy}{dt} = -x^2,$$

but that it is an unstable point of equilibrium of the linearized system.

[Hint: it may help to use a Liapounov function of the form $H(x, y) = x^m + by^n$.]

Q6.25 *An application of Liapounov's method to a class of ordinary differential systems.* Show that the null solution of the system

$$\frac{dx}{dt} = y - xf(x, y), \qquad \frac{dy}{dt} = -x - yf(x, y),$$

where $f(0, 0) = 0$ and f is continuous near the origin O of the (x, y)-plane, is

(a) stable if f is positive semidefinite at O,
(b) asymptotically stable if f is positive definite at O, and
(c) unstable if $-f$ is positive definite at O.

Hence show that the null solution of the system

$$\frac{dx}{dt} = y - x(x^4 + y^4), \qquad \frac{dy}{dt} = -x - y(x^4 + y^4)$$

is asymptotically stable.

Q6.26 *A Liapounov functional.* Consider the nonlinear diffusion equation

$$\frac{\partial u}{\partial t} = k\frac{\partial^2 u}{\partial x^2} - u^3 \tag{1}$$

for $k > 0$, together with the boundary conditions that $u(x) = 0$ at $x = 0$, π. Defining the functional $H\colon C[0, \pi] \to \mathbb{R}$ by $H(v) = \int_0^\pi \frac{1}{2}v^2\,dx$, show that $dH(u)/dt = -\int_0^\pi \{k(\partial u/\partial x)^2 + u^4\}\,dx < 0$ and hence that the null solution is stable (in a mean sense).

Show further, by the calculus of variations or otherwise, that $0 \leqslant \int_0^\pi \frac{1}{2}(\partial w/\partial x)^2\,dx - H(w)$ for all functions $w \in C^2[0, \pi]$ such that $w(x) = 0$ at $x = 0, \pi$.

Deduce that $dH(u)/dt \leqslant -H(u)$ and thence that the null solution of equation (1) is in fact asymptotically stable (in the mean).

[Liapounov's method has been generalized so as to apply to partial differential equations (Zubov 1957).]

Q6.27 *A nonlinear oscillation.* Consider the equation

$$\frac{d^2x}{dt^2} + x - \epsilon x|x| = 0.$$

Find its equilibrium points, and for what values of ϵ each exists. Find for what values of ϵ each is stable, and classify the type of the point. Sketch the bifurcation diagram in the (ϵ, x)-plane.

Show that

$$\frac{1}{2}\left(\frac{dx}{dt}\right)^2 + \frac{1}{2}x^2 - \frac{1}{3}\epsilon|x|^3 = E$$

for some constant E of integration.

Sketch the phase portraits for 'typical' values of ϵ, i.e. for some values which give rise to all topologically different portraits.

Show that there exist periodic oscillations for all ϵ, and that the period of the oscillation of amplitude a is

$$T = 4\int_0^a \frac{dx}{(a^2 - \frac{2}{3}\epsilon a^3 - x^2 + \frac{2}{3}\epsilon x^3)^{1/2}}.$$

Deduce that

$$T = 2\pi + \tfrac{8}{3}\epsilon a + O(\epsilon^2 a^2) \qquad \text{as } \epsilon a \to 0.$$

Q6.28 *Another nonlinear oscillation.* Consider the equation

$$\frac{d^2x}{dt^2} + x - \epsilon x^2 = 0$$

for an asymmetric spring. Find its equilibrium points, and for what values of ϵ each exists. Find for what values of ϵ each is stable, and classify the type of the point. Sketch the bifurcation diagram in the (ϵ, x)-plane.

Use the Lindstedt–Poincaré method to find the leading terms in the expansion of the periodic solutions. Explain the *signs* of the corrections to the period and to the time average of x for small ϵ.

Q6.29 *The precession of the perihelion of a planet in the general theory of relativity.* It is given that the equation of the orbit of a planet with plane polar coordinates (r, θ) is

$$\frac{d^2u}{d\theta^2} + u = \frac{1}{l} + \epsilon l u^2,$$

where $u = 1/r$, the sun is fixed at the origin $r = 0$, $l = h^2/GM$ and $\epsilon = 3GM/c^2 l$. Here G is the gravitational constant, c is the velocity of light, M is the mass of the sun, and h is the angular momentum per unit mass of the planet about the sun.

Show that there is a centre at $(u_N, 0)$ and a saddle point at $(u_r, 0)$ in the $(u, du/d\theta)$-plane, where $u_N(\epsilon) = 1/l + \epsilon/l + O(\epsilon^2/l)$ and $u_r(\epsilon) = 1/\epsilon l - 1/l + O(\epsilon/l)$ as $\epsilon \downarrow 0$, i.e. in the Newtonian limit. Sketch the phase portrait and identify the region of orbits representing solutions which are periodic functions of θ. (N.B. Periodicity of an orbit in *time* is a very different property.)

Define $\phi = \omega\theta$, where $2\pi/\omega$ is the period of an orbit, and assume that $\omega(\epsilon) = \omega_0 + \epsilon\omega_1 + \epsilon^2\omega_2 + \cdots$. Hence show that $\omega_0 = 1$ and $\omega_1 = -1$. Deduce that the planet is at *perihelion*, i.e. that its distance from the sun is a local minimum, at successive angles θ differing by $2\pi/\omega = 2\pi + 2\pi\epsilon + O(\varepsilon^2)$ as $\epsilon \downarrow 0$.

[This gives the precession of the perihelion of the planet by the angle $2\pi\epsilon = 6\pi GM/c^2 l$ approximately each revolution, where l is close to the mean radius of the orbit. This result is used in one of the classic tests of Einstein's general theory of relativity.]

Q6.30 *A periodic solution of Rayleigh's equation.* It is given that the equation

$$\frac{d^2x}{dt^2} + x = \epsilon\left\{\frac{dx}{dt} - \frac{1}{3}\left(\frac{dx}{dt}\right)^3\right\}$$

has a unique periodic solution for all $\epsilon \neq 0$, such that $dx/dt = 0$, $d^2x/dt^2 < 0$ at $t = 0$. Then use the Lindstedt–Poincaré method to show that the solution is given by

$$x(\epsilon, t) \sim 2\cos\omega t \qquad \text{as } \epsilon \to 0,$$

where $\omega = 1 - \frac{1}{16}\epsilon + O(\epsilon^2)$.

Q6.31 *Bendixson's criterion for the non-existence of a periodic solution.* Show that if

$$\frac{dx}{dt} = F(x, y), \qquad \frac{dy}{dt} = G(x, y),$$

for continuously differentiable functions F, G, and $\partial F/\partial x + \partial G/\partial y \neq 0$ in a domain $D \subset \mathbb{R}^2$, then there is no closed orbit in D.

[Bendixson (1901).]

Show that if

$$\frac{dx}{dt} = 2x - xy^2 + \cos y, \qquad \frac{dy}{dt} = -y - x^2y + \sin x$$

then there is no closed orbit inside the circle $x^2 + y^2 = 1$.

Q6.32 *Liapounov and Poincaré–Bendixson.* Consider the nonlinear system

$$\frac{dx}{dt} = y + x^3 - x(x^2 + y^2)^2, \qquad \frac{dy}{dt} = -x + y^3 - y(x^2 + y^2)^2.$$

(a) Show that the origin is an equilibrium point. Linearize the system at the origin, sketch the phase portrait of the linearized system, and state whether the origin is a stable or unstable point of it.

Use Liapounov's direct method to show that the origin is an unstable point of equilibrium of the nonlinear system.

(b) Show that the given system has a limit cycle.

[Bristol (1989).]

Q6.33 *Liapounov, Poincaré–Bendixson and Poincaré.* Consider the nonlinear system

$$\frac{dx}{dt} = ax + y - xf(x^2 + y^2), \qquad \frac{dy}{dt} = -x + ay - yf(x^2 + y^2),$$

where a is a real parameter, f is continuous, $f(0) = 0$ and $f(x^2 + y^2) \geqslant (x^2 + y^2)^{1/2}$.

(a) Show that the origin is the only equilibrium point. Linearize the system at the origin and find what type of point it is according to the values of a.

Use the Liapounov function $H(x, y) = \frac{1}{2}(x^2 + y^2)$ to show that the origin is a stable point of equilibrium of the nonlinear system if $a \leqslant 0$ and unstable if $a > 0$. For what values of a is it certainly asymptotically stable?

(b) Show that there exists a stable limit cycle of the system if $a > 0$.

(c) Take the special case with $f(r^2) = r$ for all $r \geqslant 0$ and with $a > 0$.

Find the limit cycle explicitly by solving the system.

A Poincaré map $\mathbf{P}: \mathbb{R}^2 \to \mathbb{R}^2$ is defined for the system in this case as follows: take $\mathbf{x}(0) = \mathbf{x}_0$, integrate the system from $t = 0$ to $t = 2\pi$ to determine $\mathbf{x}_1 = \mathbf{x}(2\pi)$, and let $\mathbf{x}_1 = \mathbf{P}(\mathbf{x}_0)$ for all $\mathbf{x}_0$, where $\mathbf{x}(t) = (x(t), y(t))$. Show that

$$\mathbf{P}(\mathbf{x}) = \frac{a\mathbf{x}}{|\mathbf{x}| + (a - |\mathbf{x}|)e^{-2\pi a}}.$$

[Bristol (1988).]

Q6.34 *Demonstration of the existence of a limit cycle.* By considering the directions in which orbits cross suitable closed curves, show that the system

$$\frac{dx}{dt} = -x - y + (x^2 + 2y^2)x, \qquad \frac{dy}{dt} = x - y + (x^2 + 2y^2)y,$$

has at least one periodic solution.

Q6.35 *Demonstration of the existence of another limit cycle.* (a) Given the equation,

$$\frac{d^2x}{dt^2} + \operatorname{sgn} x = 0,$$

find explicitly the solution such that

$$x = 0, \qquad dx/dt = \tfrac{1}{4}T \qquad \text{at } t = 0,$$

where $T > 0$. State the amplitude and period of the solution in terms of T.

(b) Sketch the phase portrait of the equation

$$\frac{d^2x}{dt^2} + \epsilon(x^2 - 1)\frac{dx}{dt} + \operatorname{sgn} x = 0$$

for $0 < \epsilon \ll 1$.

(c) Show that the equation

$$\frac{d^2x}{dt^2} + \epsilon(x^2 - 1)\frac{dx}{dt} + \tanh ax = 0,$$

for $a, \epsilon > 0$, has one and only one periodic solution (except for a translation of time). Is it stable?

Note that $\tanh ax \to \operatorname{sgn} x$ as $ax \to \pm\infty$, and comment on the relationship between the phase portraits for parts (b), (c) of this question.

Q6.36 *Demonstration of the existence of yet another limit cycle.* Show that the equation

$$\frac{d^2x}{dt^2} + \left\{\left(\frac{dx}{dt}\right)^2 + x^2 - 1\right\}\frac{dx}{dt} + x^3 = 0$$

has at least one periodic solution.

Q6.37 *Change in the topological character of the orbits near a singular point owing to nonlinearity.* Show that if

$$\frac{dx}{dt} = -x + \frac{2y}{\ln(x^2 + y^2)}, \qquad \frac{dy}{dt} = -y - \frac{2x}{\ln(x^2 + y^2)},$$

then

$$\frac{dr}{dt} = -r, \qquad \frac{d\theta}{dt} = \frac{1}{\ln r},$$

where $x = r\cos\theta$, $y = r\sin\theta$, $r > 0$. Hence find θ as a function of r and describe the character of the orbits near the origin.

Describe the character of the orbits of the linearized system, namely

$$\frac{dx}{dt} = -x, \qquad \frac{dy}{dt} = -y.$$

[Nemytskii & Stepanov (1960, p. 85).]

Q6.38 *A Takens–Bogdanov bifurcation.* Show that the system

$$\frac{dx}{dt} = y, \qquad \frac{dy}{dt} = a + by + x^2 + xy$$

has equilibrium points at $\mathbf{X}_\pm = (\pm(-a)^{1/2}, 0)$ if $a < 0$, and at $(0, 0)$ if $a = 0$, there being no equilibrium point if $a > 0$.

Show that $\mathbf{X}_+$ is a saddle point, and that $\mathbf{X}_-$ is unstable if $b > (-a)^{1/2}$ and stable if $b < (-a)^{1/2}$. Moreover, classify the types of the equilibrium point $\mathbf{X}_-$ according to the values of $a < 0, b$.

*Using the Lindstedt–Poincaré method, define $\epsilon = b - (-a)^{1/2}$, and assume that $\mathbf{x} = (x, y)$ has period $2\pi/\omega$, and

$$\mathbf{x} = \mathbf{x}_0 + \epsilon^{1/2}\mathbf{x}_{1/2} + \epsilon\mathbf{x}_1 + \epsilon^{3/2}\mathbf{x}_{3/2} + \cdots, \qquad \omega = \omega_0 + \epsilon^{1/2}\omega_{1/2} + \varepsilon\omega_1 + \cdots$$

as $\epsilon \to 0$, where $\mathbf{x}_0 = \mathbf{X}_-$. Hence verify that $\omega_0 = \{2(-a)^{1/2}\}^{1/2}$ and $\omega_{1/2} = 0$, find $\mathbf{x}_{1/2}$, and show that $\omega_1 = (10 + \omega_0^2)/6\omega_0$. Deduce that there is, for fixed $a < 0$, a *sub*critical Hopf bifurcation, in the sense that these periodic solutions occur when $0 < (-a)^{1/2} - b \ll 1$.

Deduce that there is a turning point (with one eigenvalue $s = 0$) at $a = 0$, $b \neq 0$, and a Hopf bifurcation (with eigenvalues $s = \pm \mathrm{i}(2b)^{1/2}$) at $b = (-a)^{1/2}$, $a < 0$.

[The system exhibits a coincidence of turning points and Hopf bifurcations, called a *Takens–Bogdanov bifurcation*, at $a = b = 0$. Cf. Guckenheimer & Holmes (1986, §7.3).]

Q6.39 *The method of averaging*. Show that if

$$\frac{\mathrm{d}^2x}{\mathrm{d}t^2} + x = \epsilon f(x, \mathrm{d}x/\mathrm{d}t)$$

then

$$\frac{\mathrm{d}a}{\mathrm{d}t} = \frac{\epsilon}{2\pi}\int_0^{2\pi} \sin\theta f(a\cos\theta, a\sin\theta)\,\mathrm{d}\theta + O(\epsilon^2),$$

$$\omega + 1 = \frac{\epsilon}{2\pi}\int_0^{2\pi} a^{-1}\cos\theta f(a\cos\theta, a\sin\theta)\,\mathrm{d}\theta + O(\epsilon^2) \qquad \text{as } \epsilon \to 0,$$

where $x = r\cos\theta$, $\mathrm{d}x/\mathrm{d}t = r\sin\theta$, a is the average of r, and ω is the average of $\mathrm{d}\theta/\mathrm{d}t$ around one circuit of the nearly circular orbit in the $(x, \mathrm{d}x/\mathrm{d}t)$-plane.

Q6.40 *A self-excited oscillation*. Show that the origin of the (x, v)-plane is an equilibrium point of the equation

$$\frac{\mathrm{d}^2x}{\mathrm{d}t^2} - \epsilon(1 - |x|)\frac{\mathrm{d}x}{\mathrm{d}t} + x = 0,$$

where $v = \mathrm{d}x/\mathrm{d}t$. Name the type of the equilibrium point, and state when it is stable according to the value of ϵ. What other equilibrium points are there?

You may assume that there is one and only one limit cycle when $\epsilon > 0$, and that it is stable. Show that the limit cycle is given by

$$x(\epsilon, t) \to \tfrac{3}{4}\pi\cos(t - t_0) \qquad \text{as } \epsilon \downarrow 0.$$

Show that the equation is equivalent to the system

$$\frac{\mathrm{d}w}{\mathrm{d}t} = \frac{x}{\epsilon}, \qquad \frac{\mathrm{d}x}{\mathrm{d}t} = -\epsilon\{w + F(x)\},$$

where $F(x) = \frac{1}{2}x|x| - x$. Sketch the curve $w = -F(x)$ in the (w, x)-plane and indicate the locus of the limit cycle as $\epsilon \to \infty$. Hence or otherwise show that the period T of the limit cycle is given by

$$T(\epsilon) \sim 2\{\sqrt{2} - \ln(1 + \sqrt{2})\}\epsilon \qquad \text{as } \epsilon \to \infty$$

Q6.41 *Another self-excited oscillation.* Consider the equation

$$\frac{d^2x}{dt^2} - \epsilon(1 - x^4)\frac{dx}{dt} + x = 0.$$

Show that a limit cycle $x(\epsilon, t)$ is given by

$$x(\epsilon, t) \to 2^{3/4} \cos(t - t_0) \qquad \text{as } \epsilon \downarrow 0.$$

Show that the equation is equivalent to the system

$$\frac{dw}{dt} = \frac{x}{\epsilon}, \qquad \frac{dx}{dt} = -\epsilon\{w + F(x)\},$$

where $F(x) = \frac{1}{5}x^5 - x$. Sketch the curve $w = -F(x)$ in the (w, x)-plane and indicate the locus of the limit cycle as $\epsilon \to \infty$. Hence or otherwise show that the period T of the limit cycle is given by

$$T(\epsilon) \sim 2\{\tfrac{1}{4}(\alpha^4 - 1) - \ln \alpha\}\epsilon \qquad \text{as } \epsilon \to \infty,$$

where α is the positive root of the quintic equation $a^5 - 5a - 4 = 0$.

[In fact $\alpha \approx 1.65$. You may assume that $\int_0^{2\pi} \sin^2\theta \cos^4\theta \, d\theta = \frac{1}{8}\pi$.]

Q6.42 *Yet another self-excited oscillation.* Consider the equation

$$\frac{d^2x}{dt^2} + \epsilon f(x)\frac{dx}{dt} + x = 0,$$

where $f(x) = -1$ if $-1 \leqslant x \leqslant 1$ and $f(x) = 1$ if $|x| > 1$. Write down the explicit general solutions for the four cases, $|x| > 1$ and $-1 \leqslant x \leqslant 1$, for $0 < \epsilon < 2$ and $\epsilon > 2$. Sketch the phase portraits in the (x, v)-plane for $0 < \epsilon < 2$ for $\epsilon > 2$, where $v = dx/dt$, deducing plausibly that there is a limit cycle.

Show by the method of averaging that the limit cycle $x(\epsilon, t)$ is given by

$$x(\epsilon, t) \to a_e \cos(t - t_0) \qquad \text{as } \epsilon \downarrow 0,$$

where $a_e > 1$ is the root of the equation $2 \sin 2\alpha + \pi = 4\alpha$ for a, and $0 < \alpha = \arccos(1/a) < \frac{1}{2}\pi$.

Show that the differential equation is equivalent to the system

$$\frac{dw}{dt} = \frac{x}{\epsilon}, \qquad \frac{dx}{dt} = -\varepsilon\{w + F(x)\},$$

where $F(x) = -x$ for $|x| \leqslant 1$ and $F(x) = x - 2\,\mathrm{sgn}\, x$ for $|x| > 1$. Sketch the curve $w = -F(x)$ in the (w, x)-plane and indicate the locus of the limit cycle as $\epsilon \to \infty$. Hence or otherwise show that the period T of the limit cycle is given by

$$T(\epsilon) \sim 2\epsilon \ln 3 \qquad \text{as } \epsilon \to \infty.$$

Q6.43 *Rayleigh's equation.* Apply the method of averaging directly to Rayleigh's equation,

$$\frac{d^2x}{dt^2} + \epsilon\left\{\frac{1}{3}\left(\frac{dx}{dt}\right)^3 - \frac{dx}{dt}\right\} + x = 0,$$

to find the amplitude and frequency of the limit cycle as $\epsilon \to 0$. You need only find the terms up to those of order ϵ.

Q6.44 *Simple pendulum.* Writing the equation

$$\frac{d^2\theta}{dt^2} = -gl^{-1}\sin\theta$$

in the form

$$\frac{d^2\theta}{d\tau^2} + \theta = \epsilon f(\theta),$$

where $\tau = (l/g)^{1/2}t$ and $\epsilon f(\theta) = \theta - \sin\theta$ (by introducing artificially a small positive parameter ϵ), use the method of averaging to find the frequency of the periodic solutions of small amplitude approximately, giving the correction term in the square of the amplitude of the oscillation.

Q6.45 *An approximation to a limit cycle.* Use the method of averaging to approximate the solutions of the equation

$$\frac{d^2x}{dt^2} + \epsilon\left\{\left|\frac{dx}{dt}\right| - 1\right\}\frac{dx}{dt} + x = 0$$

as $\epsilon \to 0$, finding the amplitude and frequency of the oscillations for small ϵ.

Q6.46 *An amplitude equation.* Show that if

$$\frac{d^2x}{dt^2} - \epsilon(1 - x^2)\left(\frac{dx}{dt}\right)^3 + x = 0$$

and a is the average of $\{x^2 + (dx/dt)^2\}^{1/2}$ over the cycle of an oscillation, then

$$\frac{da}{dt} \sim -\frac{1}{16}\epsilon a^3(a^2 - 6) \qquad \text{as } \epsilon \to 0.$$

Hence find the amplitude and stability of the limit cycle for small ϵ.

7

Forced oscillations

The thing that hath been, it is that which is.
Ecclesiastes i, 9.

1 Introduction

Chapter 6 is an account of many and varied properties of free oscillations which are solutions of two-dimensional autonomous systems of ordinary differential equations. We saw there that the Poincaré–Bendixson theorem prohibits chaos, but saw in Chapter 5 that a two-dimensional non-autonomous system is equivalent to a three-dimensional autonomous system, and so may have chaotic solutions: the proof of the Poincaré–Bendixson theorem becomes invalid because for a non-autonomous system orbits may have different directions at the same point of the phase plane at different times. So in this chapter we shall examine *forced oscillations*, i.e. solutions of non-autonomous second-order equations of the form

$$\frac{d^2x}{dt^2} = f(x, dx/dt) + F(t). \tag{1}$$

We call F the *forcing term*, thinking of the equation as a model of an oscillation of a particle of unit mass with a restoring and damping force f driven by an external force F which is a periodic function of time. We shall see that forced oscillations have even more and more-varied properties than do free oscillations; in particular some forced oscillations are chaotic.

To illustrate the properties of forced oscillations we shall examine in detail *Duffing's equation*, namely

$$\frac{d^2x}{dt^2} + k\frac{dx}{dt} + \alpha x + \delta x^3 = \Gamma \cos \omega t, \tag{2}$$

for real parameters $k, \alpha, \delta, \Gamma, \omega$. (Duffing (1918) used the equation to model nonlinear mechanical vibrations.) With so many parameters it will help to consider various special cases, beginning in this section with a review of some properties of the linear form of the equation. First consider

$$\frac{d^2x}{dt^2} + \alpha x = \Gamma \cos \omega t \tag{3}$$

for $\alpha > 0$, an equation which represents forced simple harmonic motion without damping. The general solution is

$$x(t) = A \cos \sqrt{\alpha}t + B \sin \sqrt{\alpha}t + \frac{\Gamma \cos \omega t}{\alpha - \omega^2} \qquad \text{if } \omega^2 \neq \alpha, \tag{4}$$

for arbitrary constants A, B. This is a quasi-periodic function in general, but a periodic function if ω is a rational multiple of $\sqrt{\alpha}$. The *response*, or *forced oscillation*, is the particular integral $\Gamma \cos \omega t/(\alpha - \omega^2)$, both being proportional to the forcing term and growing without bound as $\omega^2 \to \alpha$. If $\omega^2 = \alpha$ then

$$x(t) = A \cos \omega t + B \sin \omega t + \frac{\Gamma t \sin \omega t}{2\omega}, \tag{5}$$

and there is said to be *resonance*; it can be seen that the response is a secular term which grows without bound as t increases, but which is finite at any given value of t.

Damping diminishes this response. If

$$\frac{d^2x}{dt^2} + k\frac{dx}{dt} + \alpha x = \Gamma \cos \omega t, \tag{6}$$

and $k > 0, 0 < k^2 < 4\alpha$, then

$$x(t) = e^{-kt/2}\{A \cos(\alpha - \tfrac{1}{4}k^2)^{1/2}t + B \sin(\alpha - \tfrac{1}{4}k^2)^{1/2}t\} + \frac{\Gamma\{(\alpha - \omega^2)\cos \omega t + k\omega \sin \omega t\}}{(\alpha - \omega^2)^2 + k^2\omega^2}. \tag{7}$$

Now the response is of order $\Gamma/k\omega$ when $\omega^2 = \alpha$, so there is a resonance 'peak' with finite height $\Gamma/k\omega$. The complementary function, namely $e^{-kt/2}\{A \cos(\alpha - \frac{1}{4}k^2)^{1/2}t + B \sin(\alpha - \frac{1}{4}k^2)^{1/2}t\}$, is called the *transient*, because after a long time it decays, leaving only the response to the forcing. We shall next examine how nonlinearity, like damping, modifies resonance and renders the response finite.

2 Weakly nonlinear oscillations not near resonance: regular perturbation theory

To introduce the effects of nonlinearity on resonance, first consider an illustrative example of Duffing's equation with no damping and weak nonlinearity,

$$\frac{d^2x}{dt^2} + \Omega^2 x - \epsilon x^3 = \Gamma \cos t, \tag{1}$$

on taking $\omega = 1$ without loss of generality (because this is merely a rescaling of the unit of time), and rescaling α, δ. Let us ignore the transients, and confine our attention to *synchronous oscillations*, i.e. to those solutions $x(\epsilon, t)$ of equation (1) with the same period 2π as the forcing. Also, to approximate weak nonlinearity, we shall expand

$$x(\epsilon, t) = x_0(t) + \epsilon x_1(t) + \epsilon^2 x_2(t) + \cdots \qquad \text{as } \epsilon \to 0 \tag{2}$$

for fixed Ω, Γ. Substituting expansion (2) into equation (1), we find that

$$\frac{d^2x_0}{dt^2} + \epsilon\frac{d^2x_1}{dt^2} + \cdots + \Omega^2(x_0 + \epsilon x_1 + \cdots) - \epsilon(x_0 + \epsilon x_1 + \cdots)^3 = \Gamma \cos t.$$

Equating coefficients of successive powers of ϵ, we find

$$\frac{d^2x_0}{dt^2} + \Omega^2 x_0 = \Gamma \cos t, \tag{3_0}$$

$$\frac{d^2x_1}{dt^2} + \Omega^2 x_1 = x_0^3, \tag{3_1}$$

$$\frac{d^2x_2}{dt^2} + \Omega^2 x_2 = 3x_0^2 x_1, \tag{3_2}$$

and so on.

Also the periodicity condition

$$x(\epsilon, t + 2\pi) = x(\epsilon, t) \qquad \text{for all } t, \epsilon, \tag{4}$$

gives similarly the general condition

$$x_n(t + 2\pi) = x_n(t) \qquad \text{for all } t, n. \tag{5_n}$$

Therefore equation (3_0) gives

$$x_0(t) = A \cos \Omega t + B \sin \Omega t + \frac{\Gamma \cos t}{\Omega^2 - 1}$$

if $\Omega \neq \pm 1$. Also periodicity condition (5_0) now gives

$$x_0(t) = \frac{\Gamma \cos t}{\Omega^2 - 1} \tag{6}$$

if Ω is not an integer.

Next equation (3_1) becomes

$$\frac{d^2x_1}{dt^2} + \Omega^2 x_1 = \frac{\Gamma^3(3\cos t + \cos 3t)}{4(\Omega^2 - 1)^3}.$$

This and periodicity condition (5_1) give

$$x_1(t) = \frac{3\Gamma^3 \cos t}{4(\Omega^2 - 1)^4} + \frac{\Gamma^3 \cos 3t}{4(\Omega^2 - 1)^3(\Omega^2 - 9)}. \tag{7}$$

We can proceed to find $x_2, x_3, \ldots$ in turn to determine the unique solution consistent with the hypotheses we have made. In fact the method is successful provided that Ω is not an odd integer, because if Ω is an odd integer then some equation (3_n) has a forcing term in $\cos \Omega t$ on its right-hand side to drive a secular term in x_n. Also the expansion (2) in powers of ϵ is not uniformly valid when Ω is close to an integer, i.e. is not valid in the case of resonance which we set out to examine. At any rate, let us review the results of this theory by drawing a response diagram as in Fig. 7.1. It is informative to plot the amplitude $a_0 = \Gamma/(\Omega^2 - 1)$ of the response as a function of the natural frequency Ω of the free oscillations; however, a change of sign of a_0 merely corresponds to a phase shift of π in t, so the magnitude of a_0 is a better measure of the response to the forcing Γ than is a_0 itself.

Example 7.1: weakly forced oscillations of van der Pol's equation. There are many asymptotic methods to solve weakly nonlinear problems (cf. Kevorkian & Cole 1981), but not sufficient space to illustrate them all here. However, to illustrate another asymptotic method as well as important phenomena of forced oscillations, consider the equation

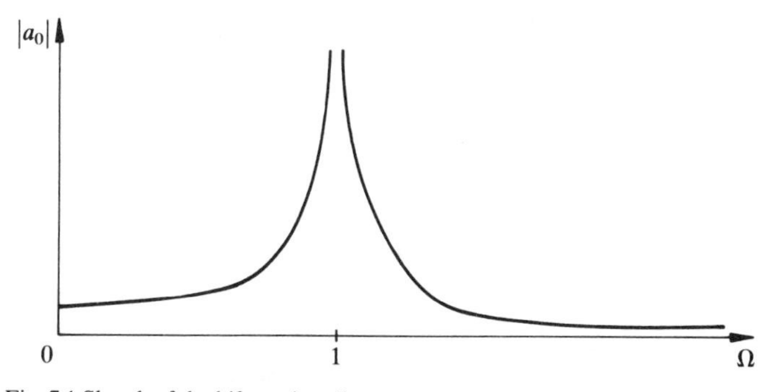

Fig. 7.1 Sketch of the bifurcation diagram in the $(\Omega, |a_0|)$-plane for fixed $\Gamma \neq 0$, i.e. of the curve $|a_0| = |\Gamma/(\Omega^2 - 1)|$.

$$\frac{d^2x}{dt^2} + x = \epsilon\left\{\gamma\cos\omega t + (1 - x^2)\frac{dx}{dt}\right\} \tag{8}$$

for $\gamma, \epsilon, \omega > 0, \omega \neq 1$. To solve the equation for small ϵ, suppose that

$$x = x_0 + \epsilon x_1 + \epsilon^2 x_2 + \cdots \qquad \text{as } \epsilon \downarrow 0, \tag{9}$$

where $x_0 = r\cos(t - \phi)$ for slowly varying functions r, ϕ of time, so that r, ϕ are functions of ϵt. Note that the limit cycle of van der Pol's equation without forcing (i.e. with $\gamma = 0$) has period 2π for small ϵ, and that the forcing has period $2\pi/\omega$.

Substitution of expansion (9) into equation (8) gives

$$\left\{\frac{d^2r}{dt^2} + 2r\frac{d\phi}{dt} - r\left(\frac{d\phi}{dt}\right)^2\right\}\cos(t - \phi) + \left(r\frac{d^2\phi}{dt^2} + 2\frac{dr}{dt}\frac{d\phi}{dt} - 2\frac{dr}{dt}\right)\sin(t - \phi)$$

$$+ \epsilon\left(\frac{d^2x_1}{dt^2} + x_1\right) + O(\epsilon^2) = \epsilon\gamma\cos\omega t + \epsilon(1 - x_0^2)\frac{dx_0}{dt} + O(\epsilon^2)$$

$$= \epsilon\gamma\cos\omega t - \tfrac{1}{4}\epsilon r(4 - r^2)\sin(t - \phi)$$

$$+ \tfrac{1}{4}\epsilon r^3 \sin 3(t - \phi) + O(\epsilon^2). \tag{10}$$

Therefore, on equating terms of order ϵ, it follows that

$$\frac{d^2x_1}{dt^2} + x_1 = \gamma\cos\omega t - \tfrac{1}{4}r(4 - r^2)\sin(t - \phi) + \tfrac{1}{4}r^3\sin 3(t - \phi)$$

$$- \frac{2r}{\epsilon}\frac{d\phi}{dt}\cos(t - \phi) + \frac{2}{\epsilon}\frac{dr}{dt}\sin(t - \phi). \tag{11}$$

Remember that r, ϕ vary slowly, so that each term above is of the same order. To ensure that x_1 is bounded, i.e. that there is no secular term like $t\sin(t - \phi)$ or $t\cos(t - \phi)$ in x_1, the coefficients of $\cos(t - \phi)$, $\sin(t - \phi)$ on the right-hand side of the above equation vanish. Therefore

$$\frac{d\phi}{dt} = O(\epsilon^2), \qquad \frac{dr}{dt} = \tfrac{1}{8}\epsilon r(4 - r^2) + O(\epsilon^2) \qquad \text{as } \epsilon \downarrow 0. \tag{12}$$

This is just what we found (equations (6.6.8)) for the limit cycle of van der Pol's equation without forcing, so that $r \to 2$ as $t \to \infty$ unless $r = 0$ at $t = 0$. It follows that

$$\frac{d^2x_1}{dt^2} + x_1 = \gamma\cos\omega t + \tfrac{1}{4}r^3\sin 3(t - \phi).$$

Therefore

$$x(\epsilon, t) = r\cos(t - \phi) - \frac{1}{32}\epsilon r^3 \sin 3(t - \phi) - \frac{\epsilon\gamma\cos\omega t}{1 - \omega^2} + o(\epsilon). \quad (13)$$

Note that the solution approaches the sum of the stable limit cycle of van der Pol's equation without forcing and a forced oscillation. However, as $\omega \to 1$ the forced oscillation of period $2\pi/\omega$ dominates the free oscillation of period 2π – this is called *entrainment*. This leads to the invalidity of the method of approximation, an issue which will be taken up in Example 7.2. □

3 Weakly nonlinear oscillations near resonance

Next we shall resolve the nonuniformity of the validity of the weakly nonlinear approximation for fixed forcing Γ, showing that the resonant response is large but finite for weakly nonlinear oscillations. This will be illustrated with the example,

$$\frac{d^2x}{dt^2} + x = \epsilon(\gamma\cos t - \beta x + x^3) \quad (1)$$

$$x(\epsilon, t + 2\pi) = x(\epsilon, t) \qquad \text{for all } t. \quad (2)$$

This equation may be used to represent the forcing of a soft spring, or of a simple pendulum when the term in $\sin x$ (i.e. $\sin\theta$) is approximated by $x - \frac{1}{6}x^3$. The terms have been chosen not only so that there is both weak nonlinearity and closeness to resonance but also so that these two effects balance as $\epsilon \to 0$ for fixed constants β, γ, i.e. that we take the *distinguished limit* as $\epsilon \to 0$.

Here substitution of expansion (2.2) in powers of ϵ into equation (1) yields first

$$\frac{d^2x_0}{dt^2} + x_0 = 0.$$

Therefore

$$x_0(t) = a_0\cos t + b_0\sin t$$

for some constants a_0, b_0. This solution satisfies the periodicity condition for all a_0, b_0.

Next, coefficients of ϵ in equation (1) give

$$\frac{d^2x_1}{dt^2} + x_1 = \gamma \cos t - \beta x_0 + x_0^3$$

$$= \gamma \cos t - \beta(a_0 \cos t + b_0 \sin t) + (a_0 \cos t + b_0 \sin t)^3$$

$$= \{\gamma - \beta a_0 + \tfrac{3}{4}a_0(a_0^2 + b_0^2)\}\cos t + b_0\{-\beta + \tfrac{3}{4}(a_0^2 + b_0^2)\}\sin t$$

$$+ \tfrac{1}{4}a_0(a_0^2 - 3b_0^2)\cos 3t + \tfrac{1}{4}b_0(3a_0^2 - b_0^2)\sin 3t.$$

The periodicity condition, $x_1(t + 2\pi) = x_1(t)$, requires that the secular terms, involving $t \sin t$ and $t \cos t$, in x_1 must vanish and therefore that the coefficients of $\cos t$ and $\sin t$ respectively on the right-hand side of the equation of x_1 must vanish, i.e. that

$$a_0\{\beta - \tfrac{3}{4}(a_0^2 + b_0^2)\} = \gamma, \qquad b_0\{\beta - \tfrac{3}{4}(a_0^2 + b_0^2)\} = 0.$$

Therefore

$$b_0 = 0, \qquad a_0(\beta - \tfrac{3}{4}a_0^2) = \gamma. \tag{3}$$

This gives

$$x_0(t) = a_0 \cos t, \tag{4}$$

$$x_1(t) = a_1 \cos t + b_1 \sin t - \tfrac{1}{32} a_0^3 \cos 3t, \tag{5}$$

where a_1, b_1 may in fact be determined uniquely in terms of a_0 by requiring that the secular terms at the next approximation be zero.

We have now found the leading terms in the approximation to the weakly nonlinear forced oscillations near resonance. Let us examine their behaviour. The leading approximation x_0 to the response is in phase (or π out of phase if a_0 has the opposite sign to $\epsilon\gamma$) with the forcing, because $b_0 = 0$; this follows from the fact that the nonlinear term is an odd function of x. The cubic (3) for a_0 has been discussed in Examples 2.2 and 2.5. It has three real roots if $\beta > (16\gamma^2/9)^{1/3}$, two if $\beta = (16\gamma^2/9)^{1/3}$ and one if $\beta < (16\gamma^2/9)^{1/3}$. It gives a pitchfork bifurcation in the (β, a_0)-plane when $\gamma = 0$ and an imperfect one for $\gamma \neq 0$, as shown in Fig. 7.2. Note that $a_0 = -(4\gamma/3)^{1/3}$ if $\beta = 0$; $a_0 = 0$ or $\pm(4\beta/3)^{1/3}$ if $\gamma = 0$; and $a_0 \sim \gamma/\beta$ as $\gamma \to 0$ for $\beta \neq 0$.

In terms of the dimensional form,

$$\frac{d^2x}{dt^2} + \alpha x + \delta x^3 = \Gamma \cos \omega t, \tag{6}$$

of Duffing's equation without damping we may choose a time scale so that $\tau = \omega t$ and deduce that

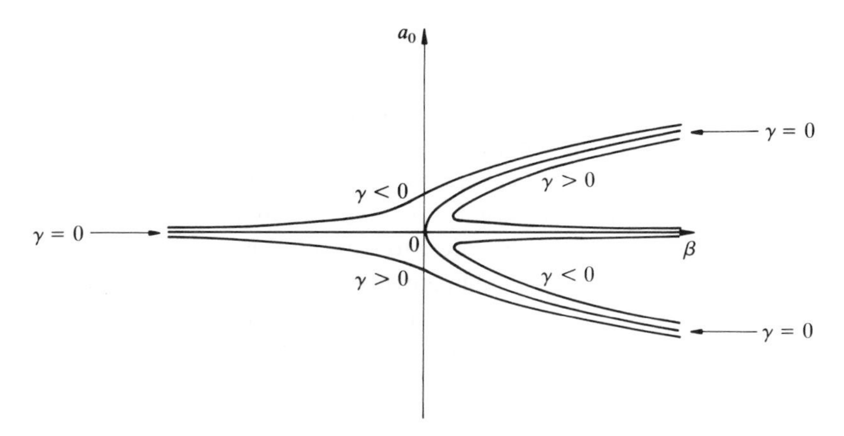

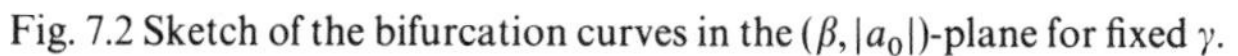

Fig. 7.2 Sketch of the bifurcation curves in the $(\beta, |a_0|)$-plane for fixed γ.

$$\frac{\mathrm{d}^2x}{\mathrm{d}\tau^2} + \frac{\alpha}{\omega^2}x + \frac{\delta}{\omega^2}x^3 = \frac{\Gamma}{\omega^2}\cos\tau,$$

i.e.

$$\frac{\mathrm{d}^2x}{\mathrm{d}\tau^2} + x = \epsilon(\gamma\cos\tau - \beta x + x^3),$$

on identifying $\Gamma/\omega^2 = \epsilon\gamma$, $\alpha/\omega^2 = 1 + \epsilon\beta$ and $\epsilon = -\delta/\omega^2$. This shows explicitly that there is weak nonlinearity and closeness to resonance when ϵ is small for fixed β, γ. Now the second of equations (3) gives $\omega^2 = \alpha + \frac{3}{4}\delta a_0^2 - \Gamma/a_0$ if $a_0 \neq 0$. If $\Gamma = 0$ then $a_0 = 0$ or $\pm\{4(\omega^2 - \alpha)/3\delta\}^{1/2}$.

We may plot the above results in the $(\omega, |a_0|)$-plane as shown in Fig. 7.3, as is conventional to give the response according to the tuning, showing, for fixed $\delta < 0$, the bifurcation curves for constant values of Γ. We see that,

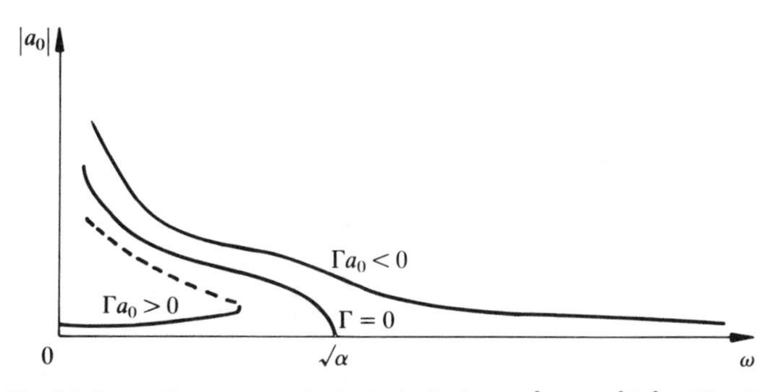

Fig. 7.3 Curves Γ = constant in the $(\omega, |a_0|)$-plane: $\omega^2 = \alpha + \frac{3}{4}\delta a_0^2 - \Gamma/a_0$ for $\delta < 0$.

for fixed Γ, the resonance peak at $\omega = \sqrt{\alpha}$ is made finite. It can be seen that near resonance there may be one periodic response in phase with the forcing or three periodic responses, of which two are π out of phase and one is in phase with the forcing. We have only considered solutions with the same period as the forcing here, but in fact they are stable as indicated by continuous curves and unstable as indicated by the broken curve in Fig. 7.3. It can be seen that if ω were to increase slowly then the solution would abruptly increase its amplitude when it became unstable, and that hysteresis might occur, just as we would anticipate for a cusp catastrophe.

If we were to include weak damping in equation (1), i.e. if we were to start instead with

$$\frac{d^2x}{dt^2} + x = \epsilon\left(\gamma \cos t - \kappa \frac{dx}{dt} - \beta x + x^3\right), \tag{7}$$

then the above asymptotic method would give the amplitude equations

$$\kappa b_0 + a_0\{\beta - \tfrac{3}{4}(a_0^2 + b_0^2)\} = \gamma, \qquad \kappa a_0 - b_0\{\beta - \tfrac{3}{4}(a_0^2 + b_0^2)\} = 0. \tag{8}$$

instead of (2). The damping leads to a phase difference between the forcing and the response, with $b_0 \neq 0$. However, it can be seen by adding the squares of equations (8) that

$$r_0^2\{\kappa^2 + (\beta - \tfrac{3}{4}r_0^2)^2\} = \gamma^2 \tag{9}$$

instead of equation (3), where the amplitude of the response is now $r_0 = (a_0^2 + b_0^2)^{1/2}$. Note the relation of this result for fixed κ to the cusp catastrophe of Example 2.2. We again plot the resulting response curves $\Gamma =$ constant in the (ω, r_0)-plane as shown in Fig. 7.4. In fact there is hysteresis owing to the instability of the solution denoted by the broken curve.

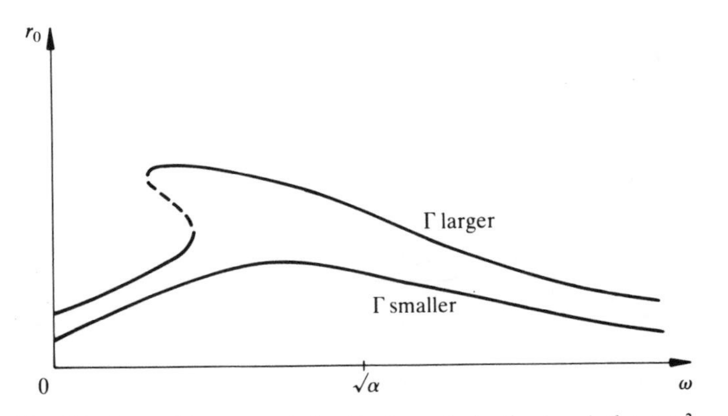

Fig. 7.4 Curves $\Gamma =$ constant, for fixed $\kappa > 0$, in the (ω, r_0)-plane: $\omega^2 = \alpha + \frac{3}{4}\delta a_0^2 - \Gamma/a_0$ for $\delta < 0$.

Example 7.2: weakly forced oscillation of van der Pol's equation near resonance. Suppose that

$$\frac{d^2x}{dt^2} + x = \epsilon\left\{\gamma\cos t + (1 - x^2)\frac{dx}{dt} - \beta x\right\} \tag{10}$$

as $\epsilon \downarrow 0$ for fixed β, γ. First note the similarity to Example 7.1, the only difference being that the limit cycle of the unforced van der Pol equation with period $2\pi/(1 + \epsilon\beta)^{1/2}$ and the forcing with period 2π are nearly in resonance here.

The equation may be solved asymptotically by using the expansion,

$$x = x_0 + \epsilon x_1 + \epsilon^2 x_2 + \cdots \qquad \text{as } \epsilon \downarrow 0, \tag{11}$$

of Example 7.1. However, we shall take the form

$$x_0 = a\cos t + b\sin t \tag{12}$$

here for slowly varying functions a, b instead of the polar variables $r = (a^2 + b^2)^{1/2}$, ϕ, assuming that da/dt, $db/dt = O(\epsilon)$ as $\epsilon \downarrow 0$, after van der Pol (1922). This will serve to display a slightly different method and also solve the problem in hand a little more conveniently. The (a, b)-plane is called the *van der Pol plane*.

Substitution of expansion (11) into equation (10) now gives

$$\left(\frac{d^2a}{dt^2} + 2\frac{db}{dt}\right)\cos t + \left(\frac{d^2b}{dt^2} - 2\frac{da}{dt}\right)\sin t + \epsilon\left(\frac{d^2x_1}{dt^2} + x_1\right) + O(\epsilon^2)$$

$$= \epsilon\left\{\gamma\cos t - \beta x_0 + (1 - x_0^2)\frac{dx_0}{dt}\right\} + O(\epsilon^2).$$

Equating terms of order ϵ, we find

$$\frac{d^2x_1}{dt^2} + x_1 = \gamma\cos t - \beta x_0 - \frac{2}{\epsilon}\frac{db}{dt}\cos t + \frac{2}{\epsilon}\frac{da}{dt}\sin t + (1 - x_0^2)\frac{dx_0}{dt}$$

$$= \gamma\cos t - \beta(a\cos t + b\sin t) - \frac{2}{\epsilon}\frac{db}{dt}\cos t + \frac{2}{\epsilon}\frac{da}{dt}\sin t$$

$$+ (1 - \tfrac{1}{4}r^2)(b\cos t - a\sin t) - \tfrac{1}{4}a(3b^2 - a^2)\sin 3t$$

$$- \tfrac{1}{4}(3a^2 - b^2)\cos 3t. \tag{13}$$

In order to ensure that x_1 is bounded, the coefficients of $\sin t$, $\cos t$ on the right-hand side of equation (13) vanish. Therefore

$$\frac{da}{dt} = \tfrac{1}{8}\epsilon(4 - r^2)a + \tfrac{1}{2}\epsilon\beta b, \quad \frac{db}{dt} = \tfrac{1}{8}\epsilon(4 - r^2)b - \tfrac{1}{2}\epsilon\beta a + \tfrac{1}{2}\epsilon\gamma. \tag{14}$$

The equilibrium points (a_0, b_0) of system (14) give periodic solutions (12) approximating solutions of equation (10):

$$\tfrac{1}{4}(4 - r_0^2)a_0 + \beta b_0 = 0, \qquad \beta a_0 - \tfrac{1}{4}(4 - r_0^2)b_0 = \gamma. \tag{15}$$

Squaring and adding, we find that

$$r_0^2\{\tfrac{1}{16}(4 - r_0^2)^2 + \beta^2\} = \gamma^2, \tag{16}$$

a familiar form of equation for the amplitude of a resonant oscillation (cf. equation (9)).

Now system (14) enables us to investigate the stability of its equilibrium points corresponding to the orbital stability of the solutions of period 2π of equation (10). As in §6.2, we calculate the Jacobian matrix at (a_0, b_0),

$$\mathbf{J} = \epsilon\begin{bmatrix} \tfrac{1}{8}(4 - r_0^2 - 2a_0^2) & \tfrac{1}{2}\beta - \tfrac{1}{4}a_0 b_0 \\ -\tfrac{1}{2}\beta - \tfrac{1}{4}a_0 b_0 & \tfrac{1}{8}(4 - r_0^2 - 2b_0^2) \end{bmatrix}.$$

Therefore

$$p = \operatorname{trace}\mathbf{J} = \tfrac{1}{2}\epsilon(2 - r_0^2),$$

$$q = \det\mathbf{J} = \tfrac{1}{4}\epsilon^2\{\tfrac{1}{16}(4 - r_0^2)(4 - 3r_0^2) + \beta^2\}.$$

Now, by using the results encapsulated in Fig. 6.4, the stability characteristics of the periodic solution (a_0, b_0) may be found according to the values of β, r_0^2 sketched in Fig. 7.5. Note that $p > 0$ below the line with equation $r_0^2 = 2$, that $q > 0$ outside the ellipse with equation

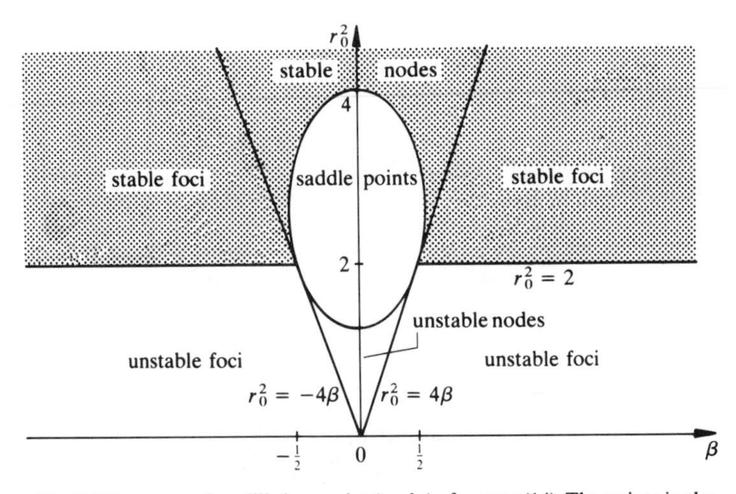

Fig. 7.5 The types of equilibrium point (a_0, b_0) of system (14). The points in the dotted region correspond to stable solutions.

$\frac{9}{16}(r_0^2 - \frac{8}{3})^2 + 3\beta^2 = 1$, and that $p^2 = 4q$ on the lines with equation $r_0^4 = 16\beta^2$.

When there is a stable point (a_0, b_0) of equilibrium of system (14) there is a stable solution of equation (10) with the period of the forcing, not the period of the natural oscillation of the unforced system. This is another example of entrainment. It is related to the synchronization of two clocks when fixed on the same board, an apparently mysterious phenomenon first reported by Huygens in the seventeenth century. □

4 Subharmonics

In solving problems of forced oscillations, we have hitherto confined our attention to one simple equation and the weakly nonlinear theory of its synchronous oscillations, i.e. its solutions with the same period as the sinusoidal forcing term. This theory may similarly be applied to different equations and to different solutions: for example, by taking a more general periodic forcing term as a Fourier series rather than a single term, by looking at the transients, and by ascertaining the stability or instability of the periodic forced oscillations. Also there may be a periodic response whose period is not the period itself, but an integral multiple of the period of the forcing term. For a forcing term $\Gamma \cos \omega t$ there often, but not always, exists a nonlinear oscillation of period $2\pi n/\omega$, for an integer n greater than one, called a *subharmonic* of order $1/n$. A brief example will serve to illustrate this point.

Consider a forced hard spring governed by the equation,

$$\frac{d^2x}{dt^2} + \alpha x + \delta x^3 = \Gamma \cos \omega t, \tag{1}$$

for $\alpha, \delta > 0$, i.e. by

$$\omega^2 \frac{d^2x}{d\tau^2} + \alpha x + \delta x^3 = \Gamma \cos \tau, \tag{2}$$

where $\tau = \omega t$. There is in fact (because the nonlinearity involves only the odd function x^3) no subharmonic of order $\frac{1}{2}$, so we shall at once seek a subharmonic of order $\frac{1}{3}$, for which

$$x(\tau + 6\pi) = x(\tau) \qquad \text{for all } \tau. \tag{3}$$

We may find x by weakly nonlinear theory for small δ, because we know that such periodic solutions exist when $\delta = 0$, $\alpha = \frac{1}{9}\omega^2$ and there is simple harmonic motion. Accordingly, we expand

$$x(\beta,\tau) = x_0(\tau) + \delta x_1(\tau) + \delta^2 x_2(\tau) + \cdots, \tag{4}$$

$$\omega(\delta) = \omega_0 + \delta\omega_1 + \delta^2\omega_2 + \cdots \qquad \text{as } \delta \to 0 \text{ for fixed } \alpha. \tag{5}$$

Therefore coefficients of δ^0 in equation (2) give

$$\omega_0^2 \frac{d^2 x_0}{d\tau^2} + \alpha x_0 = \Gamma \cos \tau. \tag{6_0}$$

Equation (6_0) has general solution

$$x_0(\tau) = a\cos(\sqrt{\alpha}\tau/\omega_0) + b\sin(\sqrt{\alpha}\tau/\omega_0) + \frac{\Gamma\cos\tau}{\alpha - \omega_0^2}$$

if $\omega_0^2 \neq \alpha$. Then the periodicity condition (3) gives

$$\omega_0 = 3\sqrt{\alpha},$$

as anticipated from the linear theory. Therefore

$$x_0(\tau) = a\cos\tfrac{1}{3}\tau + b\sin\tfrac{1}{3}\tau - \frac{\Gamma\cos\tau}{8\alpha},$$

where a and b have yet to be found.

Next, coefficients of δ^1 in equation (2) give

$$\begin{aligned}
\omega_0^2 \frac{d^2 x_1}{d\tau^2} + \alpha x_1 &= -2\omega_0\omega_1 \frac{d^2 x_0}{d\tau^2} - x_0^3 \qquad\qquad (6_1)\\
&= -2\omega_0\omega_1\left(-\tfrac{1}{9}a\cos\tfrac{1}{3}\tau - \tfrac{1}{9}b\sin\tfrac{1}{3}\tau + \frac{\Gamma\cos\tau}{8\alpha}\right)\\
&\quad - \frac{3}{4}\left\{a\left(a^2 + b^2 + \frac{\Gamma^2}{32\alpha^2}\right) - \frac{\Gamma}{8\alpha}(a^2 - b^2)\right\}\cos\tfrac{1}{3}\tau\\
&\quad - \frac{3}{4}\left\{b\left(a^2 + b^2 + \frac{\Gamma^2}{32\alpha^2}\right) + \frac{\Gamma}{4\alpha}ab\right\}\sin\tfrac{1}{3}\tau\\
&\quad + \text{terms in } \cos\tau, \sin\tau \text{ and other harmonics,}
\end{aligned}$$

because $\cos^3\frac{1}{3}\tau = \frac{1}{4}\cos\tau + \frac{3}{4}\cos\frac{1}{3}\tau$, $\sin^3\frac{1}{3}\tau = \frac{3}{4}\sin\frac{1}{3}\tau - \frac{1}{4}\sin\tau$, etc. In order that x_1 has period 6π, we require that its secular terms vanish. Therefore the coefficients of $\cos\frac{1}{3}\tau$ and $\sin\frac{1}{3}\tau$ in the above equation vanish, i.e.

$$a\left(a^2 + b^2 + \frac{\Gamma^2}{32\alpha^2} - \frac{8}{27}\omega_0\omega_1\right) - \frac{\Gamma}{8\alpha}(a^2 - b^2) = 0, \tag{7_c}$$

$$b\left(a^2 + b^2 + \frac{\Gamma^2}{32\alpha^2} - \frac{8}{27}\omega_0\omega_1\right) + \frac{\Gamma}{4\alpha}ab = 0. \tag{7_s}$$

The latter equation has one root $b = 0$, for which the former equation gives either the null solution $a = 0$ or

$$a^2 - \frac{\Gamma}{8\alpha}a + \frac{\Gamma^2}{32\alpha^2} - \frac{8}{27}\omega_0\omega_1 = 0. \quad (8_1)$$

If $b \neq 0$ then the difference of a times (7_s) and b times (7_c) implies that

$$b = \pm\sqrt{3}a,$$

so that equation (7_s) now gives

$$4a^2 + \frac{\Gamma}{4\alpha}a + \frac{\Gamma^2}{32\alpha^2} - \frac{8}{27}\omega_0\omega_1 = 0. \quad (8_2)$$

Equations (8_1), (8_2) have pairs of roots a differing by a factor of -2, because we can rewrite (8_2) as

$$(-2a)^2 - \frac{\Gamma}{8\alpha}(-2a) + \frac{\Gamma^2}{32\alpha^2} - \frac{8}{27}\omega_0\omega_1 = 0.$$

So each equation has two, one or no real roots according to the values of Γ/α and $\omega_0\omega_1 \approx (\omega^2 - 9\alpha)/\delta$.

We may invert the problem and regard ω as a given forcing frequency and find whether subharmonics of order $\frac{1}{3}$ exist according to the values of α, δ, and Γ. Also the stability of these subharmonics may be examined by use of Floquet theory and the weakly nonlinear approximation; but it will suffice here to state that some are stable.

Example 7.3: a weakly nonlinear oscillation near subharmonic parametric resonance. Consider the equation

$$\frac{d^2x}{dt^2} + \{1 + \epsilon(\beta + \cos 2t)\}x + x^2 = 0. \quad (9)$$

There is a rich variety of solutions of this equation, but here we confine our attention to subharmonic oscillations of period 2π as $\varepsilon \to 0$ for fixed β. Thus impose the condition

$$x(\epsilon, t + 2\pi) = x(\epsilon, t) \qquad \text{for all } t. \quad (10)$$

Without loss of generality, it is convenient to choose the phase, i.e. translate the origin of time, so that

$$\frac{dx}{dt} = 0 \qquad \text{at } t = 0. \quad (11)$$

Note that equation (9) is not of the form (1.1), because the coefficient of the linear term x is a function of time with period π. The subharmonic resonance of order $\frac{1}{2}$ is of a slightly different kind, called *parametric resonance*, because the period $2\pi/(1+\epsilon/\beta)^{1/2}$ of linear free oscillations is close to twice the period π of the coefficient.

The problem may be solved asymptotically by using the expansion,

$$x(\epsilon, t) = \epsilon^{1/2} x_{1/2}(t) + \epsilon x_1(t) + \epsilon^{3/2} x_{3/2}(t) + \cdots \qquad \text{as } \epsilon \to 0. \tag{12}$$

The reason for choosing powers of $\epsilon^{1/2}$ rather than ϵ will emerge later when it becomes evident that weak nonlinearity and linear resonance affect the solution at the same order, i.e. it becomes evident that the distinguished limit has been achieved. How does one choose the ansatz without foreknowledge? In practice, experience with trial and error is usually successful, at least in the long run, in finding a self-consistent asymptotic solution.

Now substitute expansion (12) into equations (9), (10) and equate coefficients of $\epsilon^{1/2}$. Therefore

$$\frac{d^2 x_{1/2}}{dt^2} + x_{1/2} = 0, \qquad \frac{dx_{1/2}}{dt} = 0 \qquad \text{at } t = 0.$$

Therefore

$$x_{1/2}(t) = r_{1/2} \cos t,$$

for some constant $r_{1/2}$ to be found later.

Next, coefficients of ϵ in equation (9) give

$$\begin{aligned} \frac{d^2 x_1}{dt^2} + x_1 &= -x_{1/2}^2, \\ &= -\tfrac{1}{2} r_{1/2}^2 (1 + \cos 2t). \end{aligned}$$

The solution of this equation and condition (11) is

$$x_1(t) = -\tfrac{1}{2} r_{1/2}^2 (1 - \tfrac{1}{3} \cos 2t) + r_1 \cos t.$$

Next, coefficients of $\epsilon^{3/2}$ give

$$\begin{aligned} \frac{d^2 x_{3/2}}{dt^2} + x_{3/2} &= -2x_{1/2}x_1 - (\beta + \cos 2t) x_{1/2} \\ &= r_{1/2}^3 (1 - \tfrac{1}{3} \cos 2t) \cos t - 2 r_{1/2} r_1 \cos^2 t \\ &\quad - r_{1/2} (\beta + \cos 2t) \cos t \\ &= r_{1/2}^3 (\tfrac{5}{6} \cos t - \tfrac{1}{6} \cos 3t) - r_{1/2} r_1 (1 + \cos 2t) \\ &\quad - r_{1/2} (\beta \cos t + \tfrac{1}{2} \cos 3t + \tfrac{1}{2} \cos t). \end{aligned}$$

To annihilate the secular terms in $r_{3/2}$, as usual we require the coefficient of $\cos t$ to be zero, i.e.

$$\tfrac{5}{6}r_{1/2}^3 - (\beta + \tfrac{1}{2})r_{1/2} = 0.$$

Therefore $r_{1/2} = \pm\{\frac{6}{5}(\beta + \frac{1}{2})\}^{1/2}$. It follows, in order that the solution be real, that if $\beta + \frac{1}{2} > 0$ then $\epsilon > 0$, and the solution of period 2π is

$$x(\epsilon, t) \sim \{\tfrac{6}{5}(\beta + \tfrac{1}{2})\epsilon\}^{1/2} \cos t \qquad \text{as } \epsilon \downarrow 0,$$

or the same solution π out of phase. If, however, $\beta + \frac{1}{2} < 0$ then $\epsilon \uparrow 0$, and if $\beta = -\frac{1}{2}$ then a different method of approximation must be used. □

Let us review the results of this chapter. Duffing's equation, and other equations of forced oscillations similarly, have solutions with a very rich structure. We have sought and found, by use of weakly nonlinear theory, solutions with the same period as the forcing and solutions whose period is an integral multiple of the period of the forcing. These solutions may also exist for strong nonlinearity, and may be stable or unstable. In short there may be many periodic solutions which attract and which repel neighbouring orbits in phase space. There may be bifurcations as the periodic solutions come and go, become stable and unstable, when the parameters vary. In particular, if a synchronous oscillation becomes unstable and is succeeded by a stable subharmonic oscillation of order $\frac{1}{2}$ as a parameter increases then there is period doubling. All this is a background which might lead us to anticipate the occurrence of chaos for strongly nonlinear solutions, and we shall see in §8.2 that chaotic solutions do indeed occur; but first in the next chapter we shall introduce chaos with an even more famous equation than Duffing's.

Further reading

§7.2 Many good books cover the theory of this chapter. In particular, Jordan & Smith (1987, Chap. 5) describe the theory of weakly nonlinear forced oscillations not near resonance.

§7.3 Jordan & Smith (1987, Chap. 5) also describe the theory of weakly nonlinear oscillations near resonance.

§7.4 Jordan & Smith (1987, Chap. 7) describe subharmonics and the stability of forced oscillations.

Problems

Q7.1 *A forced periodic oscillation.* Consider the solutions of the equation

$$\frac{d^2x}{dt^2} + x^3 = \Gamma \cos t$$

which have period 2π. Taking $x = a$ and $dx/dt = 0$ at $t = 0$ (without loss of generality), so that $x(t) = a \cos t$ + higher harmonics, and neglecting the higher harmonics, show that an approximate solution is given by $3a^3 - 4a = 4\Gamma$.

Q7.2 *Another forced periodic oscillation.* Consider the solutions of the equation

$$\frac{d^2x}{dt^2} + \operatorname{sgn} x = \Gamma \cos \omega t$$

which have period $2\pi/\omega$. Assuming that $x(t) = a \cos \omega t + b \sin \omega t$, and neglecting higher harmonics, show that an approximation to periodic solutions is given by $a = -(\pi\Gamma - 4)/\pi\omega^2$ if $\Gamma > 4/\pi$ and by $a = -(\pi\Gamma + 4)/\pi\omega^2$ if $\Gamma < -4/\pi$.

Q7.3 *A forced periodic oscillation for a nonlinear spring.* Show that if

$$\frac{d^2x}{dt^2} + \Omega^2 x + \epsilon x^2 = \Gamma \cos t$$

and x has period 2π then

$$x(\epsilon, t) = \frac{\Gamma \cos t}{\Omega^2 - 1} - \frac{\epsilon\Gamma^2}{2(\Omega^2 - 1)^2}\left(\frac{1}{\Omega^2} + \frac{\cos 2t}{\Omega^2 - 4}\right) + O(\epsilon^2)$$

as $\epsilon \to 0$ for fixed non-integral Ω.

Q7.4 *Forced oscillations for a damped nonlinear oscillator with negative linear stiffness.* Show that if

$$\frac{d^2x}{dt^2} + \delta\frac{dx}{dt} - \beta x + \alpha x^3 = 0$$

and α, β, $\delta > 0$, then there is a saddle point at the origin and, if $\delta^2 < 8\beta$, there are stable foci at $(\pm(\beta/\alpha)^{1/2}, 0)$ in the $(x, dx/dt)$-plane.

Show that if, however,

$$\frac{d^2x}{dt^2} + \delta\frac{dx}{dt} - \beta x + \alpha x^3 = \Gamma \cos \omega t$$

then there are three oscillations of period $2\pi/\omega$ such that

$$x(t) = \Gamma\{\delta\omega \sin \omega t - (\omega^2 + \beta)\cos \omega t\}/\{(\omega^2 + \beta)^2 + \delta^2\omega^2\} + O(\Gamma^2),$$

$$x(t) = \pm(\beta/\alpha)^{1/2} + \Gamma\{\delta\omega \sin \omega t - (\omega^2 - 2\beta)\cos \omega t\}/\{(\omega^2 - 2\beta)^2 + \delta^2\omega^2\} + O(\Gamma^2)$$

as $\Gamma \to 0$ for fixed $\alpha, \beta, \delta > 0$.

Q7.5 *Resonance of a nonlinear oscillation.* Find the leading approximations to the solutions of the equation

$$\frac{d^2x}{dt^2} + \Omega^2 x - \epsilon x^2 = \Gamma \cos t$$

with period 2π in the limit as $\epsilon \to 0$ (a) when Γ and non-integral Ω are fixed and (b) when $\Omega^2 = 1 + \epsilon\beta$ and $\Gamma = \epsilon\gamma$ for fixed β, γ.

[Jordan & Smith (1987, p. 132).]

Q7.6 *A subharmonic oscillation.* Supposing that

$$\frac{d^2x}{dt^2} + \Omega^2 x + \epsilon x^2 = \Gamma \cos t$$

and x has period 4π, and expanding $\Omega = \Omega_0 + \epsilon\Omega_1 + \epsilon^2\Omega_2 + \cdots$ and $x(\epsilon, t) = x_0(t) + \epsilon x_1(t) + \cdots$ as $\epsilon \to 0$, show that $\Omega_0 = \frac{1}{2}$ and $x_0(t) = a\cos\frac{1}{2}t + b\sin\frac{1}{2}t - \frac{4}{3}\Gamma\cos t$. Thence find two equations satisfied by a, b, Ω_1, and solve them.

[Hint: Lagrange's identity gives $[u\,dv/dt - v\,du/dt]_0^{4\pi} = \int_0^{4\pi} \{u(d^2v/dt^2 + \frac{1}{4}v) - v(d^2u/dt^2 + \frac{1}{4}u)\}\,dt$ for all continuously twice differentiable functions u, v. Try $v = x_1$ with $u = \cos\frac{1}{2}t$ and $\sin\frac{1}{2}t$ in turn.]

Q7.7 *An exact subharmonic solution and its stability.* Show that the equation.

$$\frac{d^2x}{dt^2} + x + \delta x^3 = \Gamma \cos 3\omega t,$$

has a periodic solution $x = X$, where $X(t) = (4\Gamma/\delta)^{1/3}\cos\omega t$, $\omega^2 = 1 + 3(\Gamma^2\delta/4)^{1/3}$.

Take $x = X + x'$ for small x', and deduce that x' satisfies a linearized equation of the form

$$\frac{d^2x'}{d\tau^2} + (a - 2q\cos 2\tau)x' = 0,$$

where $\tau = \omega t$. Evaluate a, q in terms of δ, β, ω.

*Use well-known properties of Mathieu's equation (cf. Abramowitz & Stegun 1964, Chap. 20) to deduce that the solution X is stable.

[McLachlan (1956, p. 241).]

Q7.8 *Another exact subharmonic solution and its stability.* Show that the equation,

$$\frac{d^2x}{dt^2} + \alpha x + \delta x^2 = \Gamma \cos 4t,$$

for α, δ, $\Gamma > 0$, has a solution $x = X$, where $X(t) = C + A\cos 2t$, $\alpha^2 = 16 + 4\Gamma\delta$, $A = \pm(2\Gamma/\delta)^{1/2}$, $C = (4 - \alpha)/2\delta < 0$.

Show that the linearized equation for small perturbations x' of the periodic solution X has the form

$$\frac{d^2x'}{dt^2} + (a - 2q\cos 2t)x' = 0.$$

Evaluate a, q in terms of Γ, δ.

[McLachlan (1956, p. 242).]

Q7.9 *Yet another exact periodic solution.* Given that the equation

$$\frac{d^2x}{dt^2} + (\Omega^2 - 2q\cos 2t)x + \epsilon x^3 = 0$$

has a solution X such that $X(t) = a\cos t$ for real constants a, q, ϵ, Ω, find q, Ω in terms of a, ϵ.

Expressing $x = X + x'$ and linearizing, show that

$$\frac{d^2x'}{dt^2} + (1 + \epsilon a^2 + \epsilon a^2 \cos 2t)x' = 0.$$

*Deduce that X is stable if $\epsilon > 0$.

8

Chaos

For the want of a nail the shoe was lost,
For the want of a shoe the horse was lost,
For the want of a horse the rider was lost,
For the want of a rider the battle was lost,
For the want of a battle the kingdom was lost—
And all for the want of a horseshoe-nail.

Benjamin Franklin (Expanded excerpt from *Poor Richard's Almanack*, February 1752)

1 The Lorenz system

We shall in this section introduce chaotic solutions of ordinary differential systems by detailed study of one system, which is often used as a prototype for the study of chaos. In §2 Duffing's equation with negative stiffness will be discussed briefly: its periodic solutions, their instabilities, and the onset of chaos. This leads to a quite general treatment of a common cause of the onset of chaos in §3. The last two sections bring together many parts of this book. The various sequences of bifurcations leading to chaos as a parameter of a differential system is increased are summarized in §4. Finally, the diagnosis of bifurcations and chaos on the basis of experimental measurements of an unknown dynamical system is discussed. This is a task of great importance and wide applicability; it is the inverse problem to all those considered earlier in the book, requiring induction rather than deduction.

Lorenz (1963) studied a model of two-dimensional convection in a horizontal layer of fluid heated from below:

$$\frac{\mathrm{d}x}{\mathrm{d}t} = -\sigma x + \sigma y, \tag{1_x}$$

$$\frac{\mathrm{d}y}{\mathrm{d}t} = rx - y - zx, \tag{1_y}$$

$$\frac{\mathrm{d}z}{\mathrm{d}t} = -bz + xy; \tag{1_z}$$

in which x represents the velocity and y, z the temperature of the fluid at each instant, and r, σ, b are positive parameters determined by the heating of the layer of fluid, the physical properties of the fluid, and the height of the layer. These equations, the *Lorenz system*, with three dependent variables and three parameters, have a great diversity of solutions with a complicated structure, more complicated than can be described in the few pages available here. However, we shall describe the chief properties, emphasizing the origin and nature of chaos.

*To specify the model a little more, we add that x, y, z are coefficients of Fourier components of the velocity and temperature fields which are, by use of no better than a fair approximation to solutions of the governing equations of motion and heat, assumed to vary with height and the horizontal coordinate in prescribed ways. In detail, the stream function is $\psi(x_*, z_*, t) = x(t)\sin(ax_*/d)\sin(\pi z_*/d)$ and the perturbation of the temperature field, i.e. the difference of the temperature from that of a state of rest with a uniform vertical temperature gradient, is $\theta(x_*, z_*, t) = y(t)\cos(ax_*/d)\sin(\pi z_*/d) + z(t)\sin(2\pi z_*/d)$, where x_* is the horizontal coordinate, z_* is the vertical coordinate, d is the depth of the layer of the fluid, and a is a dimensionless wavenumber. Then r is the Rayleigh number of the layer of fluid divided by its critical value for the onset of instability of the basic state of rest (i.e. of the state with $x = y = z = 0$), σ is the Prandtl number of the fluid, and $b = 4/(1 + a^2)$. The Rayleigh number is a dimensionless measure of the imposed temperature difference across the layer, representing the ratio of the destabilizing buoyancy forces to the stabilizing forces due to molecular diffusion of momentum and heat. The Prandtl number is the ratio of the coefficients of kinematic viscosity and thermal diffusion of the fluid.

First note the invariance of the system under the transformation $(x, y, z) \to (-x, -y, z)$, i.e. under reflection in the z-axis. This symmetry expresses the left–right symmetry of the thermal convection of the fluid. Also it can be seen at once that the z-axis is an orbit, i.e. if $x = y = 0$ at $t = 0$ then $x = y = 0$ for all $t > 0$; moreover, the orbits on the z-axis tend to the origin as $t \to \infty$.

We continue in the usual way by finding the equilibrium points and their linear stability, and sketching a bifurcation diagram. So, putting $\mathrm{d}\mathbf{x}/\mathrm{d}t = \mathbf{0}$, we deduce that $y = x$, $rx - y - zx = 0$, $xy - bz = 0$. A little elementary algebra now shows that an equilibrium point is either (a) the origin 0:

$$x = y = z = 0 \qquad \text{for all } r, \tag{2}$$

or (b) one of the points C, C′:

$$x = y = \pm\{b(r-1)\}^{1/2}, \qquad z = r - 1 \qquad \text{respectively for } r > 1. \quad (3)$$

Note the pitchfork bifurcation of the null solution at $r = 1$ (cf. Q5.6), if we regard σ, b as fixed and vary r. The other equilibrium points C, C′ are symmetrically placed with respect to the z-axis.

Linearization of the system for the null solution gives

$$\frac{dx'}{dt} = -\sigma x' + \sigma y', \tag{4_x}$$

$$\frac{dy'}{dt} = rx' - y', \tag{4_y}$$

$$\frac{dz'}{dt} = -bz'. \tag{4_z}$$

Therefore

$$z'(t) = z_0 e^{-bt} \tag{5}$$

and $x'(t), y'(t) \propto e^{st}$, where

$$0 = \begin{vmatrix} s+\sigma & -\sigma \\ -r & s+1 \end{vmatrix}$$
$$= s^2 + (\sigma+1)s + \sigma(1-r).$$

Therefore

$$s = -\tfrac{1}{2}(\sigma+1) \pm \tfrac{1}{2}\{(\sigma+1)^2 + 4\sigma(r-1)\}^{1/2}. \tag{6}$$

Therefore the null solution is unstable (i.e. $\text{Re}(s) > 0$ for at least one mode) if $r > 1$, and stable (i.e. $\text{Re}(s) < 0$) if $r < 1$; in fact, the three eigenvalues s_j are such that $s_1 > 0 > s_2$ and $s_3 = -b$ when $r > 1$, so that the origin is a saddle point in three-dimensions.

Next linearize about the other equilibrium points C, C′, defining perturbations

$$x' = x \mp \{b(r-1)\}^{1/2}, \qquad y' = y \mp \{b(r-1)\}^{1/2}, \qquad z' = z - (r-1)$$

for $r > 1$. Therefore

$$\frac{dx'}{dt} = \sigma y' - \sigma x', \tag{7_x}$$

$$\frac{dy'}{dt} = x' - y' \mp \{b(r-1)\}^{1/2} z', \tag{7_y}$$

$$\frac{dz'}{dt} = \pm\{b(r-1)\}^{1/2}(x' + y') - bz'. \tag{7_z}$$

Therefore $x', y', z' \propto e^{st}$, where

$$0 = \begin{vmatrix} s+\sigma & -\sigma & 0 \\ -1 & s+1 & \pm\{b(r-1)\}^{1/2} \\ \mp\{b(r-1)\}^{1/2} & \mp\{b(r-1)\}^{1/2} & s+b \end{vmatrix}.$$

$$= s^3 + (\sigma + b + 1)s^2 + b(\sigma + r)s + 2b\sigma(r-1) \tag{8}$$

$$= f(s), \qquad \text{say, for all } s.$$

To find the behaviour of the zeros s_1, s_2, s_3 of the cubic f as r increases from 1, and hence the stability of the equilibrium points, we must do a little intricate, but elementary, algebra. Note that the coefficients of the cubic are all positive, because $r > 1$ in order that the equilibrium points C, C′ exist. Therefore $f(s) > 0$ for all $s \geqslant 0$. Therefore there is instability ($\text{Re}(s) > 0$) only if there are two complex conjugate zeros of f. Now it is easy to see that when $r = 1$ the three zeros are $s = 0, -b, -(\sigma + 1)$, and therefore there is linear stability or a margin of stability. The first zero gives $s \sim -2\sigma(r-1)/(\sigma+1)$ as $r \downarrow 1$, so stability is lost in the limit as r approaches 1 from above. As r increases from 1, instability can set in only where $\text{Re}(s) = 0$, and so where two zeros are $s_1 = \text{i}\omega$, $s_2 = -\text{i}\omega$ for some real ω. But the sum of the three zeros of the cubic f is

$$s_1 + s_2 + s_3 = -(\text{coefficient of } s^2 \text{ in } f)$$
$$= -(\sigma + b + 1).$$

Therefore

$$s_3 = -(\sigma + b + 1)$$

on the margin of stability, where $s_1 = \text{i}\omega$, $s_2 = -\text{i}\omega$. Therefore, on the margin,

$$0 = f(-(\sigma + b + 1))$$
$$= \{-(\sigma + b + 1)\}^3 + (\sigma + b + 1)\{-(\sigma + b + 1)\}^2$$
$$+ b(\sigma + r)\{-(\sigma + b + 1)\} + 2b\sigma(r-1)$$
$$= r\{-b(\sigma + b + 1) + 2b\sigma\} - b\sigma(\sigma + b + 1) - 2b\sigma,$$

i.e.

$$r = \frac{\sigma(\sigma + b + 3)}{\sigma - b - 1}, = r_c, \qquad \text{say.} \tag{9}$$

So instability can arise only if σ, b are such that $r_c > 1$. Thus the points C, C′ are stable if and only if either

(a) $\sigma < b + 1$ and $1 < r$ or
(b) $\sigma > b + 1$ and $1 < r < r_c$.

In fact, if there is instability then as r increases from 1 the following happens: s_1 decreases from zero until it coalesces with s_2 (when $s_1 = s_2 < 0$), they become a complex conjugate pair, and eventually their real part increases through zero, while s_3 remains negative for all $r > 1$. We see that each of the points C, C′, when unstable, has one negative eigenvalue and two complex conjugate eigenvalues, so that neighbouring orbits spiral towards the point in the plane spanned by the eigenvectors $\mathbf{u}_1, \mathbf{u}_2$ at the same time as they leave the point in a direction parallel to $\mathbf{u}_3$; this equilibrium point is an example of what is called a *saddle-focus*.

Weakly nonlinear theory shows that in fact there is a *sub*critical Hopf bifurcation at C, C′ when $r = r_c$ if $r_c > 1$, for a range of values of σ, b. So for these values, which we shall assume in the account below, there is no attractor apparent as r increases through r_c.

Let us summarize our results in a bifurcation diagram, Fig. 8.1, for this case. Then, to prepare to understand what happens when r increases above r_c, we shall establish a few more elementary results.

We can use Liapounov's theorem to show that the null solution is globally asymptotically stable when $r < 1$ and stable if $r = 1$; for take the

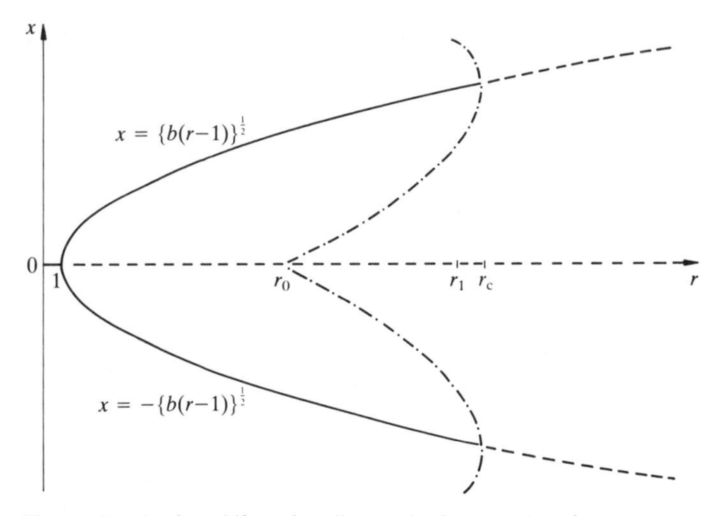

Fig. 8.1 Sketch of the bifurcation diagram in the (r, x)-plane for the Lorenz system with fixed b, σ when $r_c = \sigma(\sigma + b + 3)/(\sigma - b - 1) > 1$.

Liapounov function $H(x, y, z) = \frac{1}{2}(x^2 + \sigma y^2 + \sigma z^2)$. Then

$$\begin{aligned}\frac{dH}{dt} &= x\frac{dx}{dt} + \sigma y\frac{dy}{dt} + \sigma z\frac{dz}{dt}\\ &= \sigma x(-x + y) + \sigma y(rx - y - zx) + \sigma z(-bz + xy)\\ &= -\sigma x^2 + \sigma(1 + r)xy - \sigma y^2 - \sigma bz^2\\ &= -\tfrac{1}{2}\sigma(1 + r)(x - y)^2 - \tfrac{1}{2}\sigma(1 - r)(x^2 + y^2) - bz^2\\ &\leqslant 0\end{aligned}$$

for all x, y, z when $r \leqslant 1$, with equality if and only if $x = y = z = 0$ when $r < 1$ and $x = y, z = 0$ when $r = 1$.

We can deduce that the orbits at infinity are directed *towards* the neighbourhood of the orign as t increases, by a similar argument. We form

$$\begin{aligned}&\frac{d}{dt}\tfrac{1}{2}\{x^2 + y^2 + (z - r - \sigma)^2\}\\ &= x\frac{dx}{dt} + y\frac{dy}{dt} + (z - r - \sigma)\frac{dz}{dt}\\ &= -\sigma x(x - y) + y(rx - y - zx) - (z - r - \sigma)bz + (z - r - \sigma)xy\\ &= -\sigma x^2 - y^2 - bz^2 + b(r + \sigma)z\\ &< 0 \qquad \text{as } \mathbf{x} \to \infty.\end{aligned}$$

So $\frac{1}{2}\{x^2 + y^2 + (z - r - \sigma)^2\}$ is a positive definite function which decreases as t increases when $\mathbf{x}$ is large. Therefore the orbits there move towards $(0, 0, r + \sigma)$ in phase space.

Next, taking $d\mathbf{x}/dt = \mathbf{F}(\mathbf{x})$ for the Lorenz system (1), we form

$$\begin{aligned}\operatorname{div}\mathbf{F} &= \frac{\partial}{\partial x}\sigma(y - x) + \frac{\partial}{\partial y}(rx - y - zx) + \frac{\partial}{\partial z}(-bz + xy)\\ &= -(\sigma + b + 1) \qquad (10)\\ &< 0.\end{aligned}$$

So if $\mu(t)$ is the volume of a set of points in the phase space of (x, y, z) at time t and each point of the volume evolves according to the Lorenz system (1) then $\mu(t) \to 0$ as $t \to \infty$. It follows that the volume of any attractor must be zero. Also there can be no source of volume such as a repelling equilibrium point or limit cycle, because the sum of the real parts of the three eigenvalues is negative. Therefore there is no quasi-periodic orbit; because a quasi-periodic orbit traverses the surface of a torus, and the torus is an invariant surface, which cannot contain a diminishing volume without an interior source.

The next task should be to sketch the phase portraits for various values of r by synthesizing all the bits of information we have gleaned, just as we did for two-dimensional autonomous systems in Chapter 6. However, the task is much more difficult now because the topology of attractors is intrinsically more complicated and more variable in three-dimensional space than in two-dimensional space, and because visualization in our minds, or sketching on paper, is more difficult for three-dimensional than two-dimensional objects.

To approach the issue slowly, we next ask what the attractors are when $r > r_c$. We have found that no equilibrium point is stable; all orbits come in from infinity; volumes μ shrink as t increases; no attractor has finite volume; and there is no quasi-periodic attractor. As r increases to r_c there is a *sub*critical Hopf bifurcation. Then where do the orbits go to when $r > r_c$? Numerical experiments reveal that there is a strange attractor with fractal dimension D such that $2 < D < 3$; it seems to be topologically equivalent to the Cartesian product of a plane surface and a Cantor set. You can get an impression of it by looking at the time series in Fig. 8.2 and the orbits in

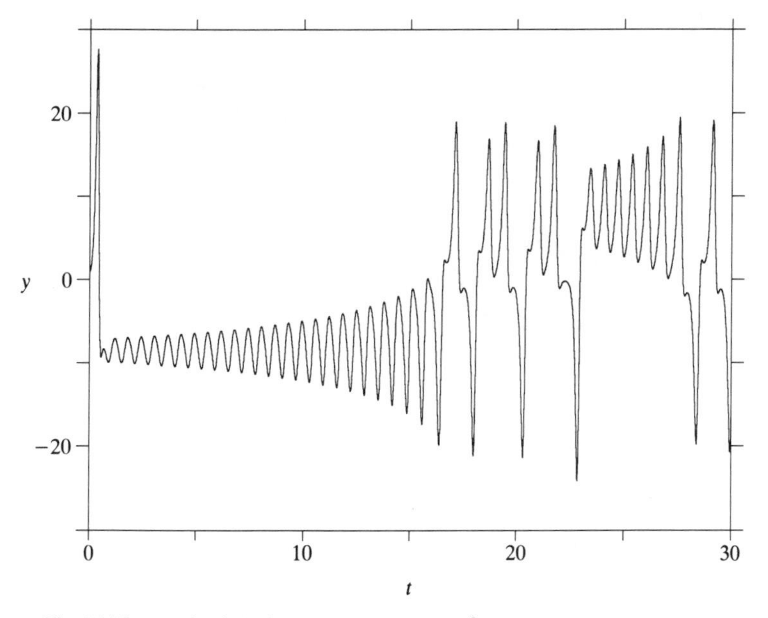

Fig. 8.2 The graph of $y(t)$ for $r = 28$, $\sigma = 10$, $b = \frac{8}{3}$, $\mathbf{x}(0) = (0, 1, 0)$ and $0 \leqslant t \leqslant 30$. Note that the y-coordinates of C, C′ in this case are ± 8.48.

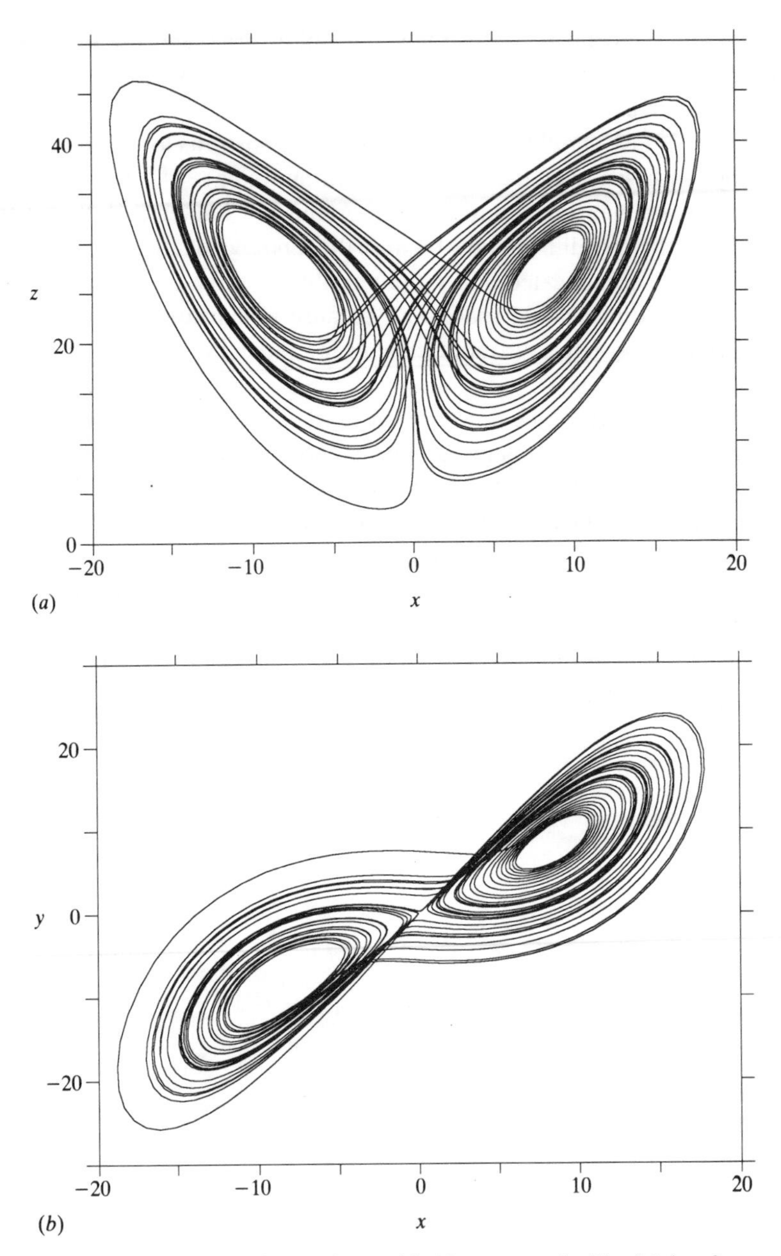

Fig. 8.3 Projections of part of an orbit (the same as in Fig. 8.2 but for $30 \leqslant t \leqslant 70$) for $r = 28$, $\sigma = 10$, $b = \frac{8}{3}$ in (*a*) the (x, z)-plane, (*b*) the (x, y)-plane, and (*c*) the (y, z)-plane. Note that the coordinates of C, C′ in this case are $(\pm 8.48, \pm 8.48, 27)$.

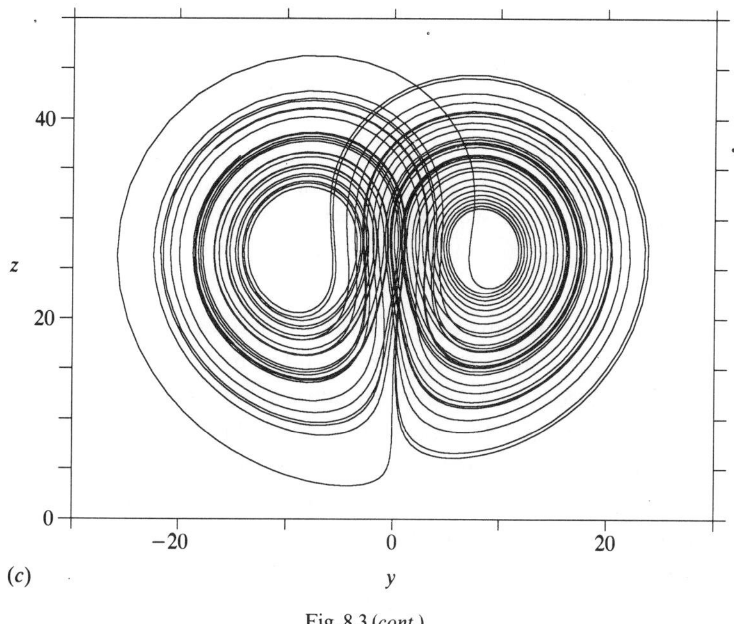

Fig. 8.3 (*cont.*)

Fig. 8.3. The results come from numerical integration of the system (1), with $\mathbf{x}(0) = (0, 1, 0)$ and $r = 28$, $\sigma = 10$, $b = \frac{8}{3}$. (You can easily verify that $r_c = 24.74$ and C, C′ $= (\pm 8.48, \pm 8.48, 27)$ with these values of σ, b.) First the orbit goes once around C and then spirals out several times around C′. It gets close to the attractor at about $t = 20$. The attractor can be seen as a number of spirals around C followed by a number of spirals around C′ and so forth; the sequence of numbers of spirals seems as if it were a random sequence. To visualize the attractor it may help to think of the groove on two inter-connected gramophone discs which are rather warped – or to think of a butterfly's wings.

Calculations for $r = 28$, $\sigma = 10$, $b = \frac{8}{3}$ also show that neighbouring orbits near the attractor separate exponentially on average, so that two orbits which start very close together soon lose all correlation with one another. Thus there is sensitive dependence on initial conditions. So if the results of Figs. 8.2, 8.3 were recalculated with a different computer or a slightly different program then the new results would probably diverge from the ones shown until the two calculated orbits $\mathbf{x}(t)$ would soon become uncorrelated, although each would traverse the same attractor.

In fact the strange attractor can be found not only for some $r > r_c$ but also for $r_1 < r \leqslant r_c$, where $r_1 > 1$. So when $r_1 < r < r_c$ the strange attractor

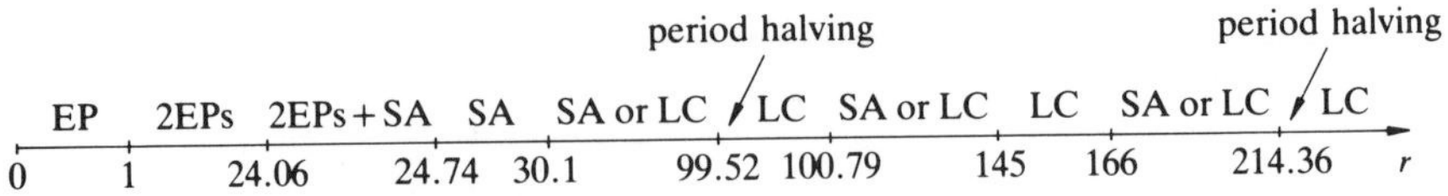

period halving period halving

EP 2EPs 2EPs+SA SA SA or LC / LC SA or LC LC SA or LC / LC

0 1 24.06 24.74 30.1 99.52 100.79 145 166 214.36 r

Fig. 8.4 Symbolic diagram specifying the attractors of the Lorenz system for $\sigma = 10, b = \frac{8}{3}$ and $0 \leqslant r < \infty$. A stable equilibrium point is denoted by 'EP', a stable limit cycle by 'LC', a strange attractor by 'SA', and the alternation of attractors (usually with a complicated sequence of bifurcations) as r increases by 'or'.

coexists with the attractors C, C'. If r were to oscillate quasi-statically in an interval including r_1 and r_c then there would be hysteresis, with attractor C or C' as r increases to r_c, strange attractor as r increases further, strange attractor as r decreases through r_c to r_1, and then attractor C or C' as r decreases further. Also note that the strange attractor originates at $r = r_1$ without being associated with the onset of instability of another attractor.

As r increases from zero to infinity there is a very complicated sequence of bifurcations and attractors. In particular, there is a period *halving* of stable limit cycles as r increases, whereby chaos ceases; this is the inverse process of period doubling and it is governed by a Feigenbaum sequence associated with the universal constant δ. Finally, as r becomes large a limit cycle becomes the unique attractor (see Q8.3 for the limit cycle as $r \to \infty$). An impression of the sequence of bifurcations is given in Fig. 8.4, based on numerical calculations for $\sigma = 10$, $b = \frac{8}{3}$. It is schematic and somewhat simplified.

There is an even richer variety of nonlinear phenomena as r varies for other pairs of values of σ, b.

The 'cause' of chaos can be seen by examining the orbits geometrically. To understand the changing topology of the orbits as r varies, try to see computer animations of the orbits (e.g. Stewart F1987). There is space here only to give an inkling of the cause. First we need to recognize that the equilibrium points and the limit cycles are fundamental to the topology, just as they are in the phase plane. However, more topologically different points of equilibrium occur in three-dimensions than occur in two-dimensions according to whether the three eigenvalues are real or complex and have positive or negative real parts.

*Let us see where the orbits near 0, C and C' go to and come from when $1 < r < r_c$. (Remember that $r_c = 24.74$ when $\sigma = 10$, $b = \frac{8}{3}$.) Then 0 is unstable, with $s_1 > 0 > s_2 > s_3$: it is a saddle point in three-dimensions. So the orbits leaving 0 are parallel and anti-parallel to an eigenvector $\mathbf{u}_1$ corresponding to s_1, and the orbits entering 0 locally lie in the space spanned by $\mathbf{u}_2$ and $\mathbf{u}_3$, i.e. a plane. Let us follow the orbits leaving 0 and trace back the orbits entering 0. It helps to make two definitions first. If a point lies on an orbit which tends to an equilibrium point X of an autonomous differential system as $t \to \infty$ then the point lies in what is called the *stable manifold* of X, denoted by $W^s(X)$; similarly, if a point lies on an orbit which tends to X as $t \to -\infty$ then the point lies in what is called the *unstable manifold* of X, denoted by $W^u(X)$. Thus if $\mathbf{x}(0) \in W^s(X)$ then the orbit $\{\mathbf{x}(t)\} \subset W^s(X)$; again, if an orbit starts at $\mathbf{x}(0)$ in $W^u(X)$ then the orbit $\{\mathbf{x}(t)\}$ remains in $W^u(X)$ for $-\infty < t < \infty$. It may help to note that if X is stable then $W^u(X)$ does not exist and $W^s(X)$ is the three-dimensional domain of attraction of X (or, more generally, the m-dimensional domain in a phase space $\mathbb{R}^m$). Thus, the unstable and stable manifolds are in a sense generalizations, suitable for an unstable point of equilibrium, of the domain of attraction of a stable point of equilibrium. It follows that, for the Lorenz system when $r > 1$, the manifold $W^u(0)$ is a curve with tangent parallel to $\mathbf{u}_1$ at 0, and $W^s(0)$ is a two-dimensional surface with tangent plane at 0 spanned by $\mathbf{u}_2$ and $\mathbf{u}_3$. (When $r = 1, \mathbf{u}_1 = (1, 1, 0), \mathbf{u}_2 = (1, -\sigma, 0)$ and $\mathbf{u}_3 = (0, 0, 1)$.) Note that no orbit can cross a stable or an unstable manifold except at an equilibrium point, so the two-dimensional surface $W^s(0)$ divides the orbits of $\mathbb{R}^3$ into two sets. When r is not too large the structure of $W^s(0)$ is fairly simple. However, we shall see that, as r increases, the structure soon develops twists and forms 'sheets', and becomes hard to visualize.

*At first for $r > 1$ the curve $W^u(0)$ leaves 0 in each of the octants of C, C′ and ends at C, C′ respectively. However, when $r = r_0$ the curve $W^u(0)$ leaves 0 in each direction and returns to 0 in the surface $W^s(0)$, as shown in Fig. 8.5(a): two homoclinic orbits between 0 and itself are established. (There are two because of the symmetry of the Lorenz system.) An orbit connecting an equilibrium point with itself, approaching the point as $t \to \pm\infty$, is called a *homoclinic orbit*, or *homoclinic connection.* It can be seen that a homoclinic orbit corresponds to a solution of infinite 'period', because although the connection is a closed curve its solution never reaches the saddle point 0 in a finite time (cf. Example 1.1). It is this special property of the Lorenz phase portrait which defines r_0 and is important in determining the changing topology of the orbits. As r increases above r_0 two unstable limit cycles (of finite period) arise from the homoclinic orbits (it is these which become the limit cycles which coalesce with the equilibrium points C, C′ as r increases to r_c). By solving the Lorenz system numerically, it can be shown that if $\sigma = 10, b = \frac{8}{3}$ then $r_0 = 13.93$ approximately. When r increases through r_0 the unstable limit cycles serve to repel $W^u(0)$ so that

J

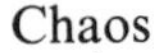

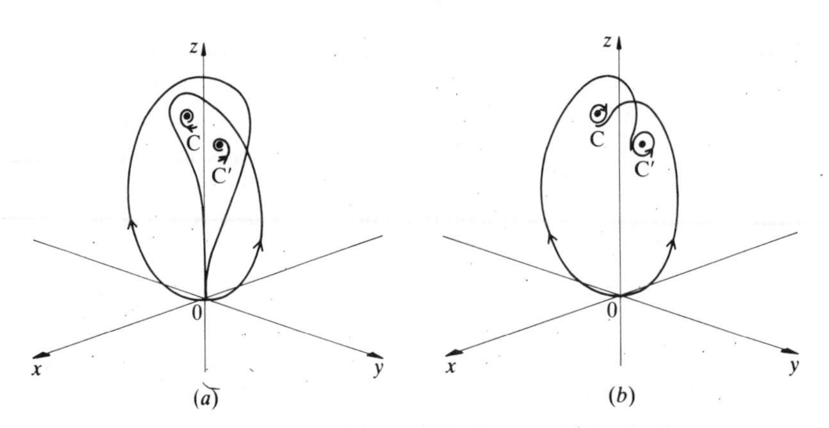

Fig. 8.5 A perspective sketch of the unstable manifold $W^u(0)$ in the phase space of the Lorenz system. (*a*) The two symmetrically located homoclinic orbits when $r = r_0$. (*b*) The two symmetrically located heteroclinic orbits when $r = r_1$.

the branch of $W^u(0)$ leaving 0 in the octant of C ends not at C but at C′, and similarly the branch leaving in the octant of C′ ends at C. As r increases further, orbits cross over from C to C′ and back an increasing number of times before finally spiralling in to either C or C′. This is sometimes called *pre-chaos* because it is difficult to distinguish an orbit that spirals around C and C′ repeatedly for a long time before eventually tending to one of those points from an orbit that spirals around them for ever.

*Almost all orbits end at either C or C′ for values of r such that these are the only attractors (the exceptions are those orbits on $W^s(0)$ and those on the stable manifolds of the two unstable limit cycles). When $r = r_1$, the curve $W^u(0)$ leaves 0 in one octant, eventually enters the unstable limit cycle about C, and leaves 0 in the opposite octant and eventually enters the unstable limit cycle about C′, as shown in Fig. 8.5 (*b*): two *heteroclinic orbits* between 0 and the limit cycles are established, a heteroclinic orbit being an orbit connecting two invariant sets as $t \to \pm\infty$. It is this property which defines r_1 and marks the onset of chaos. (The two heteroclinic orbits are established at the same value of r because of the symmetry of the system.) For $\sigma = 10$, $b = \frac{8}{3}$ it is found numerically that $r_1 = 24.06$. The chaotic attractor originates as r increases above r_1, so that for $r_1 < r < r_c$ the chaotic attractor coexists with the attractors C, C′ but for $1 < r \leqslant r_1$ the only attractors are C, C′.

*These facts are the bones of the skeleton which supports the flesh of the 'body' of the orbits. Try to picture the chaotic orbits in the three-dimensional phase space, and see the origin of chaos. It is far from easy.

**Example 8.1: perturbation of a homoclinic orbit.* Another inkling of how the break-up of a homoclinic orbit, as a parameter varies, leads to chaos is given in this example of an autonomous system in the phase *plane*. However, we find asymptotically conditions for the formation, not of chaos, but of a stable limit cycle as the parameter varies a little.

The problem is this. Suppose that

$$\frac{d\mathbf{x}}{dt} = \mathbf{F}(a, \mathbf{x}), \tag{11}$$

where $\mathbf{F}: \mathbb{R} \times \mathbb{R}^2 \to \mathbb{R}^2$ is well behaved and there is a saddle point, at $\mathbf{0}$ say. To be specific, take $\mathbf{x} = (x, y)$, $\mathbf{F}(a, \mathbf{0}) = \mathbf{0}$ and $\mathbf{J} = \begin{bmatrix} \lambda_1 & 0 \\ 0 & -\lambda_2 \end{bmatrix}$ at $\mathbf{0}$ for all a, where $\lambda_1, \lambda_2 > 0$. Further suppose that there is a homoclinic orbit through 0, in the first quadrant, say, when $a = 0$.

Then the linearized equations at 0 are

$$\frac{dx}{dt} = \lambda_1 x, \qquad \frac{dx}{dt} = -\lambda_2 y,$$

and their solutions are

$$x(t) = x_0 \exp(\lambda_1 t), \qquad y(t) = y_0 \exp(-\lambda_2 t).$$

Thus orbits near 0 resemble rectangular hyperbolae, having equations

$$\left(\frac{x}{x_0}\right)^{\lambda_2} \left(\frac{y}{y_0}\right)^{\lambda_1} = 1.$$

Consider next the orbit of system (11) which enters the small square whose sides have equations $x = \pm\epsilon$, $y = \pm\epsilon$ at (x_n, ϵ) in the first quadrant. If $0 < \epsilon \ll 1$ then the orbit leaves the square at (ϵ, y_n) approximately, where, on defining $r = \lambda_2/\lambda_1$, the above equation of the orbits gives

$$y_n = \epsilon \left(\frac{x_n}{\epsilon}\right)^r.$$

Now let us follow the orbit globally for small a. We suppose that when $a = 0$ the unstable manifold leaving 0 tangentially to the positive x-axis returns along the positive y-axis to form the homoclinic orbit. For small a we may linearize perturbations of this orbit, and deduce in principle that the orbit leaving the square at (ϵ, y_n) returns to the square at (x_{n+1}, ϵ), where

$$x_{n+1} = ka + ly_n$$

for some constants k, l. The linearization is valid because the orbit is everywhere close to the homoclinic orbit. Therefore

$$x_{n+1} = ka + mx_n^r \quad \text{for } x_n > 0 \tag{12}$$

where $m = l\epsilon^{1-r}$. This difference equation gives us the dynamics, and, in particular, shows that there is a stable limit cycle if the difference equation has a stable fixed point. The methods of Chapter 3 show that, according to the sign of $r - 1$ and value of $akm^{1/(r-1)}$, there may exist a stable fixed point (see Q3.49).

Of course, a chaotic solution cannot arise in this example of a plane autonomous system. However, now it is not difficult to conceive of a similar problem in $\mathbb{R}^3$, such that there is a cube with sides of length 2ϵ centred at a saddle point on a homoclinic orbit, and the perturbation of that orbit for small a generates a map on a face of the cube which has a chaotic attractor (cf. Glendinning & Sparrow 1984). □

2 Duffing's equation with negative stiffness

Duffing's equation offers another good illustration of an ordinary differential system with chaos. In modelling the lateral vibrations of a beam under a temporally periodic load, Holmes (1979) was led to consider the equation,

$$\frac{d^2x}{dt^2} + \delta\frac{dx}{dt} - \beta x + \alpha x^3 = f \cos \omega t \tag{1}$$

for $\beta > 0$, i.e. for *negative* stiffness. The sign of the linear stiffness $(-\beta)$ is important, because equation (1) admits no oscillations in the absence of nonlinearity and forcing, whereas the form of Duffing's equation with positive stiffness in Chapter 7 does admit them.

However, equation (1) admits *nonlinear* oscillations in the absence of damping and forcing, i.e. the equation

$$\frac{d^2x}{dt^2} - \beta x + \alpha x^3 = 0 \tag{2}$$

admits them. The solutions of equation (2) can easily be found by the methods of §1.7 and Chapter 6. The phase portrait in the (x, v)-plane is sketched in Fig. 8.6(a), where $v = dx/dt$. With the addition of positive damping, we consider

$$\frac{d^2x}{dt^2} + \delta\frac{dx}{dt} - \beta x + \alpha x^3 = 0 \tag{3}$$

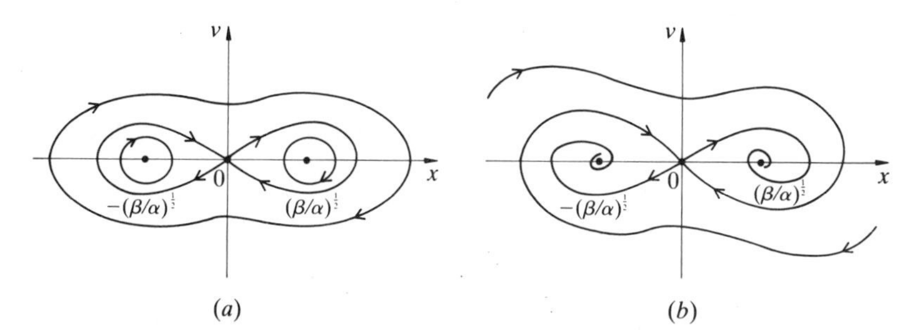

Fig. 8.6 Sketches of the phase portraits in the (x, v)-plane for $f = 0$, $\alpha > 0$, $\beta > 0$. (*a*) Equation (2), with $\delta = 0$. (*b*) Equation (3), with $\delta > 0$.

for $\delta > 0$. Then there are no oscillations, the centres at $(\pm(\beta/\alpha)^{1/2}, 0)$ in the phase plane becoming stable foci, as shown in Fig. 8.6(*b*).

*Note that the saddle point at the origin of the phase plane of Fig. 8.6(*a*) has one orbit leaving it in the first quadrant, and that this orbit coincides with the orbit approaching it in the fourth quadrant, i.e. its unstable manifold $W^u(0)$ coincides with its stable manifold $W^s(0)$ to form a separatrix enclosing closed orbits. By the invariance of equation (2) under the transformations $x \to -x$, $t \to -t$, the phase portrait is symmetric about each axis. Therefore there is a similar separatrix in the third and second quadrants. In other words, there are two symmetrically placed homoclinic connections of the saddle point with itself. (Note that the orbits form a figure of eight, whereas the two homoclinic orbits of the Lorenz system, though also symmetric, are like the outline of a butterfly's wings.) The presence of damping breaks up the homoclinic orbits, the saddle point at the origin being structurally stable but the centres at $(\pm(\beta/\alpha)^{1/2}, 0)$ becoming stable foci. If $\delta > 0$ then the unstable manifold $W^u(0)$ leaves the origin in the first quadrant but approaches the focus at $((\beta/\alpha)^{1/2}, 0)$, and the stable manifold $W^s(0)$ approaching 0 in the fourth quadrant comes from infinity. Equation (3) is symmetric under the transformation $x \to -x$, but not under time reversal because of the damping with $\delta \neq 0$.

Equations (2), (3) have no chaotic solutions because they are autonomous second-order differential equations, but, as we shall see, equation (1) is equivalent to an autonomous third-order system and does have some chaotic solutions as well as non-chaotic ones.

The weakly nonlinear theory of forced oscillations gives insight into some of the solutions. There is space enough here to consider in detail only the synchronous oscillations, i.e. those with the same period $2\pi/\omega$ as the forcing. Take $\delta = 0$ at first, so that

$$\frac{\mathrm{d}^2x}{\mathrm{d}t^2} - \beta x + \alpha x^3 = f\cos\omega t, \tag{4}$$

and approximate the forced oscillations for small f by van der Pol's method as

$$x(t) = c(t) + a(t)\cos\omega t + b(t)\sin\omega t, \tag{5}$$

where a, b, c are slowly varying functions. Then on substituting approximation (5) into equation (4), neglecting the very small terms $\mathrm{d}^2a/\mathrm{d}t^2$, $\mathrm{d}^2b/\mathrm{d}t^2$ and the higher harmonics, and equating coefficients of $\sin\omega t$, $\cos\omega t$, 1 in turn, we find

$$\left.\begin{aligned} 2\omega\frac{\mathrm{d}a}{\mathrm{d}t} &= -(\beta+\omega^2)b + 3\alpha bc^2 + \tfrac{3}{4}\alpha br^2, \\ 2\omega\frac{\mathrm{d}b}{\mathrm{d}t} &= (\beta+\omega^2)a - 3\alpha c^2a - \tfrac{3}{4}\alpha ar^2 + f, \\ \frac{\mathrm{d}^2c}{\mathrm{d}t^2} &= \beta c - \alpha c^3 - \tfrac{3}{2}\alpha cr^2, \end{aligned}\right\} \tag{6}$$

where $r = (a^2 + b^2)^{1/2}$.

This four-dimensional system has equilibrium points where

$$\left.\begin{aligned} c(ac^2 - \beta + \tfrac{3}{2}\alpha r^2) &= 0, \\ a(-\beta - \omega^2 + 3\alpha c^2 + \tfrac{3}{4}\alpha r^2) &= f, \\ b(-\beta - \omega^2 + 3\alpha c^2 + \tfrac{3}{4}\alpha r^2) &= 0. \end{aligned}\right\} \tag{7}$$

Therefore equation (4) has small synchronous oscillations (a) about the saddle point of equation (2) at the origin, given by

$$a(\tfrac{3}{4}\alpha a^2 - \beta - \omega^2) = f, \qquad b = c = 0; \tag{8}$$

and (b) about the other two equilibrium points (the centres) of equation (2), given by

$$a(2\beta - \omega^2 - \tfrac{15}{4}\alpha a^2) = f, \quad b = 0, \quad \alpha c^2 = \beta - \tfrac{3}{2}\alpha a^2 \quad \text{for } a^2 < 2\beta/3\alpha. \tag{9}$$

As seen in §7.3, the presence of damping can modify the cubic governing the amplitude of an oscillation. It can be shown similarly that if $\delta \neq 0$ then equation (1) has small synchronous oscillations (a) about the saddle point of equation (3) at the origin, given by

$$r^2\{(\tfrac{3}{4}\alpha r^2 - \beta - \omega^2)^2 + \delta^2\omega^2\} = f^2, \qquad c = 0; \tag{10}$$

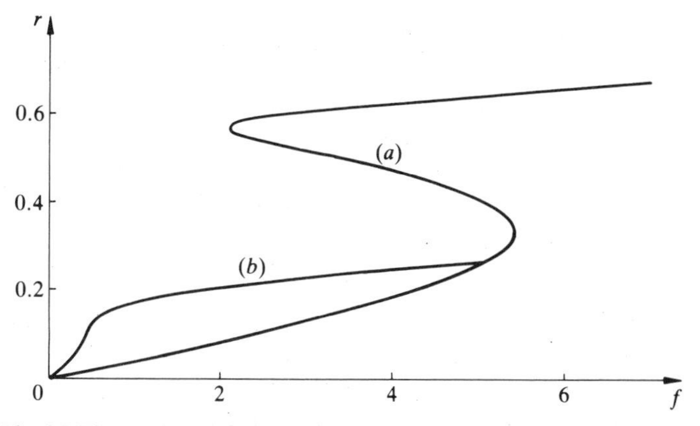

Fig. 8.7 The curves (a), (b) in the (f, r)-plane representing weakly nonlinear solution of equation (1) of period $2\pi/\omega$ for $\alpha = 100, \beta = 10, \delta = 1, \omega = 3.76$.

and (b) about the foci of equation (3), given by

$$r^2\{(2\beta - \omega^2 - \tfrac{15}{4}\alpha r^2)^2 + \delta^2\omega^2\} = f^2, \quad \alpha c^2 = \beta - \tfrac{3}{2}\alpha r^2 \quad \text{for } r^2 < 2\beta/3\alpha. \tag{11}$$

Fig. 8.7 displays the curves (10), (11) in the (f, r)-plane for 'typical' values of α, β, δ, ω. Note that the curve (b) representing oscillations about the two foci ends at $r = (2\beta/3\alpha)^{1/2}$, where it meets the curve (a) representing oscillations about the saddle point.

Of course, a periodic solution will be found after numerical integration of its differential equation for a long time only if it is stable. So to find the stability of the above weakly nonlinear solutions of equation (4) with $\delta = 0$, first linearize the fourth-order system about its equilibrium point of type (a) with $a = a_0, b = c = 0$. Then the linearized system is

$$2\omega\frac{\mathrm{d}}{\mathrm{d}t}\begin{bmatrix} a' \\ b' \end{bmatrix} = \begin{bmatrix} 0 & -\beta - \omega^2 + \tfrac{3}{4}\alpha a_0^2 \\ \beta + \omega^2 - \tfrac{9}{4}\alpha a_0^2 & 0 \end{bmatrix}\begin{bmatrix} a' \\ b' \end{bmatrix},$$

$$\frac{\mathrm{d}^2 c'}{\mathrm{d}t^2} = (\beta - \tfrac{3}{2}\alpha a_0^2)c'.$$

Therefore the point is stable if $\alpha a_0^2 > \tfrac{2}{3}\beta$ and $27\alpha^2 a_0^4 - 48(\beta + \omega^2)\alpha a_0^2 + 16(\beta + \omega^2)^2 > 0$, i.e. if $\tfrac{2}{3}\beta < \alpha a_0^2 < \tfrac{4}{9}(\beta + \omega^2)$ or $\tfrac{13}{9}(\beta + \omega^2) < \alpha a_0^2$. These results can be simply generalized for the case of damping ($\delta > 0$). It follows that in Fig. 8.7 the curve (a) represents unstable oscillations of period $2\pi/\omega$ for $r < (2\beta/3\alpha)^{1/2}$, $= 0.26$, as well as between its two turning points (at $r = 0.34, 0.56$). Also the stability of solutions of equation (4) of type (b), i.e. of the oscillations about the two centres of equation (2) without forcing, can be found similarly. This provides a framework on which an interpretation of the numerical results for the forced oscillations can be based. However, there are weakly nonlinear oscillations which are not synchronous, and the strongly nonlinear phenomena are more complicated.

When $f \neq 0$ it is often convenient to use an autonomous third-order system equivalent to (1) rather than to use (1) itself. So consider the suspended system,

$$\frac{dx}{dt} = v, \qquad \frac{dv}{dt} = \beta x - \delta v - \alpha x^3 + f \cos \omega\theta, \qquad \frac{d\theta}{dt} = 1, \tag{12}$$

where $\theta \in S^1$, i.e. where a point on the circumference of the circle S^1 of length $2\pi/\omega$ is represented by its polar angle $\omega\theta$, and $\theta + 2n\pi/\omega$ is identified with θ for $n = \pm 1, \pm 2, \ldots$; this periodicity of θ is helpful to represent the periodicity of the forcing. The structural stability of equation (3) ensures that the orbits of system (12) are topologically equivalent of those of equation (3) in the phase space $\mathbb{R}^2 \times S^1$ of (x, v, θ) for sufficiently small f. Thus there are (cf. Q7.4) two stable attracting oscillations close to $(\pm(\beta/\alpha)^{1/2}, 0)$ and an oscillating saddle-type orbit close to $(0, 0)$.

In order to help to describe the orbits, next define a Poincaré map $P_f^{t_0}$: $\Sigma \to \Sigma$, where $\Sigma = \{(x, v, \theta) \in \mathbb{R}^2 \times S^1 : \theta = t_0 \in [0, 2\pi/\omega)\}$, such that $P_f^{t_0}$ maps a point on an orbit of equation (1) at time t_0 to the point at time $t_0 + 2\pi/\omega$ on the same orbit. Thus $P_f^{t_0}$ is effectively a stroboscopic map of the (x, v)-plane. The value of t_0 is significant because the forcing term $f \cos \omega t$ varies with its phase. However, $P_0^{t_0}$ is defined and can be seen to be independent of t_0, being merely the flow $\phi_{2\pi/\omega}$ for equation (3); it has three fixed points, at the equilibrium points $(0, 0)$, $(\pm(\beta/\alpha)^{1/2}, 0)$ of equation (3). A solution of equation (1) having period $2p\pi/\omega$ for a positive integer p corresponds to a p-cycle of the plane map $P_f^{t_0}$, with different p-cycles for different values of t_0 in general.

*We can define the stable manifold $M^s(X)$ and unstable manifold $M^u(X)$ of a fixed point X of a map **F** by analogy with the stable and unstable manifolds of an equilibrium point of an autonomous differential system. Thus $M^s(X)$ is the invariant set of **F** such that if $\mathbf{x} \in M^s(X)$ then $\mathbf{F}^n(\mathbf{x}) \to X$ as $n \to \infty$, and $M^u(X)$ is the invariant set of **F** such that if $\mathbf{x} \in M^u(X)$ then $\mathbf{F}^n(\mathbf{x}) \to X$ as $n \to -\infty$. Now we have noted that a fixed point X of $P_0^{t_0}$ coincides with an equilibrium point of equation (3). Also the stable manifold $M_0^s(X)$ of the fixed point X of $P_0^{t_0}$ coincides with the stable manifold $W^s(X)$ of the equilibrium point X of equation (3); similarly the unstable manifold $M_0^u(X)$ of $P_0^{t_0}$ coincides with $W^u(X)$ of equation (3). When, however, f is small but not zero, $\mathrm{P}_f^{t_0}$ has three fixed points which are no more than close to the three equilibrium points of equation (3).

On the basis of the weakly nonlinear theory above, it can readily be appreciated that, with development of subharmonic solutions and chaos, the structure of the bifurcations of the solutions of equation (1) as f, α, β, δ and ω vary is complicated. Indeed, it is astonishingly complicated. The origin of some of the chaos is explained in Example 8.2 (other chaos arises from period doubling). The large number of parameters makes it impractical to describe here more than a few of the bifurcations. We shall describe the bifurcations only as f varies for fixed $\alpha = 100$, $\beta = 10$, $\delta = 1$, $\omega = 3.6$, after Holmes (1979). He sought the attractors by integrating the equation numerically for long times and for various values of f. He found by the method of averaging that the weakly nonlinear theory of the oscillations of period $2\pi/\omega$ is a good approximation when $f \lesssim 0.5$ or $f \gtrsim 2.5$, and that the oscillations are stable then. Thus he computed the oscillations about the origin for values of r above that at the upper turning point (i.e. for $r \gtrsim 0.56$), and the oscillations about the foci for $0 < f \lesssim 0.95$. At $f \approx 0.95$ the oscillations of type (b) start period doubling. Strange attractors appear for $1.1 \lesssim f \lesssim 2.5$, but with windows of stable periodic oscillations, including an oscillation of period $10\pi/\omega$, i.e. period 5, for $1.15 \lesssim f \lesssim 1.2$.

A few solutions are shown in Fig. 8.8 of the phase plane. In Fig. 8.8(*a*) the orbit of a large synchronous oscillation about all three fixed points is seen coexisting with a strange attractor shown up by the stroboscopic map. In Fig. 8.8(*b*) the orbit of a period-three solution circles first about the right-hand focus, then about the left-hand focus, and finally about all three equilibrium points of equation (3).

Many more details about Duffing's equation and its chaotic solutions have been given by Holmes (1979) and others (cf. Guckenheimer & Holmes (1986, §2.2)).

*3 The chaotic break-up of a homoclinic orbit: Mel'nikov's method

A common cause of chaos in ordinary differential systems, namely the formation and break-up of a homoclinic connection as a parameter is varied, is explained in this section. The essential topological aspects of the break-up will be described and some general methods will be used, but at the same time we shall take an illustrative case and show how to ascertain the presence of chaos by an asymptotic method due to Mel'nikov (1963).

Consider then the case of a Hamiltonian system of one degree of free-

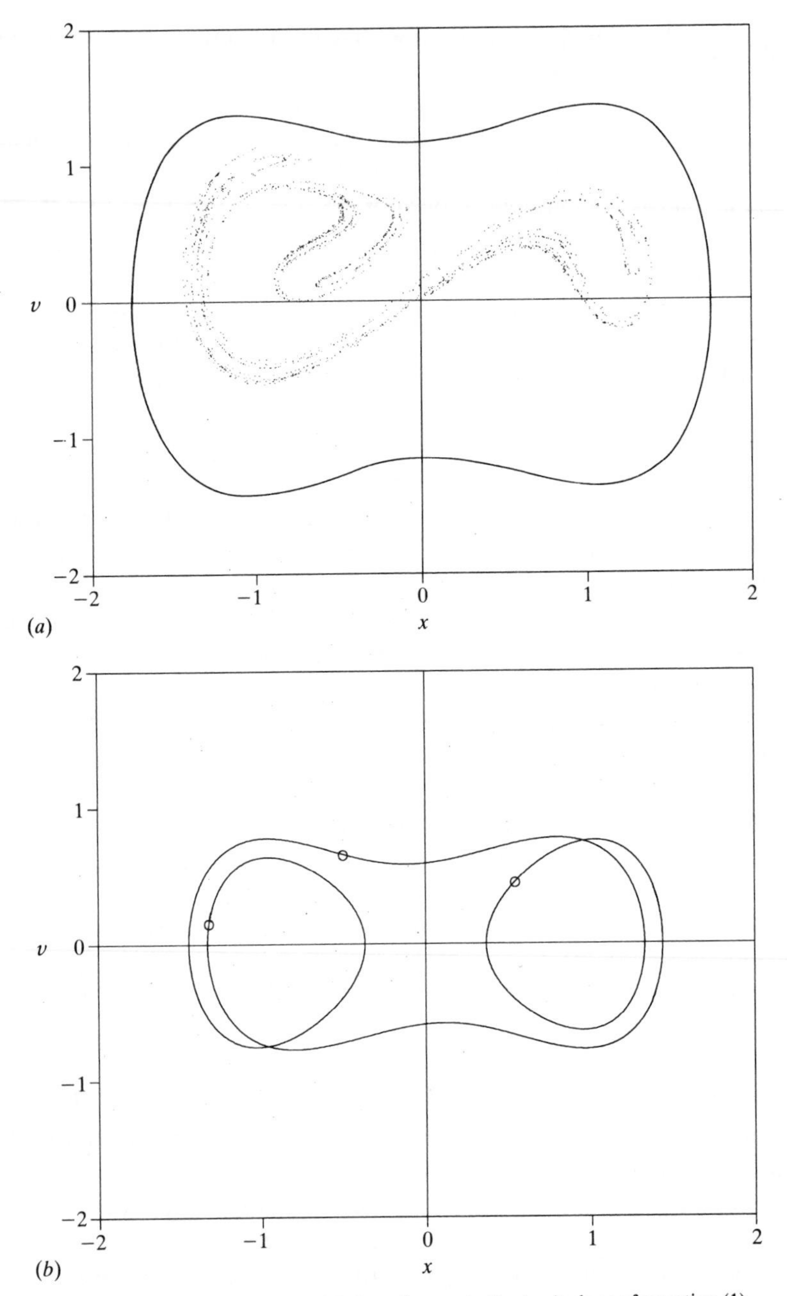

Fig. 8.8 Sketches of orbits and Poincaré maps in the (x, v)-plane of equation (1) for $\alpha = \beta = \omega = 1$, $f = 0.3$ (after Guckenheimer & Holmes (1986, Fig. 2.2.5)). (*a*) A large stable orbit of period $2\pi/\omega$ and a stroboscopic Poincaré map of a strange attractor for $\delta = 0.15$. (*b*) A large stable orbit of period $6\pi/\omega$ for $\delta = 0.22$, with three fixed points shown by circles.

dom with a periodic perturbation, namely

$$\frac{d\mathbf{x}}{dt} = \mathbf{F}(\mathbf{x}) + \epsilon\mathbf{f}(\mathbf{x}, t), \tag{1}$$

where $\mathbf{x} = [x, y]^{\mathrm{T}} \in \mathbb{R}^2$, $\mathbf{F}: \mathbb{R}^2 \to \mathbb{R}^2$ with $\mathbf{F} = [\partial H/\partial y, -\partial H/\partial x]^{\mathrm{T}}$ for some function $H(y, x)$, $\mathbf{f}: \mathbb{R}^2 \times \mathbb{R} \to \mathbb{R}^2$ with $\mathbf{f}(\mathbf{x}, t + T) = \mathbf{f}(\mathbf{x}, t)$ for all $\mathbf{x}$, t, and $\mathbf{F}$, $\mathbf{f}$ are well-behaved. Thus $\mathbf{F}$ is Hamiltonian, and $\mathbf{f}$ has period T and may or may not be Hamiltonian. The methods can be applied to perturbations of systems with non-Hamiltonian $\mathbf{F}$ and to differential systems of order higher than two, but equation (1) will serve well for illustration.

For $\epsilon = 0$ we find the basic system, i.e.

$$\frac{d\mathbf{x}}{dt} = \mathbf{F}(\mathbf{x}); \tag{2}$$

it is an autonomous Hamiltonian system, so its orbits lie on curves with equation $H(y, x) = E$ for different constants E. We shall in addition assume that the basic system has a saddle point, $\mathbf{X}_0$ say, with a homoclinic connection. Let $\{\mathbf{q}_0(t)\}$ be the homoclinic orbit, so that $\mathbf{q}_0$ satisfies equation (2) and

$$\mathbf{q}_0(t) \to \mathbf{X}_0 \qquad \text{as } t \to \pm\infty. \tag{3}$$

Note that $\mathbf{q}_0(0)$ may be chosen to be any given point of the orbit, so that $\mathbf{q}_0$ is defined uniquely only up to a translation in time. Of course, $\mathbf{q}_0(t)$ lies in both the stable $\mathrm{W}^{\mathrm{s}}(\mathbf{X}_0)$ and unstable manifolds $\mathrm{W}^{\mathrm{u}}(\mathbf{X}_0)$ of $\mathbf{X}_0$ for equation (2) for all t, and $\mathrm{W}^{\mathrm{u}}(\mathbf{X}_0) = \mathrm{W}^{\mathrm{s}}(\mathbf{X}_0)$ along the homoclinic orbit. Equation (2) having a two-dimensional phase space, these manifolds are merely plane curves.

For $\epsilon \neq 0$ it is helpful to consider the suspended system, namely

$$\frac{d\mathbf{x}}{dt} = \mathbf{F}(\mathbf{x}) + \epsilon\mathbf{f}(\mathbf{x}, \theta), \qquad \frac{d\theta}{dt} = 1. \tag{4}$$

This autonomous system, with the three-dimensional phase space of $(\mathbf{x}, \theta)$, is (cf. §5.1) equivalent to (1). We define a Poincaré map $\mathrm{P}_\epsilon^{t_0}: \Sigma^{t_0} \to \Sigma^{t_0}$ for the suspended system (4), where $\Sigma^{t_0} = \{(\mathbf{x}, \theta): \theta = t_0 \in [0, T)\}$, as follows. Take a point $(\mathbf{x}_0, t_0)$ as initial point of an orbit of the system, and integrate the system from $t = t_0$ to $t = t_0 + T$ to get the solution $(\mathrm{P}_\epsilon^{t_0}(\mathbf{x}_0), t_0 + T)$; thus $\mathrm{P}_\epsilon^{t_0}$ is again defined as a stroboscopic map. (We shall regard t_0 as a variable later, because $f \cos \omega t$ and hence $\mathrm{P}_\epsilon^{t_0}$ depend on their phases.) When ϵ is small, $\mathrm{P}_\epsilon^{t_0}$ has a fixed point, at $\mathbf{X}_\epsilon^{t_0}$, say, where $\mathbf{X}_\epsilon^{t_0} = \mathbf{X}_0 + O(\epsilon)$ as $\epsilon \to 0$. This fixed point corresponds to a periodic orbit of system (4), as illustrated

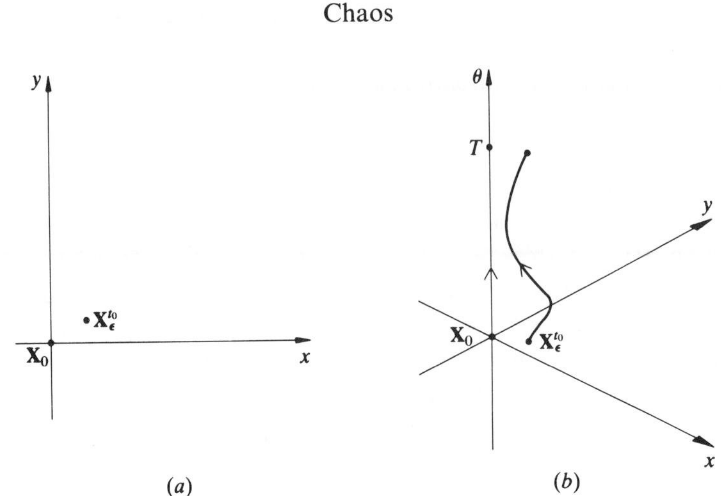

Fig. 8.9 (*a*) The fixed points $\mathbf{X}_\varepsilon^{t_0}$ of the map $\mathrm{P}_\varepsilon^{t_0}$ and $\mathbf{X}_0$ of the map $\mathrm{P}_0^{t_0}$ in the (x, y)-plane. (*b*) The periodic orbit of system (1) corresponding to the fixed point $\mathbf{X}_\varepsilon^{t_0}$ and the periodic orbit of system (2) corresponding to the fixed point $\mathbf{X}_0$. Note that, by periodicity, the plane $\theta = T$ should coincide with the plane $\theta = 0$.

in Fig. 8.9. For small ϵ, the fixed point is a saddle point of the Poincaré map, and the distance from the periodic orbit to the unperturbed orbit, i.e. the line $\mathbf{x} = \mathbf{X}_0$, is of order ϵ.

Recall that the stable and unstable manifolds of the equilibrium point $\mathbf{X}_0$ of the basic system (2) for $\epsilon = 0$ coincide (see also Fig. 8.10(*a*)). This leads us to enquire about the analogous behaviour for small $\epsilon \neq 0$. Before using perturbation theory to find this behaviour quantitatively, let us examine the possible types of behaviour geometrically. First denote the stable manifold of the fixed point $\mathbf{X}_\epsilon^{t_0}$ of the Poincaré map $\mathrm{P}_\epsilon^{t_0}$ by $\mathrm{M}_\epsilon^{\mathrm{s}}(\mathbf{X}_\epsilon^{t_0})$, and the unstable manifold by $\mathrm{M}_\epsilon^{\mathrm{u}}(\mathbf{X}_\epsilon^{t_0})$. In the present problem with a two-dimensional phase space for equation (1), these manifolds are plane curves. They depend on t_0, moving and 'waving' with period T as t_0 varies. In particular, $\mathrm{M}_0^{\mathrm{s}}(\mathbf{X}_0) = \mathrm{W}^{\mathrm{s}}(\mathbf{X}_0)$ and $\mathrm{M}_0^{\mathrm{u}}(\mathbf{X}_0) = \mathrm{W}^{\mathrm{u}}(\mathbf{X}_0)$, so $\mathrm{M}_0^{\mathrm{u}}(\mathbf{X}_0) = \mathrm{M}_0^{\mathrm{s}}(\mathbf{X}_0)$ along the homoclinic orbit. Also note that if $\mathbf{x} \in \mathrm{M}_\epsilon^{\mathrm{s}}(\mathbf{X}_\epsilon^{t_0})$ then the images $(\mathrm{P}_\epsilon^{t_0})^n(\mathbf{x}) \to \mathbf{X}_\epsilon^{t_0}$ exponentially as $n \to \infty$; similarly, if $\mathbf{x} \in \mathrm{M}_\epsilon^{\mathrm{u}}(\mathbf{X}_\epsilon^{t_0})$ then the pre-images $(\mathrm{P}_\epsilon^{t_0})^n(\mathbf{x}) \to \mathbf{X}_\epsilon^{t_0}$ as $n \to -\infty$.

The stable manifold $\mathrm{M}_\epsilon^{\mathrm{s}}(\mathbf{X}_\epsilon^{t_0})$ and the unstable manifold $\mathrm{M}_\epsilon^{\mathrm{u}}(\mathbf{X}_\epsilon^{t_0})$ either do or do not intersect. If they do not intersect, then they are as shown in Fig. 8.10(*b*), or (*c*). If they do intersect, then in general they intersect transversely, as shown in Fig. 8.10(*d*), and as first envisaged by Poincaré (1899, §397) with the prescience of a genius: 'Let us seek to visualize the pattern formed by the two curves and their infinite number of intersections ... The

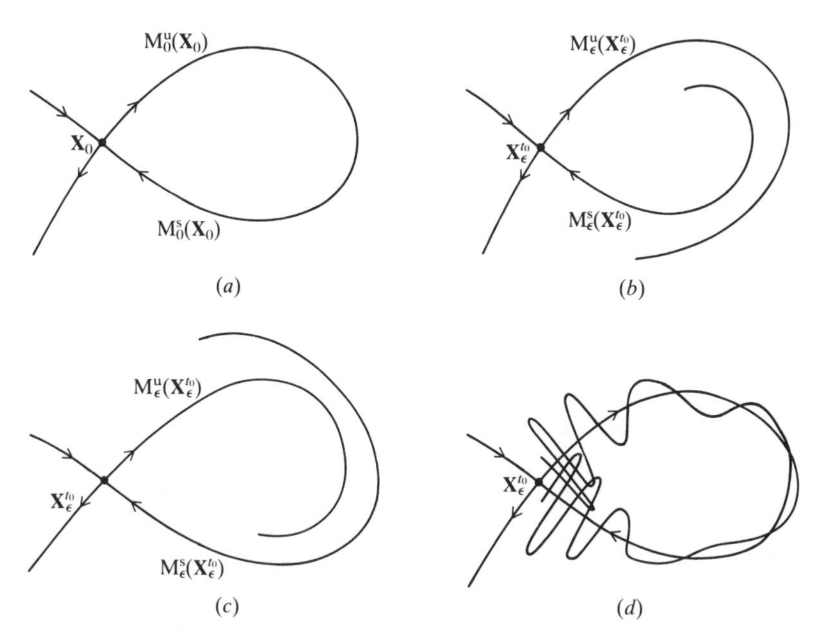

Fig. 8.10 Sketches of some possible configurations of the stable and unstable manifolds of a fixed point of a plane map. (*a*) The homoclinic connection of $\mathbf{X}_0$. (*b*), (*c*) Non-intersecting manifolds of $\mathbf{X}_\epsilon^{t_0}$. (*d*) Transversely intersecting manifolds of $\mathbf{X}_\epsilon^{t_0}$: a homoclinic tangle.

intersections form a kind of grid . . . with an infinitely tight mesh; each curve never intersects itself, but must fold upon itself in a very complicated way in order to intersect infinitely often the vertices of the grid'. This pattern is now called a *homoclinic tangle.*

Let us see why a homoclinic tangle appears as it does. If the stable and unstable manifolds of the Poincaré map intersect at one point, $\mathbf{x}$, say, then all the images and pre-images of $\mathbf{x}$ must lie in each of the manifolds, because each manifold is an invariant set of the map. Therefore the manifolds intersect an infinity of times. Moreover, because an orbit on one of the manifolds approaches or leaves the saddle point $\mathbf{X}_\epsilon^{t_0}$ of the map exponentially, successive images or pre-images get closer and closer together as they approach the saddle point. Also continuity implies that a point on one side of the stable manifold $M_\epsilon^s(\mathbf{X}_\epsilon^{t_0})$ is mapped by $P_\epsilon^{t_0}$ to a point on the same side; similarly, a point is mapped to a point on the same side of the unstable manifold. It follows that one 'lobe' between the two manifolds is mapped by $P_\epsilon^{t_0}$ into another lobe an even number of lobes ahead. So for systems (1) which are area-preserving, or approximately so, the areas of neighbouring lobes are equal, or nearly equal. Therefore the height of the lobes increases as they approach the saddle point, because their base decreases exponentially. The large height may lead to more complicated intersections of the two manifolds than those depicted in Fig. 8.10(*d*), although a manifold cannot intersect itself except at a fixed point, because an orbit of the differential system (4) is unique for given initial conditions. This is an heuristic explanation of the configuration of Fig. 8.10(*d*), which has been sketched for illustration rather than quantitative detail. It should also be borne in mind that the intersection of the two manifolds may be even more intricate if they intersect on the 'other' side of the saddle point.

Consider next the successive images of points in the interior of a small square near the saddle point, under iteration of the Poincaré map. The centre of the square will move slowly near the saddle point, while the square itself is repeatedly contracted in the direction of the stable eigenvector (i.e. in the direction parallel to the stable manifold) and stretched in the direction of the unstable eigenvector (i.e. in the direction parallel to the unstable manifold). Eventually the square will 'escape' from the neighbourhood of the saddle point, move more rapidly around (near the homoclinic orbit if ϵ is small), and then return to the neighbourhood of the saddle point if the manifolds intersect. As the square moves around, it will in general be folded. So it may return near to its original position as a horseshoe map (cf. §3.6) of its original shape, as indicated in Fig. 8.11; this property is, in fact, implied in general by the existence of the homoclinic tangle. This leads to chaos, the square returning again and again as the iterated horseshoe map of its original self.

After this qualitative description of the breaking of a homoclinic orbit

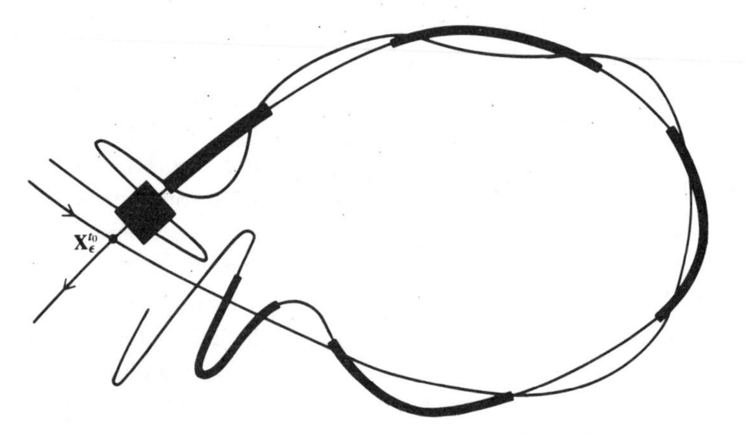

Fig. 8.11 A sketch of the contracting, stretching and folding of a square as it moves round a homoclinic tangle.

as ϵ varies from zero, let us find the details quantitatively for small ϵ. The success of the asymptotic method, due to Mel'nikov (1963), depends on knowledge of the global properties of the basic system (2); of course, in the present case this system is integrable. It is already apparent that a crucial issue is whether the stable and unstable manifolds intersect transversely, leading to a homoclinic tangle and chaos. The method gives the distance between the manifolds approximately, and hence a condition for their intersection.

Before launching into the technical details of the Mel'nikov method with an elaborate notation, let us review the essence of the method. We have already chosen arbitrarily a point on the homoclinic orbit of the basic system (2), and taken the phase of the solution $\mathbf{q}_0$ so that $\mathbf{q}_0(0)$ is the chosen point. We shall proceed to find, when ϵ is small, the distance between the stable manifold $M^s_\epsilon(\mathbf{X}^{t_0}_\epsilon)$ and the unstable manifold $M^u_\epsilon(\mathbf{X}^{t_0}_\epsilon)$ near $\mathbf{q}_0(0)$ for all t_0. Although the distance depends on both $\mathbf{q}_0(0)$ and the phase t_0, we shall, for convenience, regard it as a function of t_0 for fixed $\mathbf{q}_0(0)$. Thus we shall find whether there exists a value of t_0 such that the manifolds $M^s_\epsilon(\mathbf{X}^{t_0}_\epsilon)$, $M^u_\epsilon(\mathbf{X}^{t_0}_\epsilon)$ intersect transversely near our chosen point $\mathbf{q}_0(0)$ in the section Σ^{t_0}. This is equivalent to finding whether there exists a $\mathbf{q}_0(0)$ such that the manifolds intersect transversely in the *fixed* section Σ^{t_0}, because t_0 is merely a phase of $\mathbf{q}_0$ and a translation in time is equivalent to a translation around the homoclinic orbit. So, if there exists a value of t_0 such that the manifolds $M^s_\epsilon(\mathbf{X}^{t_0}_\epsilon)$, $M^u_\epsilon(\mathbf{X}^{t_0}_\epsilon)$ intersect transversely an infinity of times then they do so not only for this value of t_0 but for all values of t_0.

First find the equations of the orbits, say $\{\mathbf{q}^s_\epsilon(t, t_0), t\}$, $\{\mathbf{q}^u_\epsilon(t, t_0), t\}$, which satisfy equation (4) and lie in the extensions of the stable and unstable manifolds respectively in the three-dimensional phase space. There exist uniformly valid expansions of these solutions of system (1) of the form

$$\left.\begin{aligned} \mathbf{q}^s_\epsilon(t, t_0) &= \mathbf{q}_0(t - t_0) + \epsilon\mathbf{q}^s_1(t, t_0) + O(\epsilon^2) \qquad \text{for } t \geqslant t_0 \\ \mathbf{q}^u_\epsilon(t, t_0) &= \mathbf{q}_0(t - t_0) + \epsilon\mathbf{q}^u_1(t, t_0) + O(\epsilon^2) \qquad \text{for } t \leqslant t_0 \end{aligned}\right\} \text{as } \epsilon \to 0. \tag{5}$$

We can in principle find $\mathbf{q}^s_\epsilon$, and $\mathbf{q}^u_\epsilon$ likewise, by regular perturbation theory. Noting that $\mathbf{q}^s_\epsilon$ is a solution of system (1) and $\mathbf{q}_0$ of (2), linearize (1) about $\mathbf{q}_0$ for small ϵ to get

$$\frac{d\mathbf{q}^s_1(t, t_0)}{dt} = \mathbf{J}(\mathbf{q}_0(t - t_0))\mathbf{q}^s_1(t, t_0) + \mathbf{f}(\mathbf{q}_0(t - t_0), t) \tag{6^s}$$

for $t \geqslant t_0$, where $\mathbf{J}$ is the Jacobian matrix of $\mathbf{F}$. In addition, we require that $\mathbf{q}^s_1(t, t_0) \to \lim_{\epsilon\to 0}\{(\mathbf{X}^{t_0}_\epsilon - \mathbf{X}_0)/\epsilon\}$ as $t \to \infty$ in order that $\mathbf{q}^s_\epsilon(t, t_0) \to \mathbf{X}^{t_0}_\epsilon$. Also $\mathbf{q}^u_\epsilon$ satisfies a similar equation (6^u) for $t \leqslant t_0$ and similar boundary condi-

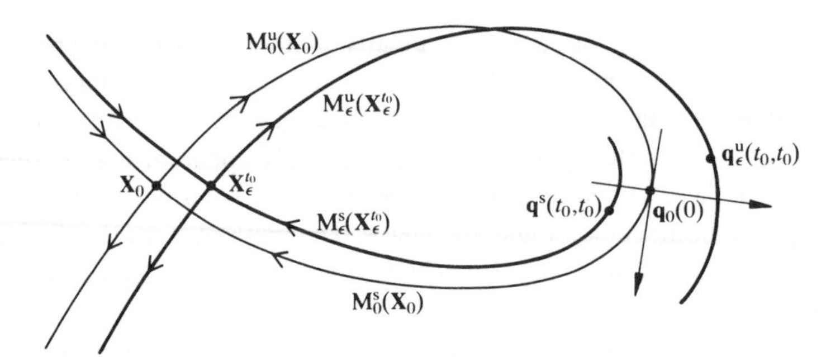

Fig. 8.12 The distance between the orbits on the stable and unstable manifolds near $\mathbf{q}_0(0)$ at time t_0.

tion as $t \to -\infty$. It will be seen that the form of equations (6^s), (6^u) but, fortunately, not their explicit solutions $\mathbf{q}_1^s$, $\mathbf{q}_1^u$, is needed to find the condition for the onset of chaos.

We can define the distance between $M_\epsilon^s(\mathbf{X}_\epsilon^{t_0})$ and $M_\epsilon^u(\mathbf{X}_\epsilon^{t_0})$ by measuring it, at least for small ϵ, in the direction normal to the homoclinic orbit of the basic system (1) at the point $\mathbf{q}_0(0)$. Thus we define the displacement

$$\mathbf{d}(t_0) = \mathbf{q}_\epsilon^u(t_0, t_0) - \mathbf{q}_\epsilon^s(t_0, t_0) \tag{7}$$
$$= \epsilon\{\mathbf{q}_1^u(t_0, t_0) - \mathbf{q}_1^s(t_0, t_0)\} + O(\epsilon^2) \qquad \text{as } \epsilon \to 0.$$

For small ϵ, the two points $\mathbf{q}_\epsilon^u(t_0, t_0)$, $\mathbf{q}_\epsilon^s(t_0, t_0)$ are in general slightly displaced from $\mathbf{q}_0(0)$, and the manifolds are nearly tangential to the homoclinic orbit at $\mathbf{q}_0(0)$, as shown in Fig. 8.12. So we may resolve the displacement $\mathbf{d}$ in the direction of the normal to the homoclinic orbit at $\mathbf{q}_0(0)$ in order to measure the distance between the manifolds. Now, if $\mathbf{F} = [F_1, F_2]^T$ then the unit outward normal vector is $\mathbf{n} = [-F_2(\mathbf{q}_0(0)), F_1(\mathbf{q}_0(0))]^T/|\mathbf{F}(\mathbf{q}_0(0))|$. So the Mel'nikov distance between the two manifolds at $\mathbf{q}_0(0)$ is defined as

$$D(t_0) = \mathbf{d}\cdot\mathbf{n} \tag{8}$$

$$= \frac{\epsilon\mathbf{F}(\mathbf{q}_0(0)) \wedge \{\mathbf{q}_1^u(t_0, t_0) - \mathbf{q}_1^s(t_0, t_0)\}}{|\mathbf{F}(\mathbf{q}_0(0))|} + O(\epsilon^2) \qquad \text{as } \epsilon \to 0, \tag{9}$$

where the *wedge product* of any given pair of vectors $\mathbf{a} = [a_1, a_2]^T$, $\mathbf{b} = [b_1, b_2]^T$ is defined as $\mathbf{a} \wedge \mathbf{b} = a_1 b_2 - a_2 b_1$.

To evaluate D, it helps to eliminate $\mathbf{q}_1^u$, $\mathbf{q}_1^s$ from expression (9) by using equations (6^u), (6^s), because $\mathbf{q}_1^u$, $\mathbf{q}_1^s$ are not known explicitly. This may be done by first defining

$$\Delta^s(t, t_0) = \mathbf{F}(\mathbf{q}_0(t - t_0)) \wedge \mathbf{q}_1^s(t, t_0). \tag{10^s}$$

It follows, on differentiating by parts, that

$$\begin{aligned}\frac{\mathrm{d}\Delta^{\mathrm{s}}(t,t_0)}{\mathrm{d}t} &= \mathbf{J}(\mathbf{q}_0(t-t_0))\mathbf{F}(\mathbf{q}_0(t-t_0)) \wedge \mathbf{q}_1^{\mathrm{s}}(t,t_0) \\ &\quad + \mathbf{F}(\mathbf{q}_0(t-t_0)) \wedge \frac{\mathrm{d}\mathbf{q}_1^{\mathrm{s}}(t,t_0)}{\mathrm{d}t} \\ &= \mathbf{J}(\mathbf{q}_0(t-t_0))\mathbf{F}(\mathbf{q}_0(t-t_0)) \wedge \mathbf{q}_1^{\mathrm{s}}(t,t_0) \\ &\quad + \mathbf{F}(\mathbf{q}_0(t-t_0)) \wedge \{\mathbf{J}(\mathbf{q}_0(t-t_0))\mathbf{q}_1^{\mathrm{s}}(t,t_0) + \mathbf{f}(\mathbf{q}_0(t-t_0),t)\} \\ &= \mathrm{trace}\{\mathbf{J}(\mathbf{q}_0(t-t_0))\}\mathbf{F}(\mathbf{q}_0(t-t_0)) \wedge \mathbf{q}_1^{\mathrm{s}}(t,t_0) \\ &\quad + \mathbf{F}(\mathbf{q}_0(t-t_0)) \wedge \mathbf{f}(\mathbf{q}_0(t-t_0),t) \\ &= \mathbf{F}(\mathbf{q}_0(t-t_0)) \wedge \mathbf{f}(\mathbf{q}_0(t-t_0),t), \end{aligned} \tag{11}$$

because $(\mathbf{Ja}) \wedge \mathbf{b} + \mathbf{a} \wedge (\mathbf{Jb}) = (\mathrm{trace}\,\mathbf{J})(\mathbf{a} \wedge \mathbf{b})$ identically, and $\mathrm{trace}\,\mathbf{J} = \partial F_1/\partial x + \partial F_2/\partial y = \partial^2 H/\partial x \partial y - \partial^2 H/\partial y \partial x = 0$. Therefore

$$\Delta^{\mathrm{s}}(t_0,t_0) = \Delta^{\mathrm{s}}(\infty,t_0) - \int_{t_0}^{\infty} \mathbf{F}(\mathbf{q}_0(t-t_0)) \wedge \mathbf{f}(\mathbf{q}_0(t-t_0),t)\,\mathrm{d}t. \tag{12^{s}}$$

Now $\Delta^{\mathrm{s}}(\infty,t_0) = \lim_{t\to\infty}\{\mathbf{F}(\mathbf{q}_0(t-t_0)) \wedge \mathbf{q}_1^{\mathrm{s}}(t,t_0)\} = 0$, because $\mathbf{q}_1^{\mathrm{s}}(t,t_0)$ is bounded and $\mathbf{F}(\mathbf{q}_0(t-t_0)) \to \mathbf{F}(\mathbf{X}_0) = \mathbf{0}$ as $t \to \infty$. Therefore

$$\Delta^{\mathrm{s}}(t_0,t_0) = -\int_{t_0}^{\infty} \mathbf{F}(\mathbf{q}_0(t-t_0)) \wedge \mathbf{f}(\mathbf{q}_0(t-t_0),t)\,\mathrm{d}t. \tag{13^{s}}$$

Similarly,

$$\Delta^{\mathrm{u}}(t_0,t_0) = \int_{-\infty}^{t_0} \mathbf{F}(\mathbf{q}_0(t-t_0)) \wedge \mathbf{f}(\mathbf{q}_0(t-t_0),t)\,\mathrm{d}t. \tag{13^{u}}$$

Therefore equation (9) gives

$$D(t_0) = \frac{\epsilon M(t_0)}{|\mathbf{F}(\mathbf{q}_0(0))|} + O(\epsilon^2) \qquad \text{as } \epsilon \to 0, \tag{14}$$

where the *Mel'nikov function* is defined as

$$M(t_0) = \int_{-\infty}^{\infty} \mathbf{F}(\mathbf{q}_0(t-t_0)) \wedge \mathbf{f}(\mathbf{q}_0(t-t_0),t)\,\mathrm{d}t. \tag{15}$$

It can be seen that in general if M has a simple zero, τ say, then D has a simple zero near τ for small ϵ, and therefore the stable and unstable manifolds intersect transversely at the point corresponding to $t_0 = \tau$. This implies that there is an infinity of intersections and a homoclinic tangle as in Fig. 8.10(d), and therefore that there is chaos. Conversely, if M has no zero then D has no zero for small ϵ, the manifolds do not intersect, as shown in Fig. 8.10(b) or (c), and there is no chaos. Note also that D and M in fact depend upon $\mathbf{q}_0(0)$ as well as t_0, but that they have zeros for all $\mathbf{q}_0(0)$ or for none.

It can easily be shown from equation (15) that M has period T, because $\mathbf{f}$ has. This also follows from the identity $\mathrm{P}_\epsilon^{t_0+T} = \mathrm{P}_\epsilon^{t_0}$ of the Poincaré maps, and verifies the result that one intersection of the stable and unstable manifold implies an infinity of intersections.

Finally, note that if $\mathbf{f}$ were Hamiltonian or were a constant then expression (15) could be simplified a little; and if $\mathbf{F}$ were not Hamiltonian then equation (11) would not follow, and the expression of D would be more complicated, but the method would still be applicable. Again, if the order of the differential system (1) were higher than two then the essence of the ideas above would be applicable but their expression more complicated.

Example 8.2: Duffing's equation. We next apply Mel'nikov's method to Duffing's equation (2.1) with negative stiffness, after Holmes (1979), and find conditions for the onset of some of the chaos described in §2. Equation (2.1) with weak damping and forcing may be written in the form

$$\frac{\mathrm{d}x}{\mathrm{d}t} = y, \qquad \frac{\mathrm{d}y}{\mathrm{d}t} = x - x^3 + \epsilon(\gamma \cos \omega t - \delta y), \tag{16}$$

where $\gamma,\ \delta,\ \omega > 0$ and $0 < \epsilon \ll 1$. On putting $\mathbf{F} = [y, x - x^3]^{\mathrm{T}}$, $\mathbf{f} = [0, \gamma \cos \omega t - \delta y]^{\mathrm{T}}$, it can be seen that equations (16) have the form (1).

For $\epsilon = 0$, there is a saddle point at 0 in the phase plane, with two symmetric homoclinic orbits (see Fig. 8.6(a)). The orbits of the basic system (2) have equation $H(y, x) = E$, i.e. $\frac{1}{2}y^2 - \frac{1}{2}x^2 + \frac{1}{4}x^4 = E$. It follows that the homoclinic orbits have equation $y^2 = x^2(1 - \frac{1}{2}x^2)$, and solutions

$$\mathbf{q}_0(t) = \pm(\surd 2 \operatorname{sech} t, -\surd 2 \operatorname{sech} t \tan t), \tag{17}$$

on taking $\mathbf{q}_0(0) = (\pm\surd 2, 0)$. Therefore the Mel'nikov function is

$$\begin{aligned} M(t_0) &= \int_{-\infty}^{\infty} \mathbf{F}(\mathbf{q}_0(t - t_0)) \wedge \mathbf{f}(\mathbf{q}_0(t - t_0), t)\,\mathrm{d}t \\ &= \int_{-\infty}^{\infty} y_0(t - t_0)\{\gamma \cos \omega t - \delta y_0(t - t_0)\}\,\mathrm{d}t \\ &= \int_{-\infty}^{\infty} y_0(s)\{\gamma \cos \omega(s + t_0) - \delta y_0(s)\}\,\mathrm{d}s, \end{aligned}$$

on substituting $s = t - t_0$ (which is an obvious, but frequently useful, substitution to simplify the Mel'nikov integral),

$$= \sqrt{2}\gamma \sin \omega t_0 \int_{-\infty}^{\infty} \operatorname{sech} s \tanh s \sin \omega s \, ds - 2\delta \int_{-\infty}^{\infty} \operatorname{sech}^2 s \tanh^2 s \, ds,$$

for the right-hand homoclinic orbit with $y_0(s) = -\sqrt{2} \operatorname{sech} s \tanh s$,

$$= \sqrt{2}\pi\gamma\omega \operatorname{sech}(\tfrac{1}{2}\pi\omega) \sin(\omega t_0) - \tfrac{4}{3}\delta. \tag{18}$$

Therefore if $2\sqrt{2}\delta \cosh(\frac{1}{2}\pi\omega) < 3\pi\gamma\omega$ then M has simple zeros and there is chaos for small $\epsilon \neq 0$, whereas if $2\sqrt{2}\delta \cosh(\frac{1}{2}\pi\omega) > 3\pi\gamma\omega$ then $M(t_0) < 0$ for all t_0 and there is no chaotic break-up of the homoclinic orbit. □

4 Routes to chaos

We have met a few sequences of bifurcations leading to chaos as a parameter, say a, of a nonlinear system varies. There are many more sequences. However, at the risk of oversimplification of research that is continuing, it may be said that this transition to chaos occurs in one of four different ways, i.e. the sequences may be divided in four classes.

(i) *Subcritical instability.* On this route a 'familiar' attractor, i.e. a point, periodic or quasi-periodic attractor, becomes unstable as a slowly increases or decreases through a critical value, a_c say, and the system then 'jumps' rapidly to a strange attractor which is not a continuous extension of the familiar attractor. An example of this occurs for the Lorenz system as r increases through r_c; there is hysteresis because the fixed points are stable if $1 < r < r_c$ and the chaotic attractor is stable if $r_0 < r$, where $1 < r_0 < r_c$. Indeed, the local theory shows that, as a increases through a value a_c where there is a *sub*critical turning point, pitchfork bifurcation or Hopf bifurcation, the attractor ceases to exist so that a solution must go somewhere else; it may therefore go abruptly either to another 'familiar' attractor or to a strange attractor.

(ii) *A sequence of bifurcations.* A route of transition to turbulent motion of a fluid was first charted by Ruelle & Takens (1971), and later revised by Newhouse, Ruelle & Takens (1978). Before the work of Ruelle & Takens, the prevailing belief was that transition to turbulence occurs after an infinity of bifurcations of increasingly complicated quasi-periodic flows as a parameter, e.g. the Reynolds number, increases. In contrast, Ruelle & Takens conjectured that a quasi-periodic solution with more than three fundamental frequencies is in general unstable, so that turbulence would ensue after only a few bifurcations.

Turbulence is another of those things which are usually easier to identify in practice than to define, but it may be said that it is a flow of a real fluid which is chaotic in space as well as time. Our concern here is with the onset of chaos rather than turbulence, and the two phenomena have been confused, so let us study this route only in terms of differential equations.

We have seen how a stable steady solution (a point attractor in phase space) becomes a periodic one (a closed curve S^1) at a Hopf bifurcation, and how the periodic solution may become unstable at another bifurcation and be succeeded by a stable quasi-periodic solution with two fundamental frequencies (on the surface of a torus $\mathbb{T}^2$) as the parameter a varies. Such bifurcations may recur, with the appearance of a quasi-periodic attractor with three, or even four, fundamental frequencies. It now seems that quasi-periodic solutions with five or more frequencies are in general unstable. Also, as a varies, the frequencies in general vary and thereby become rationally related, so that for a quasi-periodic attractor with two fundamental frequencies f_1, f_2 the relation $f_2 = pf_1/q$ becomes true for some integers p and q; then the solution has become periodic and is no longer quasi-periodic. Thereafter the ratio of the frequencies may remain constant as a varies further, in which case we say the frequencies are *locked* (cf. Q3.29, and the somewhat similar phenomenon of entrainment in Chapter 7). Then after further variation of a, the ratio f_2/f_1 may start again to vary until chaos ensues. So the graph of f_2/f_1 versus a may be like a devil's staircase (cf. Q4.6) – the ratio f_2/f_1 may vary continuously and monotonically, from its value at the bifurcation in which the quasi-periodic attractor first replaces the periodic attractor, taking fixed rational values as a increases in a sequence of intervals.

This whole route, somewhat simplified, may be summarized by the scenario:

$$\text{steady} \xrightarrow[\text{Hopf}]{} \underset{f_1;\mathrm{S}^1}{\text{periodic}} \to \underset{f_1,f_2;\,\mathbb{T}^2}{\text{quasi-periodic}} \to \underset{f_1,f_2,f_3;\,\mathbb{T}^3}{\text{quasi-periodic}} \to \text{chaos}$$

as a varies monotonically. The main properties of this route to chaos are that chaos ensues after a short sequence of bifurcations with continuous changes of the attractors.

But we have not yet specified the crucial bifurcation, namely the one which brings chaos. The onset of chaos itself may follow the break-up of a *homoclinic orbit* between an unstable point of equilibrium and itself (i.e. a limiting orbit leaving the point and returning to it) or of a *heteroclinic orbit* between two equilibrium points. The Lorenz system at $r = r_1$ offers an example of this, although the chaotic attractor originates then without continuous evolution from a quasi-periodic solution.

(iii) *Period doubling.* Feigenbaum analysed the infinite succession of period doubling *en route* to chaos for one-dimensional difference equations. This route is found in many differential systems, with the same universal scaling (*although the phenomenon is essentially two-dimensional for differential equations, because a Floquet multiplier cannot vanish (Q5.17) and a complex multiplier must be one of a complex conjugate pair, and so a multiplier has to go from 1 to -1 in the unit circle of the complex plane without passing through the origin as a varies monotonically from a_r to a_{r+1}). Period doubling at $a = a_c$ is illustrated in Fig. 8.13, where the orbits in the phase space of the variable $\mathbf{x}$ and the graphs of $x(t)$ are sketched, x being a typical component of $\mathbf{x}$.

The Lorenz system as r decreases through 99.52 or 214.36 is a differential system with period doubling. The logistic and the Hénon maps give classic examples of period doubling for difference equations.

(iv) *Intermittent transition.* On this route, first charted by Pomeau & Manneville (1980), a limit cycle becomes unstable as the parameter a increases through a critical value, a_c say. As a increases to a_c the limit cycle

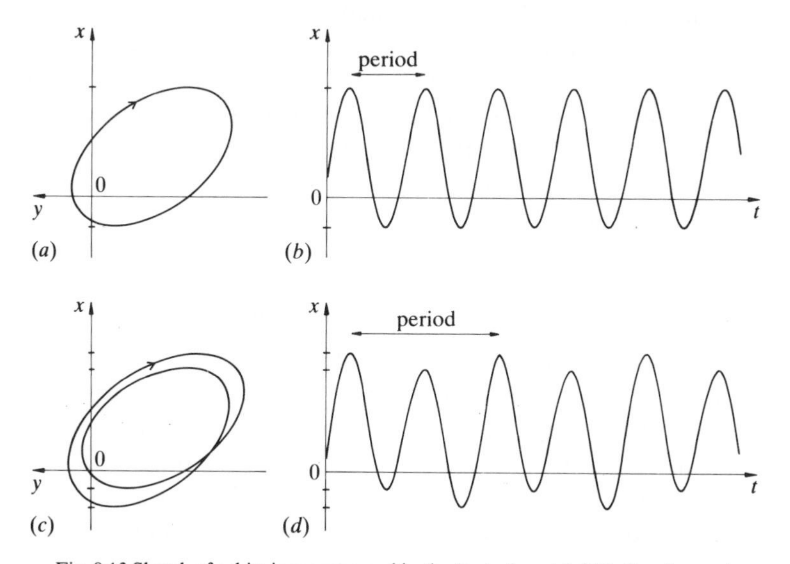

Fig. 8.13 Sketch of orbits in $\mathbf{x}$-space and in the (t, x)-plane (a), (b) before $(a \leqslant a_c)$ and (c), (d) after $(a > a_c)$ period doubling.

becomes unstable to smaller and smaller disturbances. At $a = a_c$ it coalesces with a similar unstable limit cycle, as at a turning point of a Poincaré map. For small positive $a - a_c$ the same cycle persists most of the time, but is occasionally and rarely interrupted by 'bursts' which are not small. As $a - a_c$ increases the bursts occur more frequently but do not change much in magnitude. The average time between the 'random' bursts tends to infinity like $(a - a_c)^{-1/2}$ as $a \downarrow a_c$.

An example of this occurs for the Lorenz system as r increases through 166, when $\sigma = 10$, $b = \frac{8}{3}$. Another example occurs for the logistic map $F(a, x) = ax(1 - x)$ as a *de*creases through $a_c = 1 + \sqrt{8}$. For $a > a_c$ there is a stable and an unstable three-cycle; the cycles coalesce at $a = a_c$, where F^3 has a turning point. For $0 < a_c - a \ll 1$ there is chaos of this intermittent character; although there is no three-cycle, there is a 'ghost' of the three-cycle near which x_n may linger for many iterations before eventually moving away, then returning to linger again, and so forth.

The difference equation $x_{n+1} = x_n - x_n^2 - a$ has a turning point at $a = 0$, $x = 0$, and illustrates the *non-chaotic* phase of intermittency. There are a stable fixed point $(X = (-a)^{1/2})$ and an unstable fixed point $(X = -(-a)^{1/2})$ if $a < 0$, one weakly stable fixed point $(X = 0)$ if $a = 0$, and no fixed point if $a > 0$. However, if $0 < a \ll 1$ the origin acts as a 'ghost' fixed point in the following sense. A point x_n stays very close to the origin for a long time of the order of $a^{-1/2}$ as $a \downarrow 0$, before eventually moving away from the origin (Q8.10). It is this long proximity to the ghost of a fixed point of a one-dimensional map near a turning point which represents the non-chaotic phase of intermittency when the solution of a differential system is nearly periodic, the one-dimensional map being a Poincaré map of the differential system. However, this particular difference equation is deficient as a model of the chaotic phase of intermittency because it cannot reinject x_n back near the origin once x_n has left the neighbourhood of the origin.

5 Analysis of time series

An important practical problem, with wide ramifications, is the interpretation of signals from a nonlinear system. A 'system' may be a natural phenomenon, a laboratory experiment, or a numerical experiment. Often we know little about the structure of a system, but can control some of the parameters which specify it and can measure its output, or some of its output. The 'output' is effectively the solution as a function of time. The measurements are of various quantities, either over an interval of time or at a sequence of instants. If we are ignorant of the nature of the system, i.e. if we do not know the equations which govern it, then we may seek to use the measurements to learn about the nature, the physical mechanisms, etc. In short the system may be like a 'black box', so we need to be able to use measurements of its output to find the states of the system, their stability and bifurcations as the control parameters vary. This may tell us something about what is 'inside' the black box. These ideas can be applied to phenomena ranging from sunspot cycles or weather patterns to growth of a population of foxes or stock exchange prices. Here we shall discuss only such nonlinear systems as we have met before in this book.

It is usually easy to recognize a stable steady state of a system, although sometimes the presence of 'noise' may make an equilibrium point difficult to identify unequivocally, especially if it is only weakly stable.

A periodic solution may be detectable by 'eye', i.e. merely by looking at the graph, or a list, of measurements of one state variable as a function of time. Confirmation of the periodicity may come from detecting the same period, albeit different phases, in measurements of other state variables. Again, noise may confuse matters. So may errors of measurement and inaccuracy of the numerical processing of the data.

Quasi-periodic solutions are difficult to recognize as such merely by looking at a graph of the output and chaos is even more difficult. Indeed, the distinction between the time signal of a system in a chaotic state and noise, i.e. the random errors of measurement and data processing as well as the extraneous perturbations of the system itself, is not easy in principle or practice.

We need more-objective, or, rather, less-subjective, methods to identify the attractor by examination of the output. Some common methods of diagnosis of the signal from a nonlinear system are summarized as follows. Let us suppose that the signal measured is the time series $d(t)$.

(i) Examine the time series by eye.

(ii) Examine correlations, e.g. $\overline{d(s+t)d(s)} = T^{-1}\int_0^T d(s+t)d(s)\,\mathrm{d}s$, where T is a 'long' time. The graph of $\overline{d(s+t)d(s)}$ as a function of t may be revealing. If d is periodic or quasi-periodic then so is its autocorrelation $\overline{d(s+t)d(s)}$. But if d is chaotic and has zero mean then its autocorrelation decays rapidly with t because of sensitive dependence on initial conditions and the consequent effective independence of two parts of a solution unless they occur at nearly the same time.

(iii) Plot a phase plane, with a state variable $d(t)$ as abscissa and $d(t + \tau)$ as ordinate, where t is the independent variable acting as the parameter of the orbit and τ is some suitable positive number. This is called a *scatter diagram* in statistics, and the *method of delays* in the theory of nonlinear systems. Packard *et al.* (1980) and Takens (1981) have proposed and analysed this method of constructing a vector of whatever dimension is desired from the time series of a single scalar. We need τ to be neither so small that $d(t + \tau)$ is closely correlated with $d(t)$, nor so large that $d(t + \tau)$ is independent of $d(t)$. We should be wary of aliasing in case, for example, there were a periodic solution and τ happened to be the period. It is often a good way to choose τ so that it is the least value for which the average $\overline{d(s + \tau)d(s)} = 0$, another is to use trial and error. An alternative method is to plot a phase plane with two state variables, $d_1(t)$ and $d_2(t)$, say, as abscissa and ordinate; provided, of course, that two variables can be measured. Then we have a phase plane which gives some projection of an orbit, even though we do not know the dimension of the attractor traversed by the orbit or of the phase space in which the orbit belongs. However, experience at looking at such projections helps the diagnosis of an attractor as a limit cycle (with a closed curve in the plane), a quasi-periodic or a chaotic solution.

(iv) Make a Poincaré section of the phase plane above or of other phase spaces.

(v) Take a Fourier transform of the output. Given measurements of a state variable, d say, over an interval of length T, we may compute the Fourier series of d. Fast Fourier transforms offer an efficient computational method to do this, so that a signal from an experiment may be processed 'on line'. Suppose then that a state variable $d(t)$ has been measured for $0 \leqslant t \leqslant T$. In practice there is a finite number of measurements at discrete instants; say, then, that we measure $d_n = d(nh)$ for $n = 0, 1, \ldots, 2^N$, where $h = T/2^N$ is the sampling time and N is an integer. Therefore the Fourier series may be found as an approximation to the Fourier transform by taking h quite small and T quite large. The accuracy of the spectrum of high frequencies is limited by the smallness of the sampling time h, and of low frequencies by the length T of the interval of measurement; we do not expect to resolve frequencies $f \gtrsim h^{-1}$ or $f \lesssim T^{-1}$.

It is common practice to plot the power spectrum $P_x(f)$ of a time series $x(t)$, where $P_x(f) = |\tilde{x}(f)|^2$ and $\tilde{x}$ is the complex Fourier transform of x.

For a sinusoidal function there is a single peak of the spectrum at its frequency f_1. This is characteristic of a periodic solution just after its origin at a Hopf bifurcation. But in general, for a well-behaved periodic function there are peaks of exponentially diminishing amplitude at f_1 (fundamental frequency), $2f_1$ (first harmonic), $3f_1$ (second harmonic), etc. In an ideal model with a continuum of precise measurements over an infinite interval, these peaks have the zero width and infinite height of a delta function, but noise in the system, in the measurements, and in processing of the signal gives narrow peaks of finite height in practice.

To get a glimpse at the nature of the spectrum of a quasi-periodic function, consider the example with $d(t) = a_1 \cos mf_1 t + a_2 \cos nf_2 t$ for integers m, n and incommensurate fundamental frequencies f_1, f_2. Then take a typical nonlinear term, say,

$$\begin{aligned} d^2(t) &= a_1^2 \cos^2 mf_1 t + 2a_1 a_2 \cos mf_1 t \cos nf_2 t + a_2^2 \cos^2 nf_2 t \\ &= \tfrac{1}{2}(a_1^2 + a_2^2) + \tfrac{1}{2}a_1^2 \cos 2mf_1 t + \tfrac{1}{2}a_2^2 \cos^2 2nf_2 t \\ &\quad + a_1 a_2 \{\cos(mf_1 + nf_2)t + \cos(mf_1 - nf_2)t\}. \end{aligned}$$

This, and other nonlinear interactions similarly, may generate not only peaks at $2mf_1$, $2nf_2$ but also at $\pm mf_1 \pm nf_2$. There are several examples in §6.4 and Chapter 7 of this weakly nonlinear generation. Thus nonlinearity in the equations of a system may generate not only all harmonics but also all sum and difference frequencies. These can be identified in the Fourier spectrum of output from a system, and hence lead to a diagnosis of a quasi-periodic attractor.

Chaotic solutions have spectra with broad bands rather than isolated peaks, and with a high noise level, but nevertheless may have prominent peaks and a lot of structure. Chaos is not formless, and is usually far from white noise.

*(vi) The dimension D of an attractor can be found *in principle* by measuring only a single state variable, say d, because the measurements of $d(t)$ imply the measurements of all derivatives of d in principle and thus of all the state variables $x \in \mathbb{R}^m$ of the solution. This may be better done by constructing a Takens vector $(x(\tau), x(t+\tau), \ldots, x(t+m\tau))$, because numerical differentiation is subject to error. However, an infinite time series with infinite precision is needed to specify the attractor with certainty, and in practice experiments yield a finite time series with a lot of noise, so calculations of D may be difficult unless D is 0, 1 or 2. It is usually easier, though, to calculate the fractal dimension of the attractor from the output of a numerical system because the noise is low. We see then that $D = 0$ for a steady solution, $D = 1$ for a periodic solution on S^1, $D = 2$ for a quasi-periodic solution on $\mathbb{T}^2$ with two fundamental frequencies, and so forth, but that D may have a non-integral value for a chaotic solution. Thus D

gives the effective degree of freedom of the solution or the dimension of the submanifold in which the attractor lies.

*(vii) Find the Liapounov exponents of the system. These are often a means of calculating D. They also may serve to confirm the presence of chaos.

Example 8.3: looking at time series and their analyses. To appreciate these methods it is helpful to see their application to some specific cases. First examine analyses of time series found by numerical integration of the Rössler system. This is the system of Q5.5, and we have taken parameters $a = b = 0.2$ (although these details are not important here). Now look at the results of both time series and power spectra in Fig. 8.14. They suggest that there is a periodic attractor when $c = 2.6$, that it has undergone period doubling before c increases to 3.5 and quadrupling before c reaches 4.1. The peaks of the spectra are not of infinite height and zero width because of round-off and truncation errors in both the production and analysis of the time series (of finite duration). The resultant noise dominates the signal at the level of the troughs between the peaks. However, the peaks of the fundamental and its harmonics are clearly identifiable in Fig. 8.14(*a*). Also the appearance of the subharmonic of order $\frac{1}{2}$ is clearly discernible in Fig. 8.14(*b*), and of the subharmonic of order $\frac{1}{4}$ in Fig. 8.14(*c*). This analysis is confirmed by the phase plots of Fig. 8.15.

Next look at the time series and its power spectrum in Fig. 8.16. The data come from Ryrie's (1992) numerical integration of a high-order system of ordinary differential equations devised by Curry *et al.* (1984) to model thermal convection of a layer of fluid heated from below. The solution appears to be quasi-periodic. It is not always possible to identify the fundamental frequencies f_1, f_2 unambiguously in practice. However, at the *onset* of a quasi-periodic attractor following the instability of a periodic solution as a parameter increases, we may identify f_1 as the fundamental frequency of the periodic solution and f_2 as the frequency of its linear instability (*given by the Floquet exponent), so that f_1 is the frequency of the highest peak in the spectrum of the newly created quasi-periodic solution. As the parameter increases further, it may become more difficult to identify f_1 and f_2 from the spectrum. In any event, a plausible interpretation (Ryrie 1992) of Fig. 8.16 is that $f_1 \approx 6.73$, corresponding to the highest peak, and $f_2 \approx 2.51$.

The next case is of time series and a Fourier spectrum found by numerical integration of the Lorenz system (1.1) for $r = 28$, $\sigma = 10$, $b = \frac{8}{3}$, shown

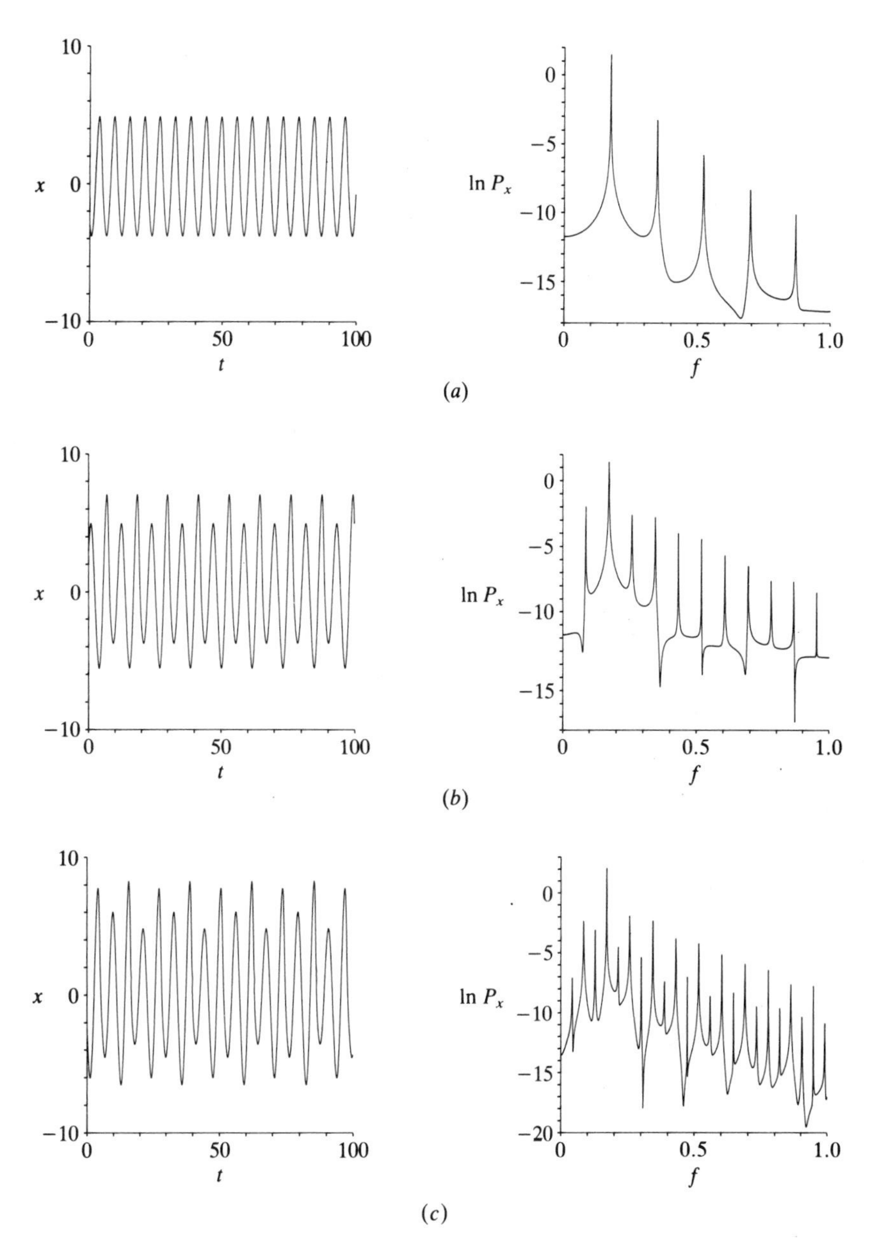

Fig. 8.14 Solutions of the Rössler system $\mathrm{d}x/\mathrm{d}t = -y - z$, $\mathrm{d}y/\mathrm{d}t = x + \frac{1}{5}y$, $\mathrm{d}z/\mathrm{d}t = \frac{1}{5} + z(x - c)$ as a time series $x(t)$ and the logarithm of its power spectrum versus frequency, $\ln P_x(f)$, where P_x is the square of the modulus of the complex Fourier transform of x, for (*a*) $c = 2.6$, a periodic, (*b*) $c = 3.5$, a period-doubled, and (*c*) $c = 4.1$, a period-quadrupled solution.

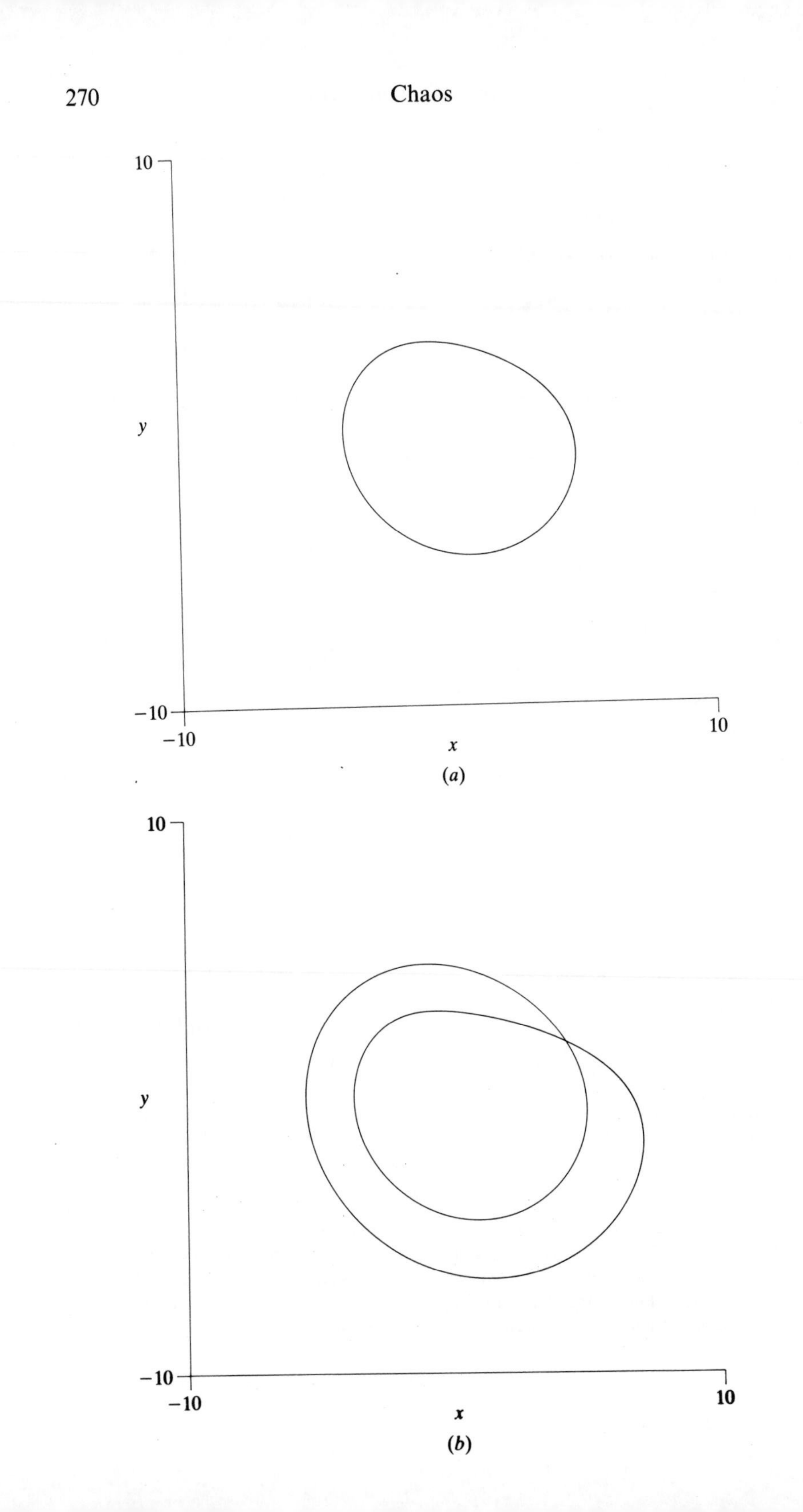

(a)

(b)

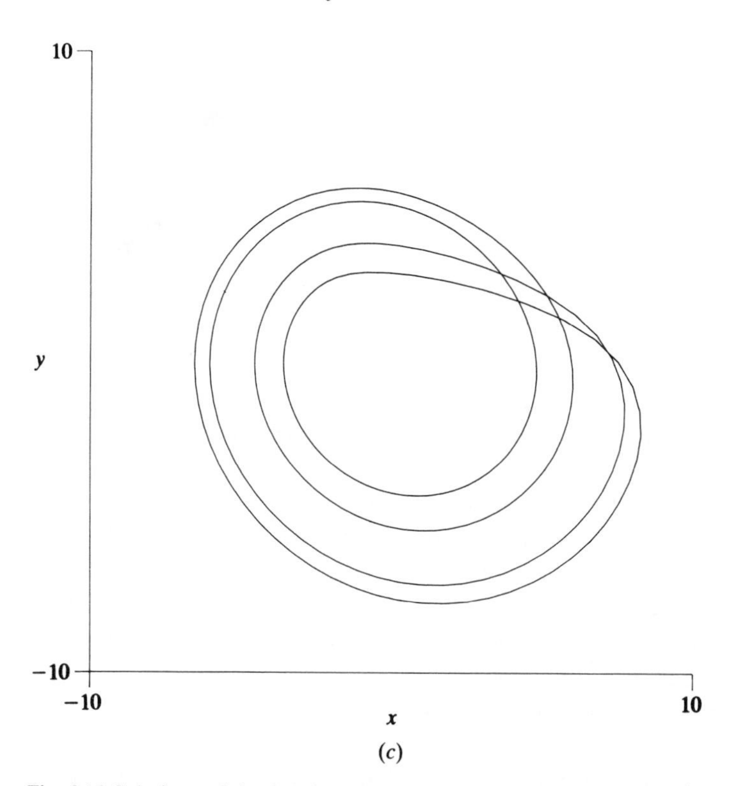

(c)

Fig. 8.15 Solutions of the Rössler system $\mathrm{d}x/\mathrm{d}t = -y - z$, $\mathrm{d}y/\mathrm{d}t = x + \frac{1}{5}y$, $\mathrm{d}z/\mathrm{d}t = \frac{1}{5} + z(x - c)$ in the (x, y)-plane for (a) $c = 2.6$, a periodic, (b) $c = 3.5$, a period-doubled, and (c) $c = 4.1$, a period-quadrupled solution. Each orbit moves clockwise as t increases.

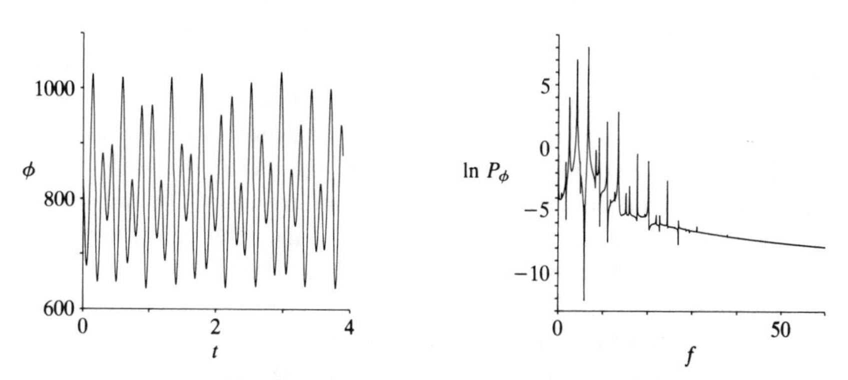

Fig. 8.16 A quasi-periodic solution found by integrating a high-order differential system which models themal convection: a time series $\phi(t)$ and the logarithm of its power spectrum versus frequency, $\ln P_\phi(f)$ (after Ryrie (1991)).

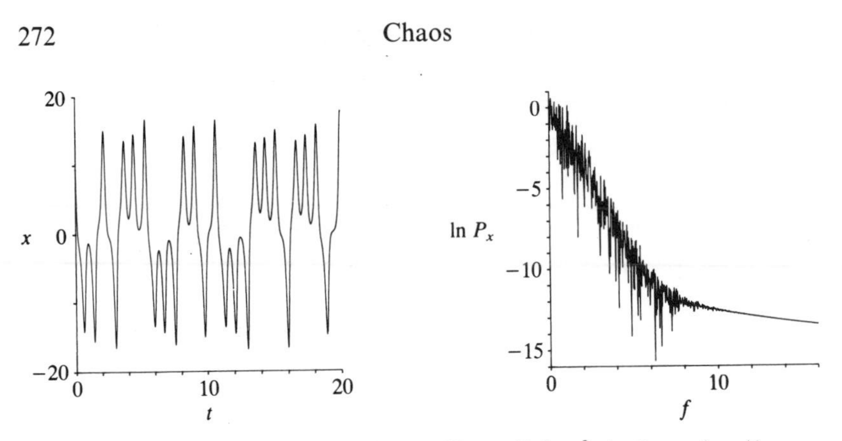

Fig. 8.17 The Lorenz system (1.1) for $r = 28$, $\sigma = 10$, $b = \frac{8}{3}$: the time series $x(t)$ and the logarithm of its power spectrum versus frequency, $\ln P_x(f)$.

in Fig. 8.17. The erratic motion about the two unstable points of equilibrium is indicated by the time series, and leads to the broad-band power spectrum with a large number of closely packed peaks which fall off rapidly as their frequency increases. This is characteristic of chaos. (The calculations are not accurate enough nor sampled often enough to resolve the high-frequency oscillations.)

These cases are of data computed by numerical solution of systems of ordinary differential equations. As is typical of such cases, there is a very low level of noise and the systems happen to be known. This is in contrast to cases with data from laboratory experiments, even careful ones. Observations of natural phenomena, such as meteorological and biological data, usually have even more noise than laboratory observations do.

So the last case is of laboratory observations. They have been made by Read *et al.* (1992) in experiments on the motion of a liquid in a differentially heated rotating annulus, a model of the large-scale motion of the Earth's atmosphere. Fig. 8.18 represents one run of the experiment. Fig. 8.18(*a*) shows a short excerpt of the time series for the heat transfer H (in watts) across the liquid as well as the temperature T (in degrees Celsius) at a point fixed in the middle of the annular container. Fig. 8.18(*b*) shows the power spectrum P_T of the temperature as a function of frequency f (in radians per second). Fig. 8.18(*c*) shows a phase portrait of the orbit constructed in a special way from the time series for T, and Fig. 8.18(*d*) shows a Poincaré section of the orbit. Read *et al.* explain the details. They interpret the series as indicating that the system is in a quasi-periodic state with two fundamental frequencies, because of the positions of the peaks in the spectrum, because Fig. 8.18(*c*) looks like the projection of an orbit winding around a torus, and because Fig. 8.18(*d*) looks like the section of a torus. Fig. 8.19 similarly represents another run. Fig. 8.19(*a*) shows some of the time series

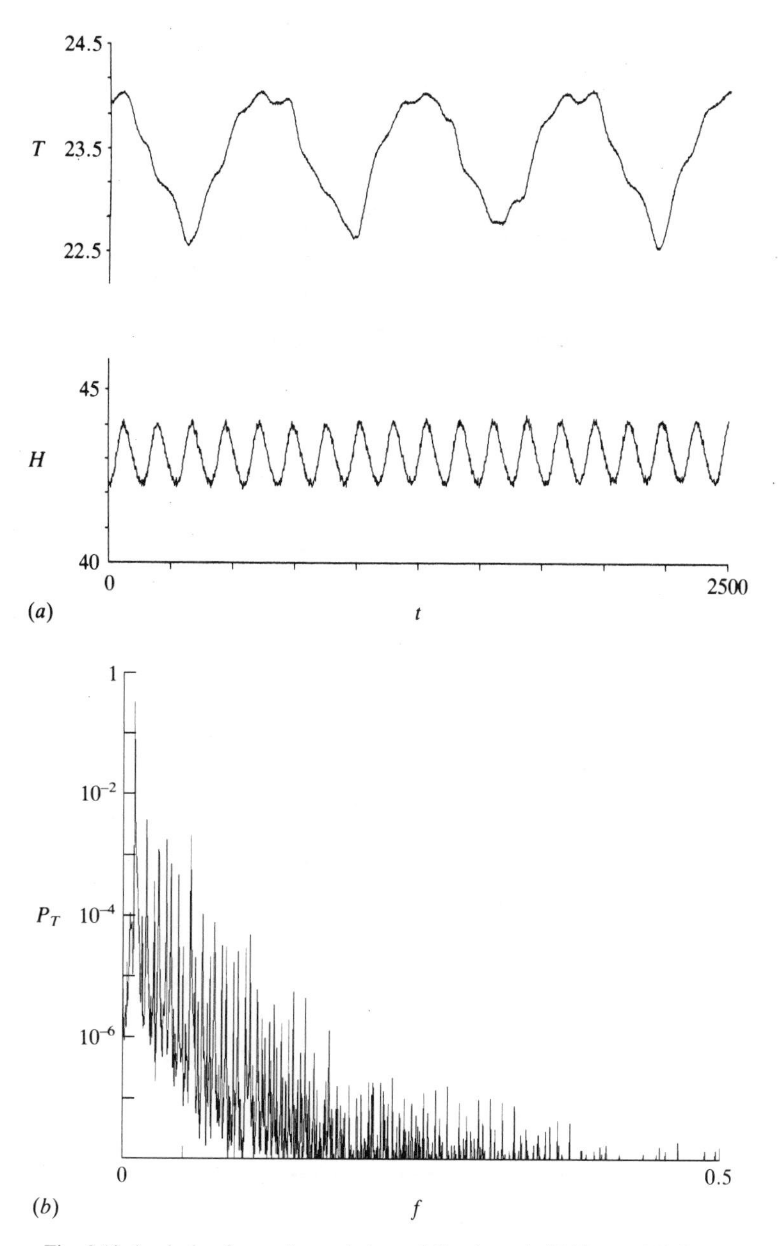

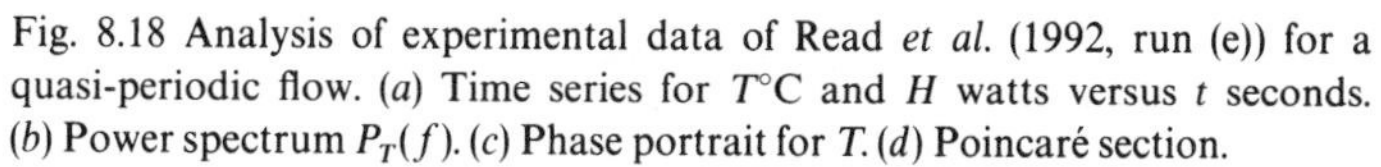
Fig. 8.18 Analysis of experimental data of Read *et al.* (1992, run (e)) for a quasi-periodic flow. (*a*) Time series for T°C and H watts versus t seconds. (*b*) Power spectrum $P_T(f)$. (*c*) Phase portrait for T. (*d*) Poincaré section.

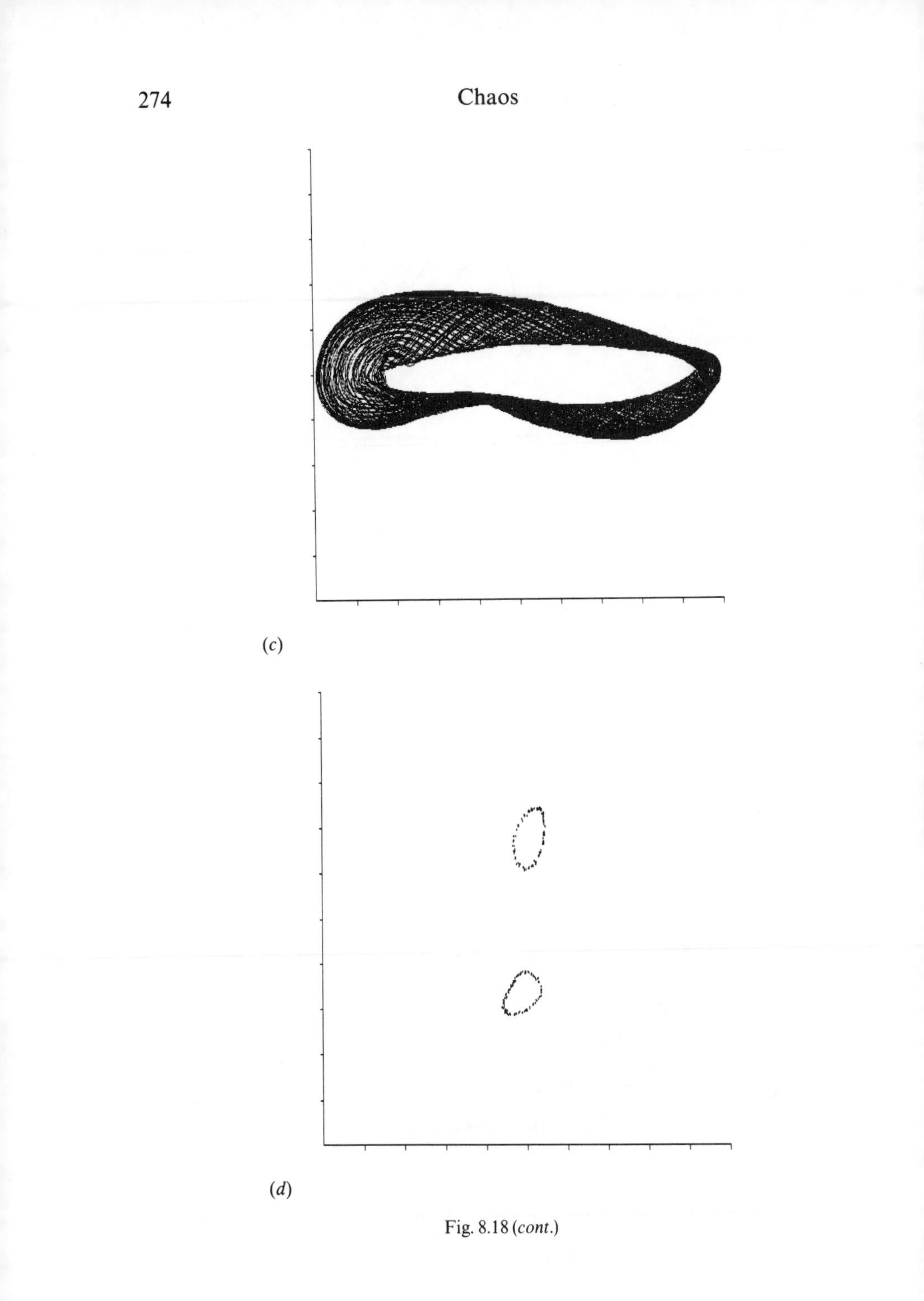

Fig. 8.18 (*cont.*)

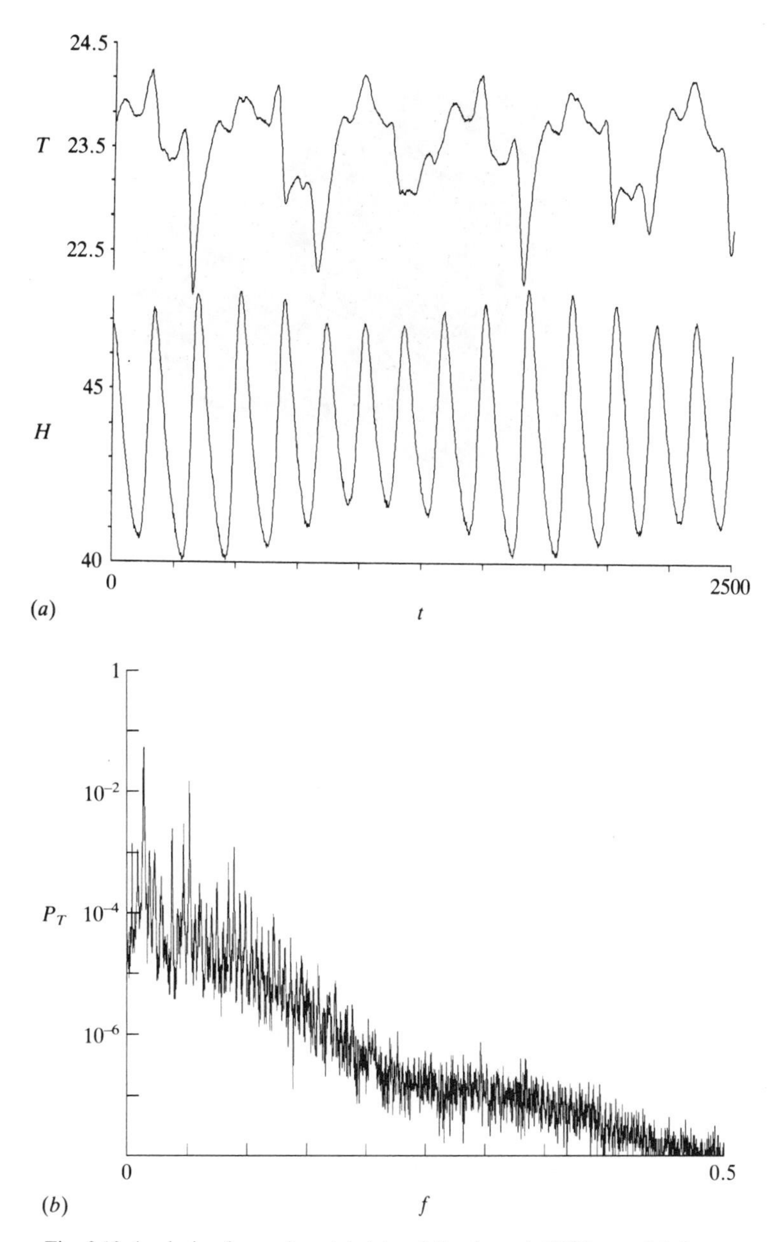

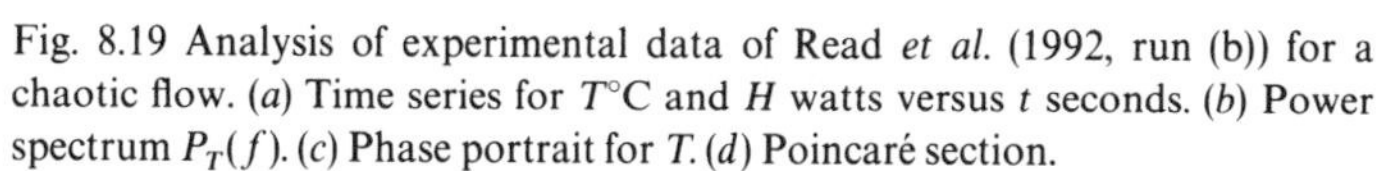
Fig. 8.19 Analysis of experimental data of Read *et al.* (1992, run (b)) for a chaotic flow. (*a*) Time series for T°C and H watts versus t seconds. (*b*) Power spectrum $P_T(f)$. (*c*) Phase portrait for T. (*d*) Poincaré section.

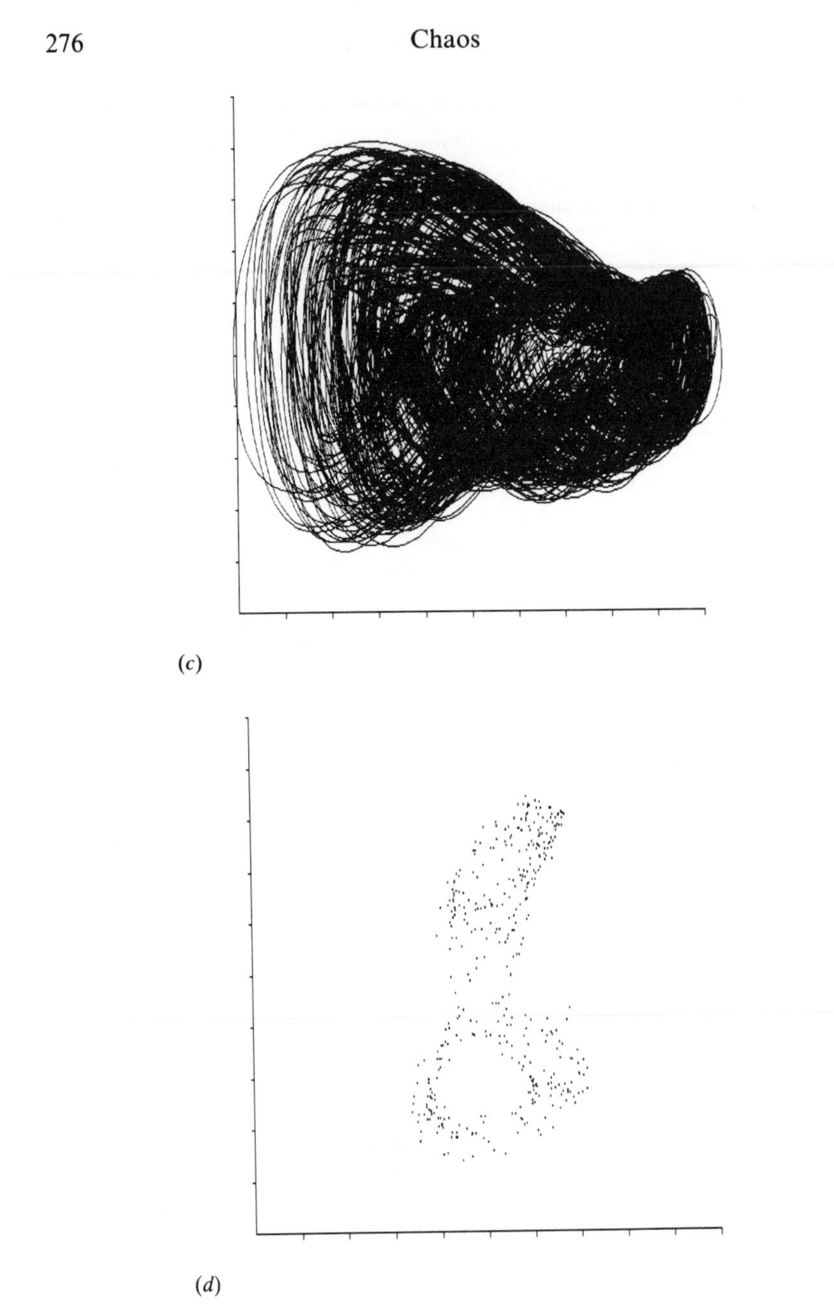

(*c*)

(*d*)

Fig. 8.19 (*cont.*)

for T and H, Fig. 8.19(b) shows P_T, Fig. 8.19(c) shows the phase portrait constructed from the series for T, and Fig. 8.19(d) shows its Poincaré section. Read *et al.* interpret the series as indicating that the system is in a chaotic state, because of the broad-band spectrum, the convoluted phase plot of the orbit, and the diffuse Poincaré section. □

It is important to remember that, even if the phenomenon observed is well-modelled by a system $d\mathbf{x}/dt = \mathbf{F}(\mathbf{a}, \mathbf{x})$ for $\mathbf{F}: \mathbb{R}^l \times \mathbb{R}^m \to \mathbb{R}^m$, in general $l, m, \mathbf{a}, \mathbf{x}$ and $\mathbf{F}$ are unknown, and we may be unable to measure $\mathbf{x}$ directly, but instead measure p data of the form $\mathbf{d}(\mathbf{a}, t) = \mathbf{g}(\mathbf{a}, \mathbf{x}(t))$ over some intervals of time for some unknown function $\mathbf{g}: \mathbb{R}^l \times \mathbb{R}^m \to \mathbb{R}^p$. This poses a formidable inverse problem: to find $\mathbf{F}$, or at least something about $\mathbf{F}$, or even about one orbit $\mathbf{x}(t)$, from knowledge of $\mathbf{d}(\mathbf{a}, t)$. It is as if we spend our lives in a cave and seek to understand the world outside merely by looking at some shadows cast upon the wall of the cave.

Further reading

§8.1 Sparrow (1982) describes the geometrical theory of the Lorenz system at length, although the existence of chaos for the system has yet to be proved rigorously.

§8.2 Guckenheimer & Holmes (1986, Chap. 2) describe solutions of Duffing's equation with negative stiffness.

§8.3 Lichtenberg & Lieberman (1983, Chap. 7) and Guckenheimer & Holmes (1986, Chap. 4) describe Mel'nikov's method with greater generality, rigour and detail.

§8.5 There is a well-developed statistical theory of the output of systems, though these are traditionally assumed to be linear. The books by Priestley (1981, 1988) are recommended to learn this theory. There is also a large literature on signal processing. King & Drazin (1992) have recently reviewed the analysis of time series.

Problems

Q8.1 *Oscillations about neutral equilibrium points of the Lorenz system.* Show that if $r = r_c, = \sigma(\sigma + b + 3)/(\sigma - b - 1) > 1$ then the eigenvalues governing the stability of the equilibrium points C, C′ of the Lorenz system are $s = \pm i\{2b\sigma(\sigma + 1)/(\sigma - b - 1)\}^{1/2}, -(\sigma + b + 1)$. Find the eigenvectors belonging to these eigenvalues.

*Q8.2 *Weakly nonlinear theory of periodic solutions of the Lorenz system.* Defining $\mathbf{x} = [x, y, z]^T$, $r_c = \sigma(\sigma + b + 3)/(\sigma - b - 1) > 1$, $X_c = Y_c = \{b(r_c - 1)\}^{1/2}$, $Z_c = r_c - 1$, $\mathbf{X}_c = [X_c, Y_c, Z_c]^T$, $\mathbf{X} = \mathbf{x} - \mathbf{X}_c$, $\omega_c =$

$\{2b\sigma(\sigma+1)/(\sigma-b-1)\}^{1/2}$, $\epsilon = |r - r_c|$, $\tau = \omega t$ for real ω, $\mathbf{J} = \begin{bmatrix} -\sigma & \sigma & 0 \\ 1 & -1 & -X_c \\ Y_c & X_c & -b \end{bmatrix}$, and $\mathbf{X}' = d\mathbf{X}/d\tau$, show that the Lorenz system may be rewritten without approximation as

$$\mathbf{JX} - \omega\mathbf{X}' = [0, ZX \mp \epsilon(X_c + X), -XY]^T,$$

according to whether $r - r_c = \pm\epsilon$.

Let $\mathbf{u} = [u, v, w]^T$ denote an eigenvector (unique up to an arbitrary multiplicative constant) of $\mathbf{J}$ and ξ of $\mathbf{J}^T$ belonging to the same eigenvalue $i\omega_c$.

Assuming that there exists a solution $\mathbf{X}$ of period 2π in τ such that

$$\omega = \omega_c + \epsilon\omega_1 + \ldots, \mathbf{X} = \epsilon^{1/2}\mathbf{X}_{1/2} + \epsilon\mathbf{X}_1 + \epsilon^{3/2}\mathbf{X}_{3/2} + \cdots \qquad \text{as } \epsilon \to 0,$$

show first that a solution for $\mathbf{X}_{1/2}$ is given by

$$\mathbf{X}_{1/2} = a_{1/2}\mathbf{u}e^{i\tau} + \bar{a}_{1/2}\bar{\mathbf{u}}e^{-i\tau}$$

for some complex constant $a_{1/2}$ to be determined later.

Show secondly that

$$\mathbf{JX}_1 - \omega_c\mathbf{X}_1' = [0, Z_{1/2}X_{1/2} \mp X_c, -X_{1/2}Y_{1/2}]^T.$$

Using the above solution for $\mathbf{X}_{1/2}$, deduce that

$$\mathbf{X}_1 = a_1\mathbf{u}e^{i\tau} + \bar{a}_1\bar{\mathbf{u}}e^{-i\tau} + \mathbf{1} + |a_{1/2}|^2\mathbf{u}_{10} + a_{1/2}^2\mathbf{u}_{12}e^{2i\tau} + \bar{a}_{1/2}^2\bar{\mathbf{u}}_{12}e^{-2i\tau}$$

for some complex constant a_1, where

$$\mathbf{Ju}_{10} = [0, w\bar{u} + \bar{w}u, -(u\bar{v} + \bar{u}v)]^T,$$

$$\mathbf{Ju}_{12} - 2i\omega_c\mathbf{u}_{12} = [0, wu, -uv]^T,$$

$$\mathbf{J1} = [0, -X_c, 0]^T.$$

Thirdly show that

$$\mathbf{JX}_{3/2} - \omega_c\mathbf{X}_{3/2}' = \omega_1\mathbf{X}_{1/2}' + [0, \mp X_{1/2} + Z_{1/2}X_1 + Z_1X_{1/2}, -(X_{1/2}Y_1 + X_1Y_{1/2})]^T.$$

Now seek only the component, say $\mathbf{q}e^{i\tau}$, of $\mathbf{X}_{3/2}$ proportional to $e^{i\tau}$ and evaluate the expression $\mathbf{Jq} - i\omega_c\mathbf{q}$. Taking the product of this expression with ξ^T, deduce the solvability condition that

$$i\omega_1\mathbf{u}^T\xi = [0, \pm u - (wl + un) - |a_{1/2}|^2(\bar{w}u_{12} + \bar{u}w_{12} + wu_{10} + uw_{10}), (um + vl) + |a_{1/2}|^2(\bar{u}v_{12} + \bar{v}u_{12} + uv_{10} + vu_{10})]\xi,$$

on using a (fairly) natural notation. Use this complex scalar condition to find both of the real quantities $|a_{1/2}|^2$ and ω_1. Deduce a condition for the periodic solutions to be subcritical, and verify numerically that they are indeed subcritical when $\sigma = 10, b = \frac{8}{3}$.

[Beware of the length of the algebraic manipulation needed to answer this question.]

Q8.3 *The Lorenz system for large r.* Transform the system

$$\frac{dx}{dt} = \sigma(y - x), \qquad \frac{dy}{dt} = rx - y - zx, \qquad \frac{dz}{dt} = xy - bz$$

into the system

$$\frac{d\xi}{d\tau} = \eta - \epsilon\sigma\xi, \qquad \frac{d\eta}{d\tau} = -\zeta\xi - \epsilon\eta, \qquad \frac{d\zeta}{\zeta\tau} = \xi\eta - \epsilon b(\zeta + \sigma)$$

without approximation, where $\epsilon = r^{-1/2}$, $\xi = \epsilon x$, $\eta = \epsilon^2\sigma y$, $\zeta = \sigma(\epsilon^2 z - 1)$, $\tau = t/\epsilon$.

Putting $\epsilon = 0$ for fixed ξ, η, ζ, show that

$$\xi^2 - 2\zeta = 2A, \qquad \eta^2 + \zeta^2 = B^2$$

for constants A, B of integration, and thence that

$$\xi'^2 = F(\xi),$$

where F is defined by $F(\xi) = C + A\xi^2 - \frac{1}{4}\xi^4$ and C is another constant. Deduce that there exists a solution $\xi(\tau)$ of period $2\int_{\xi_1}^{\xi_2}\{F(\xi)\}^{-1/2}\,d\xi$, where ξ_1, ξ_2 are simple zeros of F such that $F(\xi) > 0$ for $\xi_1 < \xi < \xi_2$.

[In fact $\xi(\tau)$ can be found in elementary terms of elliptic functions.]

Q8.4 *The Rikitake two-disc dynamo.* It is given that the self-generation of the Earth's magnetic field is modelled by the equations,

$$\frac{dx_1}{dt} = -vx_1 + yx_2, \qquad \frac{dx_2}{dt} = -vx_2 + (y - a)x_1, \qquad \frac{dy}{dt} = 1 - x_1x_2,$$

where x_1, x_2 represent the currents in two large eddies in the Earth's core, y the angular velocity of one eddy, a the positive constant difference of the angular velocities of the two eddies, and v the ratio of the mechanical time-scale to the magnetic diffusion time-scale of the dynamo formed by the two rotating eddies and their electromagnetic interaction.

Show that the two equilibrium points are

$$x_1 = \pm k, \qquad x_2 = \pm k^{-1}, \qquad y = vk^2,$$

where $v(k^2 - k^{-2}) = a$, and analyse their stability.

Show that if $\mathbf{x} = (x_1, x_2, y)$ and the system is expressed as $d\mathbf{x}/dt = \mathbf{F}$ then $\operatorname{div}\mathbf{F} = -2v$. Deduce that the volume of an attractor is zero.

Show that if $v = 0, a = 0$, then there exists a solution with $x_1 = A\cosh u$, $x_2 = A\sinh u$, $\frac{1}{2}y^2 = u - \frac{1}{4}A^2\cosh 2u + B$ for constants A, B of integration.

Use a computer to explore some chaotic solutions of the sytem for a, $v > 0$.

[Rikitake (1958), Allan (1962), Cook & Roberts (1970). The equilibrium points represent the coincidence of the geomagnetic north pole with the geographic north and south poles, and chaotic solutions the wandering of the geomagnetic pole and its reversals.]

*Q8.5 *The non-chaotic break-up of a homoclinic orbit.* Show that the equilibrium point of the equation

$$\frac{\mathrm{d}^2x}{\mathrm{d}t^2} + k\frac{\mathrm{d}x}{\mathrm{d}t} - x + x^2 = 0$$

at the origin of the $(x, \mathrm{d}x/\mathrm{d}t)$-plane has a homoclinic orbit when $k = 0$. Sketch the phase portraits for $-1 < k < 0, k = 0, 0 < k < 1$, labelling clearly the homoclinic orbit, the stable manifolds and the unstable manifolds of the origin.

*Q8.6 *The stable manifold of a map of the unit square.* The stable manifold $\mathbf{M}^s(\mathbf{X})$ of a fixed point $\mathbf{X}$ of a map $\mathbf{F}$ is defined as the set of points $\mathbf{x}$ such that $\lim_{n\to\infty} \mathbf{F}^n(\mathbf{x}) = \mathbf{X}$.

Show that if the piecewise linear map $\mathbf{F}$: $\mathbb{T}^2 \to \mathbb{T}^2$ is defined by

$$\mathbf{F}(\mathbf{x}) \equiv [y, x + y]^{\mathrm{T}} \text{ modulo } 1,$$

where $\mathbf{x} = [x, y]^{\mathrm{T}}$, then $\mathbf{0}$ is a fixed point with eigenvalues $\frac{1}{2}(1 \pm \sqrt{5})$. Find the corresponding eigenvectors. Deduce that $\mathbf{M}^s(\mathbf{0})$ is dense in $\mathbb{T}^2$.

*Q8.7 *A weakly forced simple pendulum.* It is given that the equation of motion of a forced simple pendulum is

$$\frac{\mathrm{d}^2\theta}{\mathrm{d}t^2} + \sin\theta = \epsilon(\alpha + \gamma\cos\omega t),$$

for $\alpha, \gamma > 0$.

Show that if $\epsilon = 0$ then there is a pair of heteroclinic orbits connecting saddle points at $(\pm\pi, 0)$ in the (θ, ϕ)-plane, where $\phi = \mathrm{d}\theta/\mathrm{d}t$. Deduce that one of these orbits is given by

$$\theta_0(t) = 2\arctan(\sinh t), \qquad \phi_0(t) = 2\,\mathrm{sech}\, t.$$

Show that the Mel'nikov function for the perturbation problem of the intersection near $(0, 2)$ of the unstable manifold from $(-\pi, 0)$ and the stable manifold to $(\pi, 0)$ can be expressed as

$$M(t_0) = \int_{-\infty}^{\infty} \phi_0(t - t_0)(\alpha + \gamma\cos\omega t)\,\mathrm{d}t,$$

and deduce that

$$M(t_0) = 2\pi\{\alpha + \gamma\,\mathrm{sech}(\tfrac{1}{2}\pi\omega)\cos(\omega t_0)\}.$$

Hence show that there is chaos for small ϵ if $\gamma > \alpha\cosh(\frac{1}{2}\pi\omega)$.

*Q8.8 *A Mel'nikov function of a Hamiltonian system which may be identically zero.* Show that if there is a perturbed Hamiltonian of the form $H(y, x, t) = H_0(y, x) + \epsilon h(y, x, t)$ then the system's Mel'nikov function may be expressed as

$$M(t_0) = \int_{-\infty}^{\infty} \{H_0(\mathbf{q}_0(t - t_0)), h(\mathbf{q}_0(t - t_0), t)\} \, dt,$$

on defining the *Poisson brackets* such that

$$\{H_0, h\} = \frac{\partial H_0}{\partial x}\frac{\partial h}{\partial y} - \frac{\partial H_0}{\partial y}\frac{\partial h}{\partial x}.$$

Show that if $H_0(y, x) = \frac{1}{2}(x^2 + y^2) - \frac{1}{3}x^3$, $h(y, x, t) = \frac{1}{2}x^2 \cos \omega t$ then the resultant system is

$$\frac{dx}{dt} = y, \qquad \frac{dy}{dt} = -x + x^2 - \epsilon x \cos \omega t.$$

Show that if $\epsilon = 0$ then there is a centre at $(0, 0)$ and a saddle point at $(1, 0)$ in the (x, y)-plane, with a homoclinic orbit given by

$$x_0(t) = \tfrac{1}{2}\{3 \tanh^2(\tfrac{1}{2}t) - 1\}, \qquad y_0(t) = \tfrac{3}{2}\tanh(\tfrac{1}{2}t)\,\mathrm{sech}^2(\tfrac{1}{2}t).$$

Deduce that the Mel'nikov function is

$$M(t_0) = -\tfrac{3}{4}\sin(\omega t_0) \int_{-\infty}^{\infty} \sin(\omega t) \tanh(\tfrac{1}{2}t)\,\mathrm{sech}^2(\tfrac{1}{2}t)\{3 \tanh^2(\tfrac{1}{2}t) - 1\} \, dt.$$

Define

$$I_1 = \int_{-\infty}^{\infty} \sin 2\omega s \tanh s \,\mathrm{sech}^2 s \, ds, \qquad I_2 = \int_{-\infty}^{\infty} \sin 2\omega s \tanh^3 s \,\mathrm{sech}^2 s \, ds,$$

and integrate I_1 twice by parts to deduce that $3I_2 = (2 - \omega^2)I_1$ and thence that

$$M(t_0) = \tfrac{3}{2}(1 - \omega^2)\sin(\omega t_0) I_1.$$

Now deduce that if $\omega = 1$ then $M(t_0) = 0$ for all t_0, and comment on whether chaos can be shown to occur for small ϵ in this case.

*Q8.9 *A Mel'nikov function of distance along a basic heteroclinic orbit.* Suppose that

$$\frac{dx}{dt} = \sin x \cosh y, \qquad \frac{dy}{dt} = -\cos x \sinh y + \epsilon \sin \omega t.$$

Show that if $\epsilon = 0$ then there are saddle points at $(n\pi, 0)$ in the (x, y)-plane for $n = 0, \pm 1, \pm 2, \ldots$, and that $(0, 0)$ and $(\pi, 0)$ are connected by the heteroclinic orbit $(x_0(t), 0)$, where

$$\sin\{x_0(t)\} = \mathrm{sech}[t + \ln\{\tan(\tfrac{1}{2}x_m)\}] \qquad \text{for all } t.$$

Consider, when ϵ is small, the separation near the point $(x_m, 0)$ of the unstable manifold from $(0,0)$ and the stable manifold to $(\pi, 0)$ of the Poincaré map P_ϵ^0, and show that the Mel'nikov function may be expressed as

$$M(x_m) = \int_{-\infty}^{\infty} \sin\{x_0(t)\} \sin \omega t \, dt$$

$$= -\pi \sin[\omega \ln\{\tan(\tfrac{1}{2}x_m)\}] \operatorname{sech}(\tfrac{1}{2}\pi\omega).$$

[Cox *et al.* (1990).]

Q8.10 *The non-chaotic phase of intermittency.* Suppose that $x_{n+1} = F(a, x_n)$ for $n = 0, 1, \ldots$, where $F(a, x) = x - x^2 - a$.

(a) Show that if $a < 0$ then there are two fixed points $\pm(-a)^{1/2}$, that $-(-a)^{1/2}$ is always unstable, but that $(-a)^{1/2}$ is stable if $-1 < a$ and its domain of attraction is $\mathrm{D}((-a)^{1/2}) = (-(-a)^{1/2}, 1 + (-a)^{1/2})$.

(b) Show that if $a = 0$ then there is a unique fixed point $X = 0$, and that it is (weakly) unstable.

(c) Show that if $a > 0$ then $x_n \to -\infty$ as $n \to \infty$ for all x_0. Show that if $x_0 = 0$ then

$$x_n = -na - \tfrac{1}{6}(n-1)n(2n-1)a^2 - p_3(n)n^5a^3 - p_4(n)n^7a^4 - \cdots$$

as $a \downarrow 0$ for fixed n, where p_j are certain functions such that $p_j(n) = O(1)$ as $n \to \infty$ for $j = 3, 4, \ldots$. Deduce that $x_n = O(na)$ as $a \downarrow 0$ for fixed n^2a, and thence that $x_n = o(1)$ until n is of order of magnitude of $a^{-1/2}$.

Q8.11 *Autocorrelations.* Show that if $d(t) = c + a\cos ft + b\sin ft$ then $\overline{d(s+t)d(s)} = c^2 + \tfrac{1}{2}(a^2 + b^2)\cos ft$, and that if $d(t) = c + a_1\cos f_1 t + a_2 \cos f_2 t$ for $f_2 \neq \pm f_1$ then $\overline{d(s+t)d(s)} = c^2 + \tfrac{1}{2}a_1^2 \cos f_1 t + \tfrac{1}{2}a_2^2 \cos f_2 t$.

*Appendix: Some partial-differential problems

We have seen that bifurcations and chaos for a system of difference or ordinary differential equations often occur in lower dimensions than the dimension of the system. Similarly, although a partial differential system has an infinite dimension, its bifurcations and chaos often occur in a manifold of low finite dimension. Indeed, turning points, transcritical bifurcations, pitchfork bifurcations, Hopf bifurcations, limit cycles etc. arise for partial differential systems. This can be demonstrated in many cases by use of one of a few perturbation techniques, for example the Liapounov–Schmidt reduction or centre manifold theory. The essence of these techniques is to consider perturbations of marginal stability in which the values of both the parameters and the state variables are close to those corresponding to marginal stability, and in which the effects of these two kinds of perturbations are balanced asymptotically. At the margin of stability, the number of eigenvalues whose real parts are zero is usually small, so that their eigenfunctions span a low-dimensional space; all components of an initial disturbance not in this space being strongly damped. The centre manifold of a weakly nonlinear system is tangential to this space as the margin of stability is approached.

The fact that phenomena of interest occur in a low-dimensional manifold makes the dynamics much easier to understand, but it seems that some phenomena, for example turbulent motion of a fluid, cannot be represented in a low-dimensional manifold.

In this brief introduction to nonlinear systems, we do not treat partial differential systems in detail. However, a few problems which illustrate instability and bifurcations are given below.

QA.1 *Derivation of an infinite ordinary-differential system from a partial-differential problem.* Consider the real equation

$$\frac{\partial u}{\partial t} = \frac{\partial^2 u}{\partial x^2} - \left(\frac{\partial u}{\partial x}\right)^2$$

with periodic boundary conditions,

$$u(x + 2\pi, t) = u(x, t) \qquad \text{for all } x, t.$$

Show that the problem is invariant if $x \to x + a$ for a real constant a. Taking the complex Fourier series

$$u(x, t) = \sum_{n=-\infty}^{\infty} u_n(t)\mathrm{e}^{\mathrm{i}nx},$$

show that $u_{-n} = \bar{u}_n$ and

$$\frac{\mathrm{d}u_n}{\mathrm{d}t} = -n^2 u + \sum_{m=-\infty}^{\infty} m(n-m)u_m u_{n-m}.$$

Deduce that the last equation is invariant if $u_n \to \mathrm{e}^{\mathrm{i}na}u_n$.

QA.2 *Another nonlinear diffusion equation and a pitchfork bifurcation.* (a) Consider the system

$$\frac{\partial u}{\partial t} - f(u) = \kappa\frac{\partial^2 u}{\partial z^2}, \qquad u = 0 \text{ at } z = 0, \pi,$$

for $\kappa > 0$ and some well-behaved function f. Suppose that it has a steady solution $u = U(z)$. Expressing $u = U + u'$, and linearizing for a small perturbation u', deduce that

$$\frac{\partial u'}{\partial t} - f'(U)u' = \kappa\frac{\partial^2 u'}{\partial z^2}, \qquad u' = 0 \text{ at } z = 0, \pi.$$

Taking normal modes of the form $u'(z, t) = \hat{u}(z)\mathrm{e}^{st}$, derive the eigenvalue problem

$$\frac{\mathrm{d}^2\hat{u}}{\mathrm{d}z^2} + \kappa^{-1}\{f'(U) - s\}\hat{u} = 0, \qquad \hat{u} = 0 \text{ at } z = 0, \pi.$$

(b) Supposing further that $f(0) = 0$, $f'(0) > 0$, and $U(z) = 0$ for all z, find all the eigenvalues s, and deduce that the null solution is stable if $\kappa < \kappa_c = f'(0)$ and the marginally stable mode is $\hat{u} = \sin z$.

(c) Show that if $f(u) = \sin u$ then $\kappa_c = 1$.

To perturb the marginally stable solution for this f, first define $\epsilon^2 = \kappa_c - \kappa$ for $\epsilon > 0$ and $\tau = \epsilon^2 t$, so that

$$\epsilon^2\frac{\partial u}{\partial \tau} - \sin u = (\kappa_c - \epsilon^2)\frac{\partial^2 u}{\partial z^2}, \qquad u = 0 \text{ at } z = 0, \pi.$$

Then assume that

$$u(z, t, \kappa) = \epsilon A(\tau)\sin z + \epsilon^2 u_2(z, \tau) + \epsilon^3 u_3(z, \tau) + \cdots \qquad \text{as } \epsilon \to 0,$$

and equate coefficients of $\epsilon, \epsilon^2, \epsilon^3$ in the above system to deduce that

$$\frac{\mathrm{d}A}{\mathrm{d}\tau} = A - \tfrac{1}{8}A^3.$$

[Matkowsky (1970).]

QA.3 *The Swift–Hohenberg model.* Consider the real equation

$$\frac{\partial u}{\partial t} = \left\{R - \left(\frac{\partial^2}{\partial x^2} + 1\right)^2\right\}u - u^3$$

for $-\infty < x < \infty$ and $t \geqslant 0$, where R is a positive parameter. Find the stability of the null solution by the method of normal modes and Fourier components, i.e. by linearizing the equation, assuming that $u \propto \mathrm{e}^{st+ikx}$ for real wavenumber k, and finding s as a function of R and k. Plot the values of k versus R when the condition for marginal stability is satisfied. What is the critical value R_c of R above which there is instability for at least one value of k?

Find other solutions which are independent of x and t, and identify a pitchfork bifurcation of the null solution.

Assuming that u has period 2π in x, and expressing the solution of the equation as the complex Fourier series

$$u(x,t) = \sum_{n=-\infty}^{\infty} u_n(t)\mathrm{e}^{inx},$$

show that $u_{-n} = \bar{u}_n$. Find an infinite system of ordinary differential equations for $\{u_n\}$. Verify the stability of the null solution and find when the other solutions are stable.

[Swift & Hohenberg (1977) proposed this partial differential equation as a model of nonlinear Rayleigh–Bénard convection, i.e. convection of a horizontal layer of fluid heated from below, the same flow which led Lorenz (1963) to his very different system. The null solution $u = 0$ represents the state of rest of the fluid, and the other equilibrium points represent steady convection cells.]

QA.4 *A nonlinear Schrödinger equation.* Give that

$$\mathrm{i}\frac{\partial u}{\partial t} + \frac{\partial^2 u}{\partial x^2} + |u|^2 u = 0,$$

show that there is a solution $u = U$ where

$$U(t) = a\exp\{\mathrm{i}a^2(t - t_0)\}$$

for real amplitude a and phase t_0.

Show that the linearized equation for a small perturbation u' is

$$\mathrm{i}\frac{\partial u'}{\partial t} + \frac{\partial^2 u'}{\partial x^2} + 2|U|^2 u' + U^2 \overline{u'} = 0,$$

where $u = U + u'$ and $\overline{u'}$ is the complex conjugate of u'. Taking normal modes of the form

$$u'(x,t) = U(t)f(t)\cos\{k(x - x_0)\}$$

for wavenumber $k > 0$ and real phase x_0, show that

$$f' = \mathrm{i}\{(a^2 - k^2)f + a^2\bar{f}\}.$$

By expressing this complex equation as a real pair of equations for the real and imaginary parts of f, or otherwise, deduce that $f \propto \mathrm{e}^{st}$ where $s^2 = k^2(2a^2 - k^2)$, and hence that the periodic solution U is unstable if $k < \sqrt{2}a$.

[This equation governs the behaviour of a weakly nonlinear wavepacket, and has many applications, e.g., in the theories of water waves and of plasmas. Cf. Drazin & Johnson (1989).]

QA.5 *The Ginzburg–Landau equation.* Given that

$$\frac{\partial u}{\partial t} = u - (1 + \mathrm{i}R)|u|^2 u + (1 + \mathrm{i}b)\frac{\partial^2 u}{\partial x^2}$$

and

$$\frac{\partial u}{\partial x} = 0 \qquad \text{at } x = 0, l$$

for real parameters R, b, and l, show that a solution is given by $u = U$, where

$$U(t) = \exp\{-\mathrm{i}R(t - t_0)\}$$

for an arbitrary phase t_0.

Show that the linearized equation governing the stability of the periodic solution U is

$$\frac{\partial u'}{\partial t} = u' - (1 + \mathrm{i}R)(2u' + U^2\bar{u}') + (1 + \mathrm{i}b)\frac{\partial^2 u'}{\partial x^2}.$$

Hence show that there are normal modes of the form

$$u'(x,t) = U(t)f(t)\cos(n\pi x/l)$$

for $n = 0, 1, 2, \ldots$. Find f and deduce that if $b < 0$ then the solution U is linearly stable provided that $R < R_c$, where

$$R_c = -\{1 + (1 + b^2)\pi^2/2l^2\}/b.$$

[Kuramoto & Koga (1982). The steady form of the equation arose originally in the theory of superconductivity (Ginzburg & Landau 1950), but has since been applied to the weakly nonlinear modulation of unstable waves of many kinds.]

QA.6 *The chemical basis of morphogenesis.* It is given that the concentrations u, v of two interacting species of molecules satisfy nonlinear diffusion equations and boundary equations of the dimensionless form

$$\frac{\partial u}{\partial t} = f(u, v) + \nabla^2 u, \qquad \frac{\partial u}{\partial t} = g(u, v) + d\nabla^2 v \qquad \text{in D},$$

$$\frac{\partial u}{\partial n} = \frac{\partial v}{\partial n} = 0 \qquad \text{on } \partial\text{D},$$

where D is the domain occupied by the molecules, d is the (positive) ratio of their diffusion coefficients, f, g are given well-behaved functions representing the interaction, and the Laplacian operator is $\nabla^2 = \partial^2/\partial x^2 + \partial^2/\partial y^2 + \partial^2/\partial z^2$.

Suppose that U, V are constants such that $f(U, V) = g(U, V) = 0$, and let u', v' be a small perturbation of this spatially uniform equilibrium solution. Then find the linearized problem satisfied by u', v'.

Assuming that the spatial eigenvalue problem

$$\nabla^2 p + k^2 p = 0 \qquad \text{in D}, \qquad \partial p/\partial n = 0 \qquad \text{on } \partial\text{D},$$

has known solution k, $p(\mathbf{x})$, show that the normal modes of the linearized problem have the form $u'(\mathbf{x}, t) = \hat{u}e^{st}p(\mathbf{x})$, $v'(\mathbf{x}, t) = \hat{v}e^{st}p(\mathbf{x})$, where

$$\begin{vmatrix} s + k^2 - f_u(U, V) & -f_v(U, V) \\ -g_u(U, V) & s + dk^2 - g_v(U, V) \end{vmatrix} = 0.$$

Find U, V when $f(u, v) = \gamma(a - u + u^2 v)$, $g(u, v) = \gamma(b - u^2 v)$ for constants γ, a, $b > 0$. Find the eigenvalues k and eigenvectors p for the one-dimensional problem with $\text{D} = \{x: 0 < x < \pi l\}$. Discuss the stability of the solutions in this case.

[Turing (1952) initiated such modelling, showing that diffusion can cause instability; Murray (1989, Chap. 14) gives a modern review.]

QA.7 *Spontaneous combustion of a slab.* It is given that the temperature ϕ of a one-dimensional slab of a uniform solid is governed by a dimensionless problem of the form

$$\frac{\partial \phi}{\partial t} = \frac{\partial^2 \phi}{\partial x^2} + \delta e^{\phi}, \qquad \phi = 0 \text{ at } x = \pm 1,$$

where the Frank-Kamenetskii parameter $\delta > 0$.

Seeking steady solutions of the form $\phi = F(x)$, verify that

$$F(x) = 2\ln(\cosh c \operatorname{sech} cx),$$

where c is such that $\delta = 2c^2 \operatorname{sech}^2 c$, $\cosh c = \exp(\tfrac{1}{2}\phi_m)$, $\phi_m = F(0)$. Sketch the bifurcation diagram in the (δ, ϕ_m)-plane, identifying a turning point at $(2(c_0^2 - 1), 2\ln(\cosh c_0))$, where $c_0 \approx 1.2$ is defined as the positive root of $c \tanh c = 1$.

Show that the linearized problem for small perturbations of this steady solution is

$$\frac{\partial\phi'}{\partial t} = \frac{\partial^2\phi'}{\partial x^2} + \delta e^F\phi', \qquad \phi' = 0 \text{ at } x = \pm 1.$$

Taking normal modes with $\phi'(x,t) = \hat{\phi}(x)e^{st}$, obtain the Sturm–Liouville problem,

$$\frac{d^2\hat{\phi}}{dx^2} + (2c^2\,\mathrm{sech}^2\,cx - s)\hat{\phi} = 0, \qquad \hat{\phi} = 0 \text{ at } x = \pm 1.$$

Deduce that s is real, and so s increases through zero at any margin of stability.

Defining $y(x) = \tanh cx$, verify that if $s = 0$ then the eigenfunction is

$$\hat{\phi}(x) = \tfrac{1}{2}y\ln\{(1+y)/(1-y)\} - 1$$

and $\coth c \tanh(\coth c) = 1$, i.e. $c = c_0$.

Show that $\hat{\phi}$ is an associated Legendre function of y for general values of s.

[Buckmaster & Ludford (1982, Chap. 12) describe the model of the slab and its steady solutions.]

QA.8 *The Proudman–Johnson equation.* Consider the system,

$$\frac{\partial^3 f}{\partial t\partial y^2} = \frac{1}{R}\frac{\partial^4 f}{\partial y^4} + f\frac{\partial^3 f}{\partial y^3} - \frac{\partial f}{\partial y}\frac{\partial^2 f}{\partial y^2}, \qquad f = \mp 1, \frac{\partial f}{\partial y} = 1 \quad \text{at } y = \pm 1,$$

for positive parameter R.

Supposing that $f = F$ is a steady solution of the system for a given function $F(y, R)$, taking small unsteady perturbations g so that $f = F + g$, and linearizing, show that

$$\frac{\partial^3 g}{\partial t\partial y^2} = \frac{1}{R}\frac{\partial^4 g}{\partial y^4} + F\frac{\partial^3 g}{\partial y^3} - F'\frac{\partial^2 g}{\partial y^2} - F''\frac{\partial g}{\partial y} + F'''g, \qquad g = \frac{\partial g}{\partial y} = 0 \text{ at } y = \pm 1,$$

where a prime denotes differentiation with respect to y. Then, taking normal modes with $g(y,t,R) = e^{st}G(y,R)$, derive the eigenvalue problem,

$$G^{iv} + R(FG''' - F'G'' - F''G' + F'''G) = RsG'', \qquad G = G' = 0 \text{ at } y = \pm 1.$$

Deduce that the steady solution F is unstable if there exists an eigenvalue such that $\mathrm{Re}\, s > 0$.

Show that a solution is given by $F(y) = y$ for all R. *Deduce that

$$G = b_1 y + b_2 + \exp(\tfrac{1}{4}Ry^2)\{b_3 U(s - \tfrac{7}{2}, \sqrt{R}y) + b_4 V(s - \tfrac{7}{2}, \sqrt{R}y)\}$$

for some constants b_1, b_2, b_3, b_4, where U, V are the standard parabolic cylinder functions. Hence or otherwise show that there is stability for $R < R_1$, where $V(-\tfrac{5}{2}, \sqrt{R_1}) = 0$, and so $R_1 \approx 4.51$.

[The system governs the unsteady non-parallel flow, at Reynolds number R, of a viscous incompressible fluid in a channel with suction at moving porous parallel walls (cf. Watson *et al.* 1990).]

QA.9 *Blow-up.* Consider the nonlinear diffusion equation

$$\frac{\partial u}{\partial t} = \frac{\partial^2 u}{\partial x^2} + u^2,$$

where $\partial u/\partial x = 0$ at $x = 0, 1$ and $u(x, 0) = u_0(x)$ for $0 \leqslant x \leqslant 1$ for a given smooth real function u_0.

Taking Fourier expansions

$$u = \tfrac{1}{2}a_0 + \sum_{n=1}^{\infty} a_n \cos n\pi x = \tfrac{1}{2} \sum_{n=-\infty}^{\infty} a_n e^{in\pi x} \qquad \text{for real } a_{-n} = a_n,$$

$$\partial u/\partial x = \sum_{n=1}^{\infty} b_n \sin n\pi x, \qquad \partial^2 u/\partial x^2 = \tfrac{1}{2}c_0 + \sum_{n=1}^{\infty} c_n \cos n\pi x$$

at time $t > 0$, show that $b_n = -n\pi a_n$, $c_n = -n^2\pi^2 a_n$. Deduce that

$$\frac{da_n}{dt} = -n^2\pi^2 a_n + \tfrac{1}{2} \sum_{m=-\infty}^{\infty} a_m a_{n-m} \qquad \text{for } t > 0.$$

Now show that

$$\frac{da_0}{dt} \geqslant \tfrac{1}{2}a_0^2.$$

Deduce that if $a_0(0) > 0$ then there exists a positive constant k such that

$$a_0(t) \geqslant 2/(k - t) \qquad \text{for all } t > 0$$

for which the solution exists. Hence show that if $\int_0^1 u_0(x)\,dx > 0$ then the solution $u(x, t)$ becomes singular in a finite time.

[This kind of singularity is called *blow-up*, and is described more generally for nonlinear partial differential equations by, for example, Palais (1988).]

Answers and hints to selected problems

By day and night he measured and calculated; covered enormous quantities of paper with figures, letters, computations, algebraic symbols; his face, which was the face of an apparently sound and vigorous man, wore the morose and visionary stare of a monomaniac; while his conversation, with consistent and fearful monotony, dealt with the proportional number π....

Thomas Mann (*The Magic Mountain*, Chap. VII, 'The great god dumps')

The answer, if one is given, to a problem is denoted by the prefix A: for example, the answer to Q1.1 is A1.1. In some cases a hint to the solution is given. Of course, when a reference is given in the text it is more useful than the brief answer here.

Chapter 1

A1.1 $X = 0 \; \forall \; a$ is stable for $a < c$. $X = (a - c)/ab \; \forall \; a \neq 0$ is stable if $a > c$; transcritical bifurcation at $(c, 0)$.

A1.2 $x(0) \uparrow X$ as $t \to \infty$, where $X = b[1 - \{1 - 4ac/b^2\}^{1/2}]/2c$ is least positive zero of right-hand side of equation.

A1.3 $X = 0 \; \forall \; a$ is stable if $a < 0$. $X = b + (b^2 + a)^{1/2} \; \forall \; a > -b^2$ is stable. $X = b - (b^2 + a)^{1/2} \; \forall \; a \geqslant -b^2$ is stable $\forall \; a > 0$. Hysteresis in oscillations.

A1.4 $\int_{t_0}^{t} \exp(\frac{1}{2}\epsilon s^2)\,\mathrm{d}s \sim a^{-1}\sqrt{\epsilon}\exp(\frac{1}{2}a^2/\epsilon)$ as $\epsilon \downarrow 0$ for fixed $a > 0$, $T_0 < 0$.

A1.5 $F(a, x) = 2(a^2 - x^2) - (a^2 + x^2)^2$ is symmetric in $\pm a$, & in $\pm x$. $F(a, x) = 2r^2 \cos 2\theta - r^4$ with $a = r\cos\theta$, $x = r\sin\theta$. $F(a, x) = 0$ has turning points at $(\pm\sqrt{2}, 0)$, transcritical bifurcation at $(0, 0)$.

A1.6 $F(a, x) = x^3 + a^3 - 3ax$ is antisymmetric in a, x. $F(a, x) = 0$ has asymptote $x + a = -1$, turning point at $(2^{2/3}, 2^{1/3})$, transcritical bifurcation at $(0, 0)$.

A1.8 $\mathrm{d}^2\theta/\mathrm{d}t^2 = \omega^2 f(k, \theta)$, where $k = a\omega^2/g > 0$, $0 \leqslant \theta < 2\pi$, $f(k, \theta) = \sin\theta(\cos\theta - 1/k)$. $f(k, \Theta) = 0$ implies $\Theta = 0, \pi \; \forall \; k$ or $\Theta = \pm\arccos(1/k)$

$\forall\ k > 1$. Test sign of $f_\theta(k, \Theta)$ for stability. $\Theta = 0$ stable $\forall\ k < 1$; $\Theta = \pi$ always unstable; $\Theta = \pm\arccos(1/k)$ stable $\forall\ k > 1$.

A1.9 $\mathrm{d}^2\theta/\mathrm{d}t^2 = f(k, \theta)$, where $f(k, \theta) = (am)^{-1}\{k(2a\cos\frac{1}{2}\theta - l)\sin\frac{1}{2}\theta - mg\sin\theta\}$, $0 \leqslant \theta < 2\pi$, $a, g, k, l, m > 0$. $f(k, \Theta) = 0$ implies $\Theta = 0\ \forall\ k$ or $\Theta = \pm\alpha\ \forall\ k > mg/(a - \frac{1}{2}l)$, where $\alpha = 2\arccos\{kl/2(ak - mg)\}$. $\Theta = 0$ stable $\forall\ k < mg/(a - \frac{1}{2}l)$, $\Theta = \pm\alpha$ stable $\forall\ k > mg/(a - \frac{1}{2}l)$.

A1.10 $\Theta = \alpha$ or $\pi - \alpha\ \forall\ k < 2/\pi$, where $0 \leqslant \alpha = \arcsin(\frac{1}{2}\pi k) \leqslant \frac{1}{2}\pi$; former equilibrium is always stable, latter unstable, turning point at $(2/\pi, \frac{1}{2}\pi)$.

A1.12 $\mathrm{d}r/\mathrm{d}t = 1 - r^2$, $\mathrm{d}\theta/\mathrm{d}t = 1$.

A1.13 $\mathrm{d}r/\mathrm{d}t = r(1 - r^2)^2$, $\mathrm{d}\theta/\mathrm{d}t = 1$.

A1.14 $\mathrm{d}r/\mathrm{d}t = rf(r)$, $\mathrm{d}\theta/\mathrm{d}t = 1$, $f(r_j) = 0$, r_j stable only for even j.

A1.15 $\mathrm{d}r/\mathrm{d}t = r(1 - f/a)$, $\mathrm{d}\theta/\mathrm{d}t = 1$. $\mathrm{d}r/\mathrm{d}t = r\{1 - g(r)/a\}$, $g(r_0) = a$. $(0,0)$ is stable for $0 < a < \frac{3}{2}$, unstable for $a > \frac{3}{2}$, with Hopf bifurcation at $(\frac{3}{2}, 0)$. For $\frac{3}{2} < a < 2$ $\exists$ a stable $(r_0 < 1)$ & an unstable $(r_0 > 1)$ limit cycle; for $1 < a < \frac{3}{2}$ $\exists$ an unstable $(r_0 > 1)$ limit cycle.

A1.16 $T = 4\int_0^E 1/\{2(E - x)\}^{1/2}\,\mathrm{d}x$.

A1.17 $\frac{1}{2}(\mathrm{d}x/\mathrm{d}t)^2 + \frac{1}{2}x^2 - \frac{1}{4}\epsilon x^4 = \frac{1}{2}a^2 - \frac{1}{4}\epsilon a^4$. Substitute $x = a\cos\phi$ in $T = \int_0^a \mathrm{d}x/[(a^2 - x^2)\{1 - \frac{1}{2}\epsilon(a^2 + x^2)\}]^{1/2}$.

A1.18 $T = 4c^{-1}\int_0^{\pi/2}(1 + \epsilon a^2\cos^2\theta)a\cos\theta/(\epsilon a\cos\theta + \frac{1}{2}\epsilon^2 a^4\cos^4\theta)^{1/2}\,\mathrm{d}\theta$ if $x = a\cos\theta$.

A1.19 $A = B = 0$ for a solitary wave.

A1.21 (a) Cf. Example 1.2. (b) Linearized equation is $\psi'' + \omega^2\psi = 0$, so $\psi = \sin\omega s$. $\omega s = \int_0^\psi \mathrm{d}\psi/\{4(\cos\psi - \cos\alpha)\}^{1/2}$.

A1.23 Express $[a, b] = \mathrm{I}$, $[m, M] = F(\mathrm{I})$. Then $F(\mathrm{I}) \supset \mathrm{I}$ implies that $m \leqslant a < b \leqslant M$. Therefore $\exists\ c, d \in \mathrm{I}$ such that $F(c) = a$, $F(d) = b$. Therefore $F(c) - c \leqslant 0 \leqslant F(d) - d$, & intermediate-value theorem gives at least one zero of $F(x) - x$ in $[c, d]$.

A1.25 By 2nd mean-value theorem, $\exists\ \xi \in (a, b)$ such that $0 = f(X) = f(x_n) + (X - x_n)f'(x_n) + \frac{1}{2}(X - x_n)^2 f''(\xi)$. But $f(x_n) = (x_n - x_{n+1})f'(x_n)$. Therefore $|x_{n+1} - X| = \frac{1}{2}f''(\xi)|x_n - X|^2/f'(x_n)$.

A1.27 $F(x) = r + \delta/(r + x)$. $X = \pm\sqrt{a}$. $F'(\pm\sqrt{a}) = (r \mp \sqrt{a})/(r \pm \sqrt{a})$. $\sqrt{a}$ is stable, $-\sqrt{a}$ is unstable.

A1.28 X is stable fixed point of G if $-2 < a\{F'(X) - 1\} < 0$.

A1.29 Define $F(x) = x + hf(x)$. $|F'(0)| < 1$ iff $-2 < hf'(0) < 0$.

A1.32 To estimate x_n as $n \to \infty$ heuristically, assume that $x_n \sim bn^p$ as $n \to \infty$, test this ansatz for consistency, & deduce that $p, b = 1$.

A1.35 $s_1, s_2 = -b \pm (b^2 - c)^{1/2}$, $\mathbf{u}_j = [1, s_j]^{\mathrm{T}}$. $b = -\frac{1}{2}$, $c = -1$, $x_0 = 0$, $x_1 = 1$, $s_1, s_2 = \frac{1}{2}(1 \pm \sqrt{5})$, $\xi_1, \xi_2 = \pm 1/\sqrt{5}$.

A1.36 On differentiating a function of a function, $G'(X) = F'(X)F'(F(X)) = F'(X)F'(Y)$, $= G'(Y)$ similarly.

A1.37 $X = \{-1 \pm (1 + a)^{1/2}\}/a$; former point is (just) stable, the latter unstable. $F^2(x) = x\ \forall\ x \neq -1/a$, so only cycles are the 2-cycles $\{x, (1 - a)/(1 + ax)\}\ \forall\ x \neq -1/a, X$.

Chapter 2

A2.2 Use implicit-value theorem. A zero of H is either a fixed point of G or a member of a 2-cycle of G.

A2.3 $A = 0$ or $a = A + n^2\pi^2$. Pitchfork bifurcations at $(n^2\pi^2, 0)$.

A2.4 (i) Imperfect transcritical bifurcation. (ii) Imperfect transcritical bifurcation & turning point, with hysteresis for $\delta < 0$. (iii) Isola for $\delta > 0$, isolated point for $\delta = 0$, no solution for $\delta < 0$.

A2.5 $F = 0$ gives $x = 0$ or parabola $(x + \frac{1}{2}\delta)^2 = a + \frac{1}{4}\delta^2$ with nose at $a = -\frac{1}{4}\delta^2$, $x = -\frac{1}{2}\delta$; i.e. turning point at $(-\frac{1}{4}\delta^2, -\frac{1}{2}\delta)$, transcritical bifurcation at $(0,0)$ if $\delta \neq 0$.

A2.6 $F(1,b,0,x) = 0$ gives $x = 0$ or parabola $(x - 1)^2 = b - 2$, with transcritical bifurcation at $b = 3, x = 0$ & turning point at $b = 2, x = 1$. $F = F_x = 0$ gives $3\{2 + 3c - \frac{2}{3}b - \frac{4}{3}a(3 - \frac{4}{3}a - b)\}^2 = 4(b + \frac{4}{3}a^2 - 3)^3$, a curve in (b,c)-plane with cusp at $b = 3 - \frac{4}{3}a^2$, $c = -\frac{8}{27}a^3$. On eliminating a, $(b - 3)^3 = -27c^2$.

A2.7 $x = X$, $\mathrm{d}x/\mathrm{d}t = \mathrm{d}^2x/\mathrm{d}t^2 = 0$ gives $gf'(X) - \omega^2(X - \delta a) = 0$. Linearization gives $[1 + \{f'(X)\}^2]\mathrm{d}^2x'/\mathrm{d}t^2 + \{gf''(X) - \omega^2\}x' = 0$, & so stability according to sign of $gf''(X) - \omega^2$. For the given f, using dimensional analysis, define $u = X/a$ so $F(u,b,\delta) = 0$, where $F(u,b,\delta) = u^3 - \frac{1}{2}(b - 1)u + \frac{1}{2}\delta b$. Imperfect pitchfork bifurcation at $(1,0)$ in the (b,u)-plane. Eliminate x from $F = F_u = 0$ to find cusp. Cf. Examples 2.2 & 2.5.

A2.8 The triple root occurs when the cubic has the form $(v - v_c)^3 = 0$. $F = F_\phi = 0$ gives the equation of curve in (θ, π)-plane with cusp at origin.

A2.11 If $\epsilon > 0$ then $(a - r^2)r \pm \epsilon = 0$ & $\theta = 0, \pi$ respectively, & so there is half $(r \geqslant 0)$ of an imperfect pitchfork in the (a,r)-plane for $\theta = 0$, & half for $\theta = \pi$; however $\theta = 0$ is stable & $\theta = \pi$ unstable $\forall a$, so the turning point for $\theta = \pi$ separates 2 unstable solutions.

A2.12 $\mathbf{f}(\mathbf{x}) = \mathbf{f}(\mathbf{X}) + \mathbf{J}(\mathbf{X})(\mathbf{x} - \mathbf{X})$ + nonlinear terms. The Jacobian matrix of $\mathbf{F}$ at $\mathbf{X}$ is $\mathbf{0}$.

A2.13 Equation of intersection (with x as a parameter) is $b = -4x^3 - 2ax$, $c = 3x^4 + ax^2$.

Chapter 3

A3.4 Use mathematical induction.

A3.5 X_1 is a stable fixed point of $G = F^p$ if $1 > |G'(X_1)|$.

A3.6 $X = 0 \;\forall a$, stable if $-1 < a < 1$; or $X = \pm(a - 1)^{1/2} \;\forall a > 1$, stable if $1 < a < 2$. The three 2-cycles are $\{(a + 1)^{1/2}, -(a + 1)^{1/2}\} \;\forall a > -1$, always unstable; $\{\pm [\frac{1}{2}\{a + (a^2 - 4)^{1/2}\}]^{1/2}, \pm [\frac{1}{2}\{a - (a^2 - 4)^{1/2}\}]^{1/2}\} \;\forall a > 2$, stable if $2 < a < \sqrt{5}$. Pitchfork bifurcation at $(1,0)$, flip bifurcations at $(-1,0)$, $(2, \pm 1)$ in (a,x)-plane.

A3.7 $X = 0 \ \forall\, a$, stable if $-1 < a < 1$; or $X = \pm(1-a)^{1/2} \ \forall\, a < 1$, always unstable. The three 2-cycles are $\{(-1-a)^{1/2}, -(-1-a)^{1/2}\} \ \forall\, a < -1$, stable if $-2 < a < -1$; $\{\pm[\frac{1}{2}\{-a + (a^2-4)^{1/2}\}]^{1/2}, \pm[\frac{1}{2}\{-a-(a^2-4)^{1/2}\}]^{1/2}\}$ $\forall\, a < -2$, stable if $-\sqrt{5} < a < -2$.

A3.8 (a) Assume $\exists$ p-cycle with $F(X_j) = X_{j+1}$, $F(X_p) = X_1$ & use *reductio ad absurdum*. Take $X_1 < X_2$ without loss of generality. Then monotonic property implies that $X_2 < X_3 < \cdots < X_p < X_1$, a contradiction. (b) $F(x) - x$ decreases more rapidly than $-x$ & so has unique zero. Note that if $G(x) = F^q(x)$ then $G'(x) > 0 \ \forall\, x$ if q is even & $G'(x) < 0$ if q is odd, & use previous parts of question.

A3.10 $F^n(a,x) = x/[a/\{x^2 + (a-x^2)\mathrm{e}^{-4\pi na}\}]^{1/2}$. $\mathrm{D}(0) = (-\infty,\infty)$ if $a < 0$; $\mathrm{D}(\sqrt{a}) = (0,\infty)$, $\mathrm{D}(-\sqrt{a}) = (-\infty, 0)$ if $a > 0$.

A3.11 $F^8(r) \equiv 2^8 r = (5 \times 51 + 1)r \equiv r$ modulo 51.

A3.12 Use rotation map.

A3.13 $r \to r, \theta \to \theta + \psi(r)$.

A3.14 X_1 is unstable, X_2 stable; $\mathrm{D}(X_2) = (X_1, \infty)$. $\exists$ turning point at (e, e) in the bifurcation diagram in the (a, x)-plane.

A3.15 $N(x) = \frac{1}{2}(x + a/x)$, $X = \pm\sqrt{a}$, $N'(\pm\sqrt{a}) = 0$ so (super-)stability, $\mathrm{D}(-\sqrt{a}) = (-\infty, 0)$, $\mathrm{D}(\sqrt{a}) = (0,\infty)$. $F(x) = x + h(a - x^2)$, $X = \pm\sqrt{a}$, $F'(\pm\sqrt{a}) = 1 \mp 2\sqrt{a}h$, so $\sqrt{a}$ is stable (if $0 < h < 1/\sqrt{a}$) & $-\sqrt{a}$ is unstable, $\mathrm{D}(\sqrt{a}) = (-\sqrt{a}, \sqrt{a} + 1/h)$, $\mathrm{D}(-\infty) = (-\infty, -\sqrt{a}) \cup (\sqrt{a} + 1/h, \infty)$.

A3.17 Eliminate Y, Z from $Y = aX(1-X)$, $Z = aY(1-Y)$, $X = aZ(1-Z)$ to get octic, and then discard factors $X(aX - a + 1)$. $t_1 t_2 = (a^2 + a + 1)/a^6$, $v_1 + v_2 = (3a+1)/a$, $u_1 + u_2 + v_1 + v_2 = (3a+1)(a+1)/a^2$, $t_1 + t_2 + u_1 v_2 + u_2 v_1 = (a^3 + 5a^2 + 3a + 1)/a^3$, $t_j - u_j + v_j = (a^3 - 1)/a^3$, $v_1 v_2 = 2(a^2 + a + 1)/a^2$.

A3.18 $F(a, \frac{1}{2}) = \frac{1}{4}a$, $G_x(a,x) = a^2(1-2x)(1 - 2ax + 2ax^2)$, so if $a > 2$ then G has two equal maxima at $x = \frac{1}{2}\{1 \pm (1 - 2/a)^{1/2}\}$ & a minimum at $x = \frac{1}{2}$. $G(A, \frac{1}{16}A^2(4-A)) = G(A, 1/A) = (A-1)/A$, $G(A, (A-1)/A) = (A-1)/A$. $G(A, \frac{1}{2}) = 1/A$. $F(A, 1/A) = (A-1)/A$. $F(A, \frac{1}{4}A) = 1/A$. $F(A, (A-1)/A) = (A-1)/A$. $F(A, x) \leqslant F(A, \frac{1}{2}) = \frac{1}{4}A$. $1/A < \frac{1}{2} < (A-1)/A$.

A3.19 Let $P(\theta) = \frac{1}{2} + \cos 2\pi\theta = x$. Therefore $P(2\theta) = F(x)$ by elementary trigonometry.

A3.20 $\bar{x} = \int_0^1 x f(x)\,\mathrm{d}x = \int_0^1 \sin^2(\pi\theta)\,\mathrm{d}\theta = \frac{1}{2}$, $\overline{x^2} = \int_0^1 x^2 f(x)\,\mathrm{d}x = \int_0^1 \sin^4(\pi\theta)\,\mathrm{d}\theta = \frac{3}{8}$.

A3.22 $X = 0 \ \forall\, a$, stable $\forall\, a \leqslant 1$; $X = a/(a+1) \ \forall\, a > 1$, unstable; or $0 < X \leqslant \frac{1}{2}$ for $a = 1$, always stable. $\{\frac{2}{5}, \frac{4}{5}\}$.

A3.23 The 2-cycle, $\{a/(1+a^2), a^2/(1+a^2)\}$, is unstable $\forall\, a > 1$.

A3.24 $H^{-1}(x) = \sin^2(\frac{1}{2}\pi x)$, $FH^{-1}(x) = \sin^2(\pi x)$.

A3.25 (a) $x_{n+1} = \mathrm{sn}^2(2K(m)\theta_{n+1}|m) = \mathrm{sn}^2(4K(m)\theta_n|m) = F(m, x_n)$. (b) $F(m,x) = \frac{1}{4}\mathrm{sn}^2(4K(m)\theta|m)$, where $x = \mathrm{sn}^2(2K(m)\theta_n|m)$.

A3.26 $X = 0$ is unstable, $X = \pm 1$ are (super-)stable. $\mathrm{D}(1) = (0, \infty)$, $\mathrm{D}(-1) = (-\infty, 0)$.

A3.27 $X = 0 \ \forall\, a$, stable for $-1 < a < 1$; $X = U/\pi \ \forall\, a > 1$, marginally stable if $a \cos U = -1$. Note analogy with logistic map, & incompleteness of analogy because, e.g., F is odd function of x. Pitchfork bifurcation at $(1,0)$ in (a,x)-plane, period-doubling etc. as a increases to π. If $a=\pi$ then F maps $[0,1]$ onto itself, & unstable periodic orbits are dense in $[0,1]$. Also succession of turning points as a increases.

A3.28 Graphs show uniqueness of $X \leqslant 0$. $X = \ln(-X) - \ln(-a) \sim -\ln(-a)$ as $a \to -\infty$. By chain rule, $G_x(a,x) = F_x(a,F(a,x))F_x(a,x) = F(a,F(a,x))F(a,x) = G(a,x)F(a,x)$. Therefore $G_x(a,X) = X^2$, etc. Therefore $G=-1$, $G_x = 1$, $G_{xx}=0$, $G_{xxx} = -1$ at $a=-\mathrm{e}$, $X=-1$, so line $y=x$ has triple intersection with curve $y=G(a,x)$ at $x=-1$ if $a=-\mathrm{e}$, with onset of 3 intersections for $a<-\mathrm{e}$. X_1, X_2 are roots of $\{x-G(a,x)\}/(x-X)=0$, where the Taylor series $G(a,x) = X + (x-X)X^2 + \frac{1}{2}(x-X)^2X^2(1+X) + \frac{1}{6}(x-X)^3X^2(1+3X+X^2)+\ldots$, so $\exists$ subcritical flip bifurcation at $(-\mathrm{e},-1)$ in (a,x)-plane.

A3.29 (a) See Example 3.9. (b) $G(X)=X$ iff $a+(2\pi)^{-1}\sin 2\pi X = 0$, i.e. $X = 1-(2\pi)^{-1}\sin(2\pi a/b)$ or $\frac{1}{2}+(2\pi)^{-1}\sin(2\pi a/b)$. $Y_0 = X_0 + \frac{1}{2}$ as in part (a). $a_2 = (8\pi)^{-1}\sin(4\pi X_0)$, so $-1/8\pi < a_2 < 1/8\pi$.

A3.30 $X=0\ \forall\ a$, unstable for $a>0$; $X=1\ \forall\ a$, stable for $0\leqslant a\leqslant 2$. Stable 2-cycle for $a=2.3$, 4-cycle for $a=2.6$, chaos for $a=2.9$.

A3.31 Cf. Q3.6. Note analogy with logistic map, although F is odd function of x so bifurcation diagram is symmetric in $\pm x$, with pitchfork bifurcation at $(1,0)$, flip bifurcation at $(-1,0)$ in (a,x)-plane.

A3.34 $\mathbf{K} = \mathbf{J}(\mathbf{X}_1)\mathbf{J}(\mathbf{X}_2)$. $q_1q_2 = \det\mathbf{K} = b^2$, $q_1+q_2 = \operatorname{trace}\mathbf{K} = 4(1-b)^2 - 4a + 2b$. $q_1 = 1$, $q_2 = b^2$ when $a=a_1$; $q_1=-1$, $q_2=-b^2$ when $a=a_2$. Note also that $q_2 = \bar{q}_1$, $|q_1| = |b|$ when $(1-b)^2 + \frac{1}{4}b(1-b) \leqslant a \leqslant (1-b)^2 + b(1+b)$.

A3.36 $a_1(b) = 1-b$. $X_+ = (a+1-b)/\{a^2+(1-b)^2\}$,
$X_- = -(a+b-1)/\{a^2+(1-b)^2\}\ \forall\ a > 1-b$.

A3.37 $\mathrm{V} = ((1-b+aq)/q(1+a-b), b(1-q)/(1+a-b))$.

A3.38 $\mathbf{0}$ is always stable, because $q=\mathrm{e}^{\pm i\alpha}$. Other point is always unstable, because q has form $b\pm(b^2-c)^{1/2}$ with $b>1$, real c. Fixed points of $\mathbf{T}^2$ are these two, & two complex points.

A3.41 If X, Y are rational, express $X=a/r$, $Y=b/r$ for integers a, b, r, not necessarily coprime. Then $\mathbf{F}^n(\mathbf{X}) = (c/r, d/r)$ for some integers c, d; but $\exists$ only r^2 points of the form $(c/r,d/r)$; use the pigeon-hole principle.

A3.42 $(0, f(0)/(a^b-1))\ \forall\ a\neq 1$, $(X,Y)\ \forall\ Y$, where X is a zero of f, for $a=1$. $F(x,g(x)) = (ax, g(ax))$. $\mathbf{J} = \begin{bmatrix} a & 0 \\ -a^bf'(x) & a^b \end{bmatrix}$. $q_1 = a$, $q_2 = a^b$, so $|q_1|$, $|q| > 1$, & there is instability. $\mathbf{F}(\mathbf{x}) = (ax \bmod 1, a^b(y-f(x))$ gives a continuous map of the cylinder because $f(1)=f(0)$.

A3.44 $Z = Z_\pm, = \frac{1}{2}\{1\pm(1-4a)^{1/2}\} = \frac{1}{2}(1\pm r^{1/2}\mathrm{e}^{\mathrm{i}\theta/2})$, say, where $a=\frac{1}{4}+r\mathrm{e}^{\mathrm{i}\theta}$ for $r\geqslant 0$, $-\pi<\theta\leqslant\pi$. Stability if $1>|F'(Z)|^2$, where $F'(Z_\pm) = 1\pm(1-4a)^{1/2}$. Therefore Z_+ is unstable $\forall\ a$, Z_- is stable if $r<\cos^2\frac{1}{2}\theta$, i.e. a lies in a cardioid in the complex plane. 2-cycle $\{-\frac{1}{2}\pm(-a-\frac{3}{4})^{1/2}\}$ is stable if $|a+1|<\frac{1}{4}$, a circle. Cf. Mandelbrot set.

A3.45 Use Cauchy–Riemann relations. $q_1, q_2 = F'(z), \overline{F'(z)}$.

A3.47 Cf. Q3.44 & Fig. 3.13(*a*). $a \in \mathrm{M}$ iff $F^n(a,0) \nrightarrow \infty$ iff $0 \notin \mathrm{D}(\infty)$ for *real* map $F(a,x) = x^2 + a$ (which is equivalent to logistic map). F has fixed points $X_\pm = \frac{1}{2}\{1 \pm (1-4a)^{1/2}\}\ \forall\ a \leqslant \frac{1}{4}$. X_+ is always unstable; X_- is stable & $0 \in \mathrm{D}(X_-)$ for $-\frac{3}{4} \leqslant a < \frac{1}{4}$. $\{X_1, X_2\}$ is a 2-cycle of F for $a < -\frac{3}{4}$, where X_1, $X_2 = \frac{1}{2}\{1 \pm (-3-4a)^{1/2}\}$. $\{X_1, X_2\}$ is stable & $0 \in \mathrm{D}(\{X_1, X_2\})$ for $-\frac{5}{4} \leqslant a < -\frac{3}{4}$. As a decreases there is period doubling.

Chapter 4

A4.2 (i) $\frac{1}{4} = 0.020202\ldots$. (ii) $x = 0.200020\ldots$.

A4.3 If $x \in \mathrm{S} \cap [0, \frac{1}{4})$ then $4x \in \mathrm{S}$ because the point of the quaternary expression of x is moved one place to the right. Similarly, if $y \in \mathrm{S}$ then $\frac{1}{4}y \in \mathrm{S} \cap [0, \frac{1}{4}]$, etc.

A4.4 $X = 0$, $a/(a+1)$ both unstable. F maps $(\frac{1}{3}, \frac{2}{3})$ to $(1, \frac{3}{2}]$, & $[0, \frac{1}{3}]$, $[\frac{2}{3}, 1]$ each onto $[0, 1]$ repeatedly.

A4.6 (a) Piecewise linear function F_n increases by 2^{-n} over each of 2^n subintervals of length 3^{-n}, $l_n = (\frac{2}{3})^n$. (b) Length of C_n = total horizontal + total sloping length $= 1 - l_n + (1 + l_n^2)^{1/2}$. (c) $|F_n(y) - F_n(x)| < 2|y - x|^{(\ln 3)/(\ln 2)}\ \forall\ x$, $y \in [0, 1]$, $\forall\, n$.

A4.7 $N(\epsilon) = 2^n$ when $\epsilon = 4^{-n}$.

A4.9 $N(\epsilon) \propto R^n$, $\epsilon \propto r^n$, $D = \lim_{n\to\infty}\{\ln N/\ln(\epsilon^{-1})\}$. For K, $R = 2$, $r = \frac{1}{3}$; for von Koch curve, $R = 4$, $r = \frac{1}{3}$. For decimal set, $R = 5$, $r = 10^{-1}$, $D = \ln 5/\ln 10 = \log_{10} 5$. Same method works in $\mathbb{R}^m$ if ϵ = length of sides of hypercubes.

A4.11 S_n has $N = 4^n$ subsquares with sides of length $\epsilon = 4^{-n}$. Subsquares of S_1 project onto 4 equal subintervals AC_1, $\mathrm{C}_1\mathrm{C}_2$, $\mathrm{C}_2\mathrm{C}_3$, $\mathrm{C}_3\mathrm{B}$, where $\mathrm{C}_1 = (-\frac{1}{10}, \frac{1}{20})$, $\mathrm{C}_2 = (\frac{1}{5}, -\frac{1}{10})$, $\mathrm{C}_3 = (\frac{1}{2}, -\frac{1}{4})$, & so forth.

A4.12 $D = m - 1 + \ln 2/\ln 3$.

A4.13 $F^n(\mathrm{S})$ has 2^n strips of unit length (in x-direction) & height $(\frac{1}{2}a)^n$ in y-direction. Therefore $N \sim \{1/(\frac{1}{2}a)^n\} \times 2^n$ as $n \to \infty$ if $\epsilon = (\frac{1}{2}a)^n$.

A4.15 $F(X_2) - X_2 = -\{F(X_1) - X_1\}$, so continuous function $F(x) - x$ has at least one zero between X_1, X_2, by intermediate-value theorem.

A4.16 Only fixed point of F is $X = \frac{10}{3}$. F^3 has no fixed point in $[1, 3] \cup [4, 5]$. $F^3[3, 4] = F^2[2, 4] = F[2, 5] = [1, 5]$, being monotonically decreasing map of interval in each case. Therefore F^3 has a unique fixed point in $[3, 4]$, which can only be X.

A4.19 $D = 1 + \lambda_1/|\lambda_2|$ in this case with $m = 2, k = 1, \lambda_1 > 0 > \lambda_2$.

A4.20 $\lambda_1, \lambda_2 = \ln\{\frac{1}{2}(3 \pm \sqrt{5})\}$.

Chapter 5

A5.1 $z(t) = \{z_0 - \frac{1}{2}\epsilon(\cos\theta_0 + \sin\theta_0)\}\exp(\theta_0 - t) + \frac{1}{2}\epsilon(\cos t + \sin t)$ if $z = z_0$ at $t = \theta_0$. $Z = \frac{1}{2}\epsilon(\cos\theta_0 + \sin\theta_0)$ is always stable.

A5.2 $\forall\, \epsilon > 0\ \exists\ \{t_n\}$ such that $|\mathbf{x}(t_n) - \mathbf{x}_\infty| < \epsilon$ & $t_n \uparrow \infty$ as $n \to \infty$. Therefore, $\forall\ \epsilon' > 0$, $|V(\mathbf{x}(t_n)) - V(\mathbf{x}_\infty)| < \epsilon'$ as $n \to \infty$. Therefore grad $V(\mathbf{x}(t_n)) \to \mathbf{0}$ as $n \to \infty$.

A5.4 Linearized system is $\mathrm{d}\omega_1'/\mathrm{d}t = 0$, $\mathrm{d}\omega_2'/\mathrm{d}t = (C - A)n\omega_3'/B$, $\mathrm{d}\omega_3'/\mathrm{d}t = (A - B)\omega_2'/C$. A rigid body is stable if it spins steadily about its greatest or least principal axis of inertia, but unstable if it spins about its intermediate principal axis of inertia.

A5.5 $X = \frac{1}{2}\{c \pm (c^2 - 4ab)^{1/2}\}$, $Y = -Z = -\frac{1}{2}\{c \pm (c^2 - 4ab)^{1/2}\}/a$ if $c^2 > 4ab$. Turning points at $(\pm 2(ab)^{1/2}, \pm(ab)^{1/2})$ in (c, x)-plane.

A5.6 See §8.1. (a) $r_2 = 1/b$, $\mathbf{x}_2 = [0, 0, 1/b]^{\mathrm{T}}$. N.B. *Exact* solution is $\mathrm{x} = [\delta, \delta, \delta^2/b]^{\mathrm{T}}, r = 1 + \delta^2/b$. (b) $\mathbf{x}_2 = [0, 0, A^2/b]^{\mathrm{T}}$.

A5.12 Use Euler–Lagrange equation. $\partial L/\partial q = (\mathrm{d}/\mathrm{d}t)(\partial L/\partial \dot{q}) = \mathrm{d}p/\mathrm{d}t$.

A5.13 For linearized problem of instability, $s = -A/\sqrt{2}$ or $\sqrt{2A}$ twice. Poincaré map is identity map for x & rotates y by $2\pi \tan x$ (see Example 3.9).

A5.15 (a) General solution $x(t) = (A + \alpha)\cos(t + \tau)$ is close to X at $t = 0$ for small constants α, τ. $|x(t) - X(t)| = O(\alpha, \tau)\ \forall\, t$ as $\alpha, \tau \to 0$. (b) Use Example 1.2.

A5.16 $\mathrm{d}r'/\mathrm{d}t = -(2ab\cos^2 t)r'$, $\mathrm{d}\theta'/\mathrm{d}t = (\sqrt{ab}\sin 2t)r'$. $P_a(\sqrt{a} + r') = \sqrt{a} + r'\exp(-2\pi ab) + O(r'^2)$ as $r' \to 0$.

A5.17 $\mathbf{P}(t + T) = \Phi(t + T)\mathrm{e}^{-(t+T)\mathbf{R}} = \Phi(t)\mathrm{e}^{T\mathbf{R}}\mathrm{e}^{-(t+T)\mathbf{R}} = \mathbf{P}(t)$. If $\mathbf{R}\mathbf{u} = s\mathbf{u}$ & f is a polynomial then $f(\mathbf{R})\mathbf{u} = f(s)\mathbf{u}$. $\mathrm{d}(\det\Phi)/\mathrm{d}t = (\mathrm{trace}\,\mathbf{A})\det\Phi$.

A5.18 Let x_1, x_2 be solutions of Meissner's equation such that $x_1 = 1, \dot{x}_1 = 0, x_2 = 0, \dot{x}_2 = 1$ at $t = 0$. Then $\Phi = \begin{bmatrix} x_1 & x_2 \\ \dot{x}_1 & \dot{x}_2 \end{bmatrix}$. Solving piecewise for $0 \leqslant t \leqslant \frac{1}{2}, \frac{1}{2} \leqslant t \leqslant 1$, find $x_1(1) = \cos\frac{1}{2}a\cosh\frac{1}{2}a - \sin\frac{1}{2}a\sinh\frac{1}{2}a$ and also $\dot{x}_1(1) = a(\cos\frac{1}{2}a\sinh\frac{1}{2}a - \sin\frac{1}{2}a\cosh\frac{1}{2}a)$. Similarly find $x_2(1) = (\sin\frac{1}{2}a\cosh\frac{1}{2}a + \cos\frac{1}{2}a\sinh\frac{1}{2}a)/a$, $\dot{x}_2(1) = \sin\frac{1}{2}a\sinh\frac{1}{2}a + \cos\frac{1}{2}a\cosh\frac{1}{2}a$. Therefore $0 = \det(\Phi(1) - q\mathbf{I}) = q^2 - 2bq + 1$, where $b = \frac{1}{2}\{x_1(1) + \dot{x}_2(1)\} = \cos\frac{1}{2}a\cosh\frac{1}{2}a$. Therefore $q_1, q_2 = b \pm (b^2 - 1)^{1/2}$. N.B. $q_1 = q_2 = \pm 1$ for values of a such that $\cos\frac{1}{2}a\cosh\frac{1}{2}a = \pm 1$.

A5.19 (a) If $\mathbf{x} = \begin{bmatrix} x \\ \mathrm{d}x/\mathrm{d}t \end{bmatrix}$ then $\mathbf{A} = \begin{bmatrix} 0 & 1 \\ -P(t) & 0 \end{bmatrix}$. $\Phi(1)$ is a real 2×2 matrix so its eigenvalues q_1, q_2 are real or a complex conjugate pair. trace $\mathbf{A}(t) = 0\ \forall\, t$. Therefore $q_1 q_2 = \det\Phi(\pi) = \text{constant} = \det\Phi(0) = 1$. Therefore $0 = \det(\Phi(\pi) - q\mathbf{I}) = q^2 - \{x_1(\pi) + \dot{x}_2(\pi)\}q + 1$, & $\{x_1(\pi) + \dot{x}_2(\pi)\}^2 < 4$ implies stability. At margin of stability $q_1 = q_2 = \pm 1$ (for period π, 2π respectively). (b) If $x_0(t) = \sin t$ then $a_1 = -1, x_1(t) = -\frac{1}{8}\sin 3t$.

Chapter 6

A6.1 If $\mathrm{P} = (x, y)$ then $\mathrm{Q} = (x, F(x)), \mathrm{R} = (0, F(x))$, $\overrightarrow{\mathrm{RP}} = (x, y - F(x))\,\mathrm{d}\mathbf{x}/\mathrm{d}t = (y - F(x), -x)$, $\overrightarrow{\mathrm{RP}}\cdot\mathrm{d}\mathbf{x}/\mathrm{d}t = 0$.

A6.2 Orbits are alternately semiellipses in upper & lower halves of $(x, \mathrm{d}x/\mathrm{d}t)$-plane, with centres at $(-F/\lambda, 0), (F/\lambda, 0)$ respectively, but end where $\mathrm{d}x/\mathrm{d}t = 0$, $-F/\lambda \leqslant x \leqslant F/\lambda$.

A6.3 $(-\sqrt{a}, 0)$ is a saddle point, $(\sqrt{a}, 0)$ is a stable node.

A6.4 $y = -x + bc^{-1}\ln x + \text{constant}$.

A6.5 Linearized equations at $(0, 0)$ are $\mathrm{d}x'/\mathrm{d}t = ax'$, $\mathrm{d}y'/\mathrm{d}t = -by'$; & at $(b/c, a/c)$ are $\mathrm{d}x'/\mathrm{d}t = -by'$, $\mathrm{d}y'/\mathrm{d}t = ax'$, so $\mathrm{d}^2x'/\mathrm{d}t^2 = -abx'$. Separate variables, integrate, & take exponents to find $x^b y^a = k\mathrm{e}^{c(x+y)}$ for constant k of integration on each orbit.

A6.7 Equilibrium point $(n\pi, 0)$ is a saddle point if n is even; it is a stable focus for $gl > k^2$, a stable node for $gl < k^2$ if n is odd.

A6.8 Equilibrium points are at $(X, 0)\ \forall\ X$, such that each orbit ends where it meets x-axis. Equation is piecewise linear in upper & lower half planes.

A6.9 $(a, b/a)$ is stable if $b < a^2 + 1$ or unstable if $b > a^2 + 1$, & node if $(a^2 + 1 - b)^2 > 4a^2$ or focus if $(a^2 + 1 - b)^2 < 4a^2$.

A6.10 (i) Equilibrium points are $(0, 0)\ \forall\ a$, saddle point $\forall\ a > 0$, stable node $\forall\ a \leqslant 0$; and $(\pm\sqrt{a}, 0)\ \forall\ a > 0$, stable nodes. Supercritical pitchfork bifurcation at $x = y = a = 0$.

(ii) $(0, 0)$ is a centre, $(a, 0)$ a saddle point $\forall\ a \neq 0$. Transcritical bifurcation at $(0, 0)$ in (a, x)-plane. System is invariant if $(x, a) \to (-x, -a)$ or $(y, t) \to (-y, -t)$. $x = -a$ is an invariant line. If $a = 0$, exact integral is $y^2 = x^2(\ln x + C)$.

(iv) Equilibrium points are $(0, 0)\ \forall\ a$, $(a, 0)$, $(-a, -a)$, $(-a, 2a)\ \forall\ a \neq 0$. $y = 0$ is invariant line. System is independent of a if $(x, y, t) \to (x/a, y/a, at)$, so take $a = 0, 1$ without loss of generality. If $a = 0$, exact integral is $y^2(y^2 - 2x^2) = C$. If $a = 1$, $(0, 0)$ is unstable node & $(1, 0)$, $(-1, -1)$, $(-1, 2)$ are saddle points.

(vii) Equilibrium points are $(0, 0)\ \forall\ a$ & $(0, 5a)\ \forall\ a \neq 0$. System is invariant if $(x, y, t, a) \to (-x, -y, -t, -a)$, so take $a \geqslant 0$ without loss of generality. If $a = 0$, exact integral is $x^2 = 2y^2(C - \ln y)$. $(0, 0)$, $(0, 5a)$ are unstable foci if $a > 0$.

A6.11 (a) $\mathbf{0}$ is stable focus if $a < -1$, stable node if $-1 < a < 0$, saddle point if $a > 0$. (b) $(-2, 0)$ is unstable if $a < 0$. $(-1 \pm (1 + 4a)^{1/2}, a)$ exist for $a > -\frac{1}{4}$, former is unstable for $-\frac{1}{4} < a < 0$, latter for $a > 0$. (c) Transcritical bifurcations at $(0, 0)$, $(0, -2)$, turning point at $(-\frac{1}{4}, -1)$ in (a, x)-plane. (d) Equate coefficients of a in the *three* equations, solve for $\mathbf{x}_1$. Equate coefficients of a^2 to find F_0, $\mathbf{x}_2 = A(1 - A)[1, -2]^{\mathrm{T}}/50$.

A6.12 $\phi = \ln(2/\delta r^2)$. Saddle point at $(0, 2)$.

A6.13 $0 \geqslant \mathbf{x}^{\mathrm{T}}\mathbf{F}(\mathbf{x}) = \frac{1}{2}\mathrm{d}(\mathbf{x}^{\mathrm{T}}\mathbf{x})/\mathrm{d}t$. Therefore $|\mathbf{x}(t)| \leqslant |\mathbf{x}(0)|$. Taking $\mathbf{F}(\mathbf{x}) = \mathbf{A}\mathbf{x}$, deduce $\mathbf{x}^{\mathrm{T}}\mathbf{F}(\mathbf{x}) = \frac{1}{2}\mathbf{x}^{\mathrm{T}}(\mathbf{A} + \mathbf{A}^{\mathrm{T}})\mathbf{x} \leqslant 0\ \forall\ \mathbf{x}$.

A6.14 Try *reductio ad absurdum*. It may help to find elsewhere a proof of Liapounov's theorem (a) on p. 179 & adapt it to a difference equation.

A6.16 (a) $x(t) = x_0\mathrm{e}^{-t}$, $y(t) = r_0\mathrm{e}^{-t}\cos(\theta_0 - 2t)$, $z(t) = -r_0\mathrm{e}^{-t}\sin(\theta_0 - 2t)$. (b) $H(\mathbf{x}) = x^2 + y^2 + z^2$ will do.

A6.17 Verify results & use uniqueness of solutions of differential equations.

A6.20 $K(x, y) = \frac{1}{2}r^2\{a^2 + b^2 + c^2 + d^2 + (a^2 - b^2 + c^2 - d^2)\cos 2\theta +$

$2(ab + cd)\sin 2\theta\}$. $\mathrm{d}K/\mathrm{d}\theta = 0$ if $\tan 2\theta = 2(ab + cd)/(a^2 - b^2 + c^2 - d^2)$. $\mathrm{d}H/\mathrm{d}t = 2pK + O(|\mathbf{x}|\cdot|\xi|) > pK$ as $\mathbf{x} \to \mathbf{0}$.

A6.21 $\mathrm{d}H/\mathrm{d}t = -2k(\mathrm{d}x/\mathrm{d}t)^2$.

A6.22 Take $a = 2$, say, so that $\mathrm{d}H/\mathrm{d}t = -2(x^2 - \frac{7}{5})^2 - 2y^2/25$.

A6.23 Try $H(\mathbf{x}) = x^2 + axy + by^2$ & find suitable constants a, b.

A6.24 $m = 4, n = 2, b = 2$ will do.

A6.25 Try $H(\mathbf{x}) = x^2 + y^2$.

A6.26 If $\delta\int L(w, w_x)\,\mathrm{d}x = 0$ & $L(w, w_x) = \frac{1}{2}(w_x^2 - w^2)$, then Euler–Lagrange equation is $w_{xx} + w = 0$. Minimizing function is $w(x) = \sin x$. Deduce inequality & use it to show $\mathrm{d}H(u)/\mathrm{d}t < -kH(u)$ unless $u(x, t) = 0$ almost everywhere.

A6.27 Centre at $(0, 0)\ \forall\ \epsilon$ is stable, saddle points at $(\pm 1/\epsilon, 0)\ \forall\ \epsilon \neq 0$ are unstable.

A6.28 Centre at $(0, 0)\ \forall\ \epsilon$ is stable, saddle point at $(1/\epsilon, 0)\ \forall\ \epsilon \neq 0$ is unstable. $x(\epsilon, t) = a\cos\omega t + \epsilon a^2(\frac{1}{6}\cos 2\omega t + \frac{1}{3}\cos\omega t - \frac{1}{2}) + O(\epsilon^2 a^3)$, $\omega = 1 - \frac{5}{12}\epsilon^2 a^2 + O(\epsilon^3 a^3)$ as $\epsilon a \to 0$. $2\pi/\omega > 2\pi$, $(\omega/2\pi)\int_0^{2\pi/\omega} x(\epsilon, t)\,\mathrm{d}t < 0$ because periodic orbit for $\epsilon \neq 0$ goes nearer saddle point, where motion is slow, than it would for $\epsilon = 0$.

A6.29 u_{N}, $u_{\mathrm{r}} = \{1 \mp (1 - 4\epsilon)^{1/2}\}/2\epsilon l$. Linearization gives a centre, saddle point respectively if $\epsilon < \frac{1}{4}$. Problem is essentially same as Q6.28 with $x = u - u_{\mathrm{N}}$: $\omega^2\mathrm{d}^2x/\mathrm{d}\phi^2 = \epsilon l x^2 - (1 - 4\epsilon)^{1/2}x$.

A6.30 Cf. Kevorkian & Cole (1981, §3.1.2).

A6.31 Suppose a closed orbit C encloses a domain $\mathrm{S} \subset \mathrm{D}$ & use *reductio ad absurdum*. $\Delta = \partial F/\partial x + \partial G/\partial y$ is of the same sign in D because it is non-zero & continuous. Therefore $0 \neq \int_{\mathrm{S}}\Delta\,\mathrm{d}x\mathrm{d}y = \int_{\mathrm{C}}(F\mathrm{d}y - G\mathrm{d}x)$, by divergence theorem, $= 0$ because C is an orbit.

A6.32 (a) Origin is centre, so stable. Take $H = r^2$. Therefore $\mathrm{d}H/\mathrm{d}t = 2r^4(1 - \frac{1}{2}\sin^2 2\theta - r^2) \geqslant 0$ if $r^2 < \frac{1}{2}$. (b) Take $\mathrm{D} = \{\mathbf{x}: \frac{1}{2} < r < 1\}$. Therefore $\mathrm{d}r/\mathrm{d}t > 0$ on outside circle of annulus D & $\mathrm{d}r/\mathrm{d}t < 0$ on inside circle of D. $\mathrm{d}\theta/\mathrm{d}t = -(1 + \frac{1}{4}r^2\sin^2 4\theta) \neq 0$ in D.

A6.33 (a) 0 is only equilibrium point because $\mathrm{d}\theta/\mathrm{d}t \neq 0$. $\mathrm{d}H/\mathrm{d}t = 2H\{a - f(r^2)\}$. (b) Define ρ as least zero of $f(r) = a > 0$, & take $\mathrm{D} = \{\mathbf{x}: \epsilon < r < \rho - \epsilon\}$ for ϵ so small that $\mathrm{d}r/\mathrm{d}t < 0$ on $r = \epsilon$ & $\mathrm{d}r/\mathrm{d}t > 0$ on $r = \rho - \epsilon$. (c) $\mathrm{d}r/\mathrm{d}t = r(a - r)$, $\mathrm{d}\theta/\mathrm{d}t = -1$ implies that $x(t) = a\cos(\theta_0 - t)$, $y(t) = a\sin(\theta_0 - t)$ is stable limit cycle.

A6.34 $\mathrm{d}r/\mathrm{d}t = -r\{1 - r^2(1 + \sin^2\theta)\}$, $\mathrm{d}\theta/\mathrm{d}t = 1$. Take $\mathrm{D} = \{\mathbf{x}: 1/\sqrt{2} < r < 1\}$. In fact, explicit solution is $\theta = t - t_0$, $r(t) = r_0/[\mathrm{e}^{2\theta} + r_0^2\{\frac{1}{4}(\mathrm{e}^{2\theta} + \sin 2\theta - \cos\theta) - \frac{3}{2}(\mathrm{e}^{2\theta} - 1)\}]^{1/2} \sim 1/\{\frac{1}{4}(\sin 2\theta - \cos\theta) + \frac{3}{2}\}^{1/2}$ as $\theta \to -\infty$, giving the unstable limit cycle.

A6.35 (a) Cf. Q1.16. $x(t) = \frac{1}{4}Tt + \frac{1}{2}t^2$ for $-\frac{1}{2}T \leqslant t \leqslant 0$, $x(t) = \frac{1}{4}Tt - \frac{1}{2}t^2$ for $0 \leqslant t \leqslant \frac{1}{2}T$, where period $T = 4(2a)^{1/2}$. (b) Unstable focus at $(0, 0)$. Orbits spiral anticlockwise out from $(0, 0)$ & in from infinity to stable limit cycle. (c) Use theorem of §6.6.

A6.36 Define $H(\mathbf{x}) = \frac{1}{4}x^4 + \frac{1}{2}y^2$, where $y = \mathrm{d}x/\mathrm{d}t$, & show that $\mathrm{d}H/\mathrm{d}t > 0$ for small $|\mathbf{x}|$ & $\mathrm{d}H/\mathrm{d}t > 0$ for large $|\mathbf{x}|$. Bound 0 by small and large level curves of H.

A6.37 $\theta = \ln|\ln r|$ gives focus at origin with tight spiral, whereas linearized system gives sink at origin with radial straight orbits.

A6.38 $x_{1/2} = \alpha \cos \omega t$, $y_{1/2} = -\alpha\omega_0 \cos \omega t$, where $\alpha^3 = 4\omega_0^2$.

A6.40 $da/dt \sim \frac{1}{2}\epsilon a(1 - 4a/3\pi)$, $\omega = -1 + O(\epsilon^2)$.

A6.41 $da/dt \sim \frac{1}{2}\epsilon a(1 - \frac{1}{8}a^4)$, $\omega = -1 + O(\epsilon^2)$.

A6.42 $da/dt = O(\epsilon^2)$ if $a < 1$, $da/dt = \epsilon a(\frac{1}{2} + \pi^{-1}\sin 2\alpha - 2\alpha/\pi)$ if $a > 1$, where $\alpha = \arccos(1/a)$, $\omega = -1 + O(\epsilon^2)$.

A6.44 $da/dt = O(a^2)$, $\omega = -1 + \frac{1}{16}a^2 + O(a^3)$ as $a \to 0$. In agreement with equation (1.7.13), this gives $T = 2\pi(l/g)^{1/2}/|\omega| = 2\pi(l/g)^{1/2}\{1 + \frac{1}{16}a^2 + O(a^3)\}$.

A6.45 $da/dt \sim -\frac{1}{2}\epsilon a(1 - 8a/3\pi)$, $\omega = -1 + O(\epsilon^2)$. Limit cycle is $x(\epsilon, t) \to \frac{3}{8}\pi \cos(t - t_0)$.

A6.46 $a \to \sqrt{6}$, $|\omega| \to 1$ as $\epsilon \to 0$.

Chapter 7

A7.1 $d^2x/dt^2 + x^3 = (-a + \frac{3}{4}a^3)\cos t$ + h.h. if $x(t) = a\cos t$ + h.h.

A7.2 Express $x(t) \approx a\cos\omega t + b\sin\omega t = r\cos(\omega t - \tau_0)$. Therefore $\operatorname{sgn} x(t) = \frac{1}{2}a_0 + \sum(a_n \cos n\omega t + b_n \sin n\omega t)$, where $a_n \approx 4(n\pi)^{-1}\cos(n\tau_0)$, $b_n \approx 4(n\pi)^{-1}\sin(n\tau_0)$. Therefore $\operatorname{sgn} x(t) \approx 4(a\cos\omega t + b\sin\omega t)/\pi r$.

A7.4 Expand $x = x_0 + \Gamma x_1 + \ldots$, assume x_n has period $2\pi/\omega$. Therefore $x_0 = 0, \pm(\beta/\alpha)^{1/2}$.

A7.5 Expand $x = x_0 + \epsilon x_1 + \ldots$, assume x_n has period 2π. (a) $x_0 = \Gamma\cos t/(\Omega^2 - 1)$, $x_1 = \frac{1}{2}\Gamma^2(\Omega^2 - 1)^{-2}\{\Omega^{-2} + (\Omega^2 - 4)^{-1}\cos 2t\}$. (b) $x_0 = a_0\cos t + b_0 \sin t$, where annihilation of secular terms in x_1 gives $b_0 = 0$, $a_0 = \gamma/\beta$.

A7.6 $a(\frac{1}{2}\Omega_1 - \frac{2}{3}\Gamma) = 0$, $b(\frac{1}{2}\Omega_1 + \frac{2}{3}\Gamma) = 0$.

A7.7 $a = 2 - 1/\omega^2$, $q = -1 + 1/\omega^2$.

A7.8 $a = 4$, $q = \mp(2\Gamma/\delta)^{1/2}$.

A7.9 $q = \frac{1}{4}\epsilon a^2$, $0 \leqslant \Omega^2 = 1 - \epsilon a^2$. Therefore $1 + \epsilon a^2 < 2$, $q < \frac{1}{4}$, & Abramowitz & Stegun (1964, Fig. 20.1) give stability for $\epsilon > 0$.

Chapter 8

A8.3 See §1.7.

A8.5 Saddle point at $(0, 0)$ & centre at $(1, 0)$ for $k = 0$; saddle point at $(0, 0)$ & focus at $(1, 0)$ for $0 < k^2 < 1$, the focus being stable for $0 < k < 1$ & unstable for $-1 < k < 0$.

A8.8 If $M(t_0)$ is identically zero then we can deduce only that $D(t_0) = O(\epsilon^2)$ as $\epsilon \to 0$, & cannot ascertain whether D has a simple zero without proceeding to a higher approximation.

A8.9 Note that $\ln\{\tan(\frac{1}{2}x_m)\} \to \mp\infty$ as $x_m \to 0, \pi$ respectively, & so the separation of the zeros of M tends to zero near the saddle points.

Appendix

AA.3 $R_c = \min(1 - k^2)^2 = 1$. $du_n/dt = \{R - (n^2 - 1)^2\}u_n - \sum_{p,q=-\infty}^{\infty} u_p u_q u_{n-p-q}$.

AA.5 $df/dt = -(1 + R)(f + \bar{f}) - (1 + ib)k^2 f$, where $k = n\pi/l$. Let $f = g + ih$, & get real pair of equations for g, h. Take $g, h \propto e^{st}$; thence $s = -1 - k^2 \pm (1 - 2Rbk^2 - b^2k^4)^{1/2}$.

AA.6 $u_t' = f_u(U, V)u' + f_v(U, V)v' + \nabla^2 u'$, $v_t' = g_u(U, V)u' + g_v(U, V)v' + d\nabla^2 v'$, $\partial u'/\partial n = \partial v'/\partial n = 0$. $U = a + b$, $V = b/(a + b)^2$. $p(x) = \cos kx$ for $k = 0, 1/l, 2/l, \ldots$. $s^2 + \{(1 + d)k^2 + \gamma(a + b)^2 - \gamma(b - a)/(a + b)\}s + dk^4 + \gamma k^2\{(a + b)^2 - (b - a)d/(a + b)^2\} + \gamma^2(a + b)^2 = 0$.

AA.7 $F''(x) = -2c^2 \operatorname{sech}^2 cx = -\delta e^F$. Lagrange's identity gives $(s - \bar{s})\int_{-1}^{1} \hat{\phi}\bar{\hat{\phi}}\, dx = [\bar{\hat{\phi}} d\hat{\phi}/dx - \hat{\phi} d\bar{\hat{\phi}}/dx]_{-1}^{1} = 0$ & thence real s. $(1 - y^2)d^2\hat{\phi}/dy^2 - 2y d\hat{\phi}/dy + \{2 - sc^{-2}/(1 - y^2)\}\hat{\phi} = 0$ is the associated Legendre equation of degree 1 & order $\sqrt{s}/c$.

Bibliography and author index

The numbers in square brackets following each entry give the pages of this book on which the entry is cited. I have been unable to inspect copies of a few of the works cited, so their citations are second- (or even third-) hand.

Abramowitz, M. & Stegun, I. A. 1964 *Handbook of Mathematical Functions.* Washington, DC: Natl Bureau Standards. [231, 299]

Allan, D. W. 1962 On the behaviour of systems of coupled dynamos. *Proc. Camb. Phil. Soc.* **58**, 671–93. [280]

Andronow, A. A. & Chaikin, C. E. 1949 *Theory of Oscillations.* Princeton University Press. English version of Russian book of 1937. [4, 40, 55]

Andronow, A. A. & Pontryagin, L. 1937 Systèmes grossiers. *Dokl. Akad. Nauk SSSR* **14**, 247–50 [55]

Arnol'd, V. I. 1961 Small denominators. I. Mappings of the circumference onto itself. *Izvest. Akad. Nauk Ser. Mat.* **25**, 21–86. English transl. in *Amer. Math. Soc. Transl.* (2) **46** (1965), 213–84. [118]

Arnol'd, V. I. & Avez, A. 1968 *Ergodic Problems of Classical Mechanics.* New York: Benjamin. [121]

Bender, C. M. & Orszag, S. A. 1978 *Advanced Mathematical Methods for Scientists and Engineers.* New York: McGraw-Hill. [38, 46]

Bendixson, I. 1901 Sur les courbes définies par des équations différentielles. *Acta Math.* **24**, 1–88. [208]

Benjamin, T. B. 1978 Bifurcation phenomena in steady flows of a viscous fluid I. Theory. *Proc. Roy. Soc. Lond.* **A359**, 1–26. [64]

Bernoulli, J. 1713 *Ars Conjectandi.* Basel: Impensis Thurnisorum, fratrum. [78]

Bountis, T. C. 1981 Period doubling bifurcations and universality in conservative systems. *Physica* D**3**, 577–89. [121]

Bradbury, R. 1953 The Sound of Thunder. On pp. 100–13 of *The Golden Apples of the Sun.* Garden City, NY: Doubleday. [38]

Bristol 1983, 1987–90 Examination questions on final-year mathematics course *Nonlinear Systems* at University of Bristol. [116, 201, 208, 209]

Buckmaster, J. D. & Ludford, G. S. S. 1982 *Theory of Laminar Flames.* Cambridge University Press. [288]

Budd, C. J. 1989 Applications of Shilnikov's theory to semilinear elliptic equations. *SIAM J. Math. Anal.* **20**, 1069–80. [203]

Campbell, L. & Garnett, W. 1982 *Life of James Clerk Maxwell*. London: Macmillan. [6]

Cantor, G. 1883 Ueber unendliche, lineare Punktmannichfaltigkeiten. *Math. Ann.* **21**, 545–91. Also *Gesammelte Abhandlungen* (1932), ed. E. Zermelo, pp. 165–209, Berlin: Teubner. [87, 125]

Cayley, A. 1879 The Newton–Fourier imaginary problem. *Amer. J. Math.* **2**, 97. Also *Collected Math. Papers* **10** (1896), pp. 405–06, Cambridge University Press. [109]

Chandrasekhar, S. 1969 *Ellipsoidal Figures of Equilibrium*. New Haven, CT: Yale University Press. [38]

Coddington, E. A. & Levinson, N. 1955 *Theory on Ordinary Differential Equations*. New York: McGraw-Hill. [38, 161, 168, 177, 199]

Collet, P. & Eckmann, J.-P. 1980 *Iterated Maps of the Interval as Dynamical Systems*. Boston, MA: Birkhauser. [109, 144]

Cook, A. E. & Roberts, P. H. 1970 The Rikitake two-disc dynamo system. *Proc. Camb. Phil. Soc.* **68**, 547–69. [280]

Courant, R. 1936 *Differential and Integral Calculus*, vol. 2. London: Blackie. [63]

Cox, S. M., Drazin, P. G., Ryrie, S. C. & Slater, K. 1990 Chaotic advection of irrotational flows and of waves in fluids. *J. Fluid Mech.* **214**, 517–34. [282]

Curry, J. H., Herring, J. R., Loncaric, J. & Orszag, S. A. 1984 Order and disorder in two- and three-dimensional Bénard convection. *J. Fluid Mech.* **147**, 1–38. [268]

Cvitanović, P. (ed.) 1984 *Universality in Chaos*. Bristol, England: Adam Hilger. [135]

Devaney, R. L. 1989 *An Introduction to Chaotic Dynamical Systems*. 2nd edn. Redwood City, CA: Addison-Wesley. [38, 109, 111, 132]

Dombre, T., Frisch, U., Greene, J. M., Hénon, M., Mehr, A. & Soward, A. M. 1986 Chaotic streamlines in the ABC flows. *J. Fluid Mech.* **167**, 353–91. [166]

Drazin, P. G. & Johnson, R. S. 1989 *Solitons: an Introduction*. Cambridge University Press. [43, 286]

Drazin, P. G. & Reid, W. H. 1981 *Hydrodynamic Stability*. Cambridge University Press. [38, 40, 65]

Duffing, G. 1918 *Erzwungene Schwingungen bei Veränderlicher Eigenfrequenz*. Braunschweig: Vieweg. [214]

Eggleston, H. G. 1953 On closest packing by equilateral triangles. *Proc. Camb. Phil. Soc.* **49**, 26–30. [145]

Euler, L. 1744 *De Curvis Elasticis*. Appendix on pp. 245–310 of *Methodus Inveniendi Lineas Curvas*, Lausanne: Bousquet. [4]

Falconer, K. J. 1985 *Geometry of Fractal Sets*. Cambridge University Press. [145, 147]

Falconer, K. J. 1990 *Fractal Geometry*. Chichester, England: Wiley. [109, 122, 143, 144, 148]

Fatou, P. 1906 Sur les solutions uniformes de certaines équations fonctionnelles. *Comptes Rendus* (Paris) **143**, 546–8. [109, 144]

Fatou, P. 1919 Sur les équations fonctionnelles. *Bull. Soc. Math.* (France) **47**, 161–271. [105, 109]

Feigenbaum, M. J. 1978 Quantitative universality for a class of nonlinear transformations. *J. Statist. Phys.* **19**, 25–52. [87, 135, 137]

Feigenbaum, M. J. 1979 The universal metric properties of nonlinear transformations. *J. Statist. Phys.* **21**, 669–706. [137]

Feigenbaum, M. J. 1980 The transition to aperiodic behavior in turbulent systems. *Commun. Math. Phys.* **77**, 65–86. [140]

Floquet, G. 1883 Sur les équations différentielles linéaires à coefficients périodiques. *Ann. Sci. École Norm. Sup.* **12**, 47–82 [160]

Ginzburg, V. L. & Landau, L. D. 1950 On the theory of superconductivity. *Zhur. Eksper. Teor. Fiz.* **20**, 1064–82 (in Russian). English transl. in *Collected Papers of L. D. Landau* (1965), ed. D. ter Haar, pp. 546–68, Oxford: Pergamon. [286]

Glendinning, P. & Sparrow, C. 1984 Local and global behavior near homoclinic orbits. *J. Statist. Phys.* **35**, 645–96. [246]

Golubitsky, M. & Schaeffer, D. G. 1985 *Singularities and Groups in Bifurcation Theory* I. New York: Springer-Verlag. *Appl. Math. Sci.* **51**. [63]

Griffel, D. H. 1981 *Applied Functional Analysis.* Chichester, England: Ellis Horwood. [64]

Guckenheimer, J. & Holmes, P. J. 1986 *Nonlinear Oscillations, Dynamical Systems, and Bifurcations of Vector Fields.* 2nd edn. New York: Springer-Verlag. *Appl. Math. Sci.* **42**. [211, 251, 252, 277]

Haberman, R. 1979 Slowly varying jump and transition phenomena associated with algebraic bifurcation problems. *SIAM J. Appl. Math.* **37**, 69–105. [39]

Hardy, G. H. & Wright, E. M. 1979 *An Introduction to the Theory of Numbers.* 5th edn. Cambridge University Press. [109]

Hausdorff, F. 1919 Dimension and äusseres Mass. *Math. Ann.* **79**, 157–79. [128, 129]

Hénon, M. 1966 Sur la topologie des lignes de courant dans un cas particulier. *Comptes Rendus* (Paris) **262**, 312–14. [166]

Hénon, M. 1969 Numerical study of quadratic area-preserving mappings. *Quart. Appl. Math.* **27**, 291–312. [120]

Hénon, M. 1976 A two-dimensional mapping with a strange attractor. *Commun. Math. Phys.* **50**, 69–77. [98]

Hénon, M. & Heiles, C. 1964 The applicability of the third integral of motion: some numerical experiments. *Astron. J.* **69**, 73–79. [164]

Hénon, M. & Pomeau, Y. 1976 Two strange attractors with a simple structure. On pp. 29–68 of *Turbulence and Navier–Stokes Equations*, ed. R. Temam. Berlin: Springer-Verlag. *Lecture Notes in Math.* **505**. [98, 99, 100, 119]

Holmes, P. J. 1979 A nonlinear oscillator with a strange attractor. *Phil. Trans. Roy. Soc. Lond.* A**292**, 419–48. [246, 251, 260]

Hopf, E. 1942 Abzweigung einer periodischen Lösung von einer stationären Lösung eines Differentialsystems. *Ber. Math.-Phys. Klasse Sachs. Akad. Wiss. Leipzig* **94**, 1–22. English transl. on pp. 163–93 of *The Hopf Bifurcation and its Applications*, ed. Marsden, J. E. & McCracken, M. (1976), New York: Springer-Verlag. [21]

Iooss, G. & Joseph, D. D. 1990 *Elementary Stability and Bifurcation Theory*. 2nd edn. New York: Springer-Verlag. [35, 63, 201]

Jordan, D. W. & Smith, P. 1987 *Nonlinear Ordinary Differential Equations*. 2nd edn. Oxford: Clarendon Press. [161, 197, 199, 229, 231]

Julia, G. 1918 Mémoire sur l'itération des fonctions rationelles. *J. Math. Pures Appl.* (7) **4**, 47–245. Also *Oeuvres* (1968), 121–319, Paris: Gauthier-Villars. [105, 109]

Kaplan, J. L. & Yorke, J. A. 1979 Chaotic behavior of multidimensional difference equations. On pp. 204–27 of *Functional Differential Equations and Approximation of Fixed Points*, ed. Peitgen, H.-O. & Walther, H.-O., Berlin: Springer-Verlag. *Lecture Notes in Math.* **730**. [143]

Kermack, W. O. & McKendrick, A. G. 1927 A contribution to the mathematical theory of epidemics. *Proc. Roy. Soc. Lond.* A**115**, 700–21. [200]

Kevorkian, J. & Cole, J. D. 1981 *Perturbation Methods in Applied Mathematics*. New York: Springer-Verlag. *Appl. Math. Sci.* **34**. [197, 217, 298]

King, G. P. & Drazin, P. G. 1992 Interpretation of time series from nonlinear systems. *Physica* D (in press). [277]

Kittel, C. 1986 *Introduction to Solid State Physics*. 6th edn. New York: Wiley. [66]

Koch, H. von See von Koch, H.

Kolmogorov, A. N. 1959 Entropy per unit time as a metric invariant of automorphisms. *Dokl. Akad. Nauk SSSR* **124**, 754–55. [141]

Korteweg, D. J. & De Vries, G. 1985 On the change of form of long waves advancing in a rectangular canal, and on a new type of long stationary waves. *Phil. Mag.* (5) **39**, 422–43. [43]

Kryloff, N. M. & Bogoliuboff, N. N. 1943 *Introduction to Non-linear Mechanics*. Princeton University Press. *Annals Math. Studies* **11**. English version of papers, especially 'Introduction à la mécanique non-linéaire: les méthodes approchées et asymptotiques'. *Ukrainska Akad. Nauk Inst. Mec., Chaire de Phys. Math. Annales* **1**–**2** (1937). [192]

Kuramoto, Y. & Koga, S. 1982 Anomalous period-doubling bifurcations leading to chemical turbulence. *Phys. Lett.* A**92**, 1–4. [286]

Lagrange, J. L. 1788 *Mécanique Analytique*. Paris: Courcier. [180]

Landau, L. D. 1944 On the problem of turbulence. *Comptes Rendus Acad. Sci. U.S.S.R. (Doklady)* **44**, 311–14. Also *Collected Papers* (1965), ed. D. ter Haar, pp. 445–60, Oxford: Pergamon. [13]

Landau, L. D. & Lifshitz, E. M. 1980 *Statistical Physics, Part 1*. 3rd edn. London: Pergamon. [66]

Landau, L. D. & Lifshitz, E. M. 1987 *Fluid Dynamics*. 2nd edn. London: Pergamon. [148]

Lebovitz, N. R. & Schaar, R. J. 1975 Exchange of stabilities in autonomous systems. *Stud. Appl. Math.* **54**, 229–60. [39]

Li, T.-Y. & Yorke, J. A. 1975 Period three implies chaos. *Amer. Math. Monthly* **82**, 985–92. [87, 133]

Liapounov, A. M. 1892 The general problem of the stability of motion. *Commun. Soc. Math. Kharkov*. Transl. from Ukrainian to French in 'Problème général de la stabilité du mouvement', *Ann. Fac. Sci. Toulouse* (2) **9**, (1907) 203–475; reprinted in *Annals Math. Studies* **17** (1947), Princeton University Press. [70, 178, 180]

Liapounov, A. M. 1906 Sur les figures d'équilibre peu différentes des ellipsoides d'une masse liquide homogène donnée d'un mouvement de rotation. 1: Étude générale du problème. *Zap. Akad. Nauk* (St Petersburg) **1**, 1–225. [6]

Lichtenberg, A. J. & Lieberman, M. A. 1983 *Regular and Stochastic Motion.* New York: Springer-Verlag. *Appl. Math. Sci.* **38**. [277]

Liénard, A. 1928 Étude des oscillations entretenues. *Rév. Gén d'Élect.* **23**, 901–12, 946–54. [199]

Lindstedt, A. 1883 Beitrag zur Integration der Differentialgleichungen der Störungstheorie. *Mém. Acad. Imp. Sci. St Petersbourg*, (7), **33**, no. 4, 1–20. [183]

Lorenz, E. N. 1963 Deterministic nonperiodic flow. *J. Atmos. Sci.* **20**, 130–41. [233, 285]

Lotka, A. J. 1920 Undamped oscillations derived from the law of mass action. *J. Amer. Chem. Soc.* **42**, 1595–9. [200]

Lozi, R. 1978 Un attracteur étrange du type attracteur de Hénon. *J. Phys.* (Paris) **39** (C5), 9–10. [119, 120]

Mandelbrot, B. B. 1975 *Les Objets Fractals: Forme, Hasard et Dimension.* Paris: Flammarion. [131]

Mandelbrot, B. B. 1982 *The Fractal Geometry of Nature.* San Francisco, CA: Freeman. [105, 109, 144, 146]

Matkowsky, B. 1970 A simple nonlinear dynamical stability problem. *Bull. Amer. Math. Soc.* **76**, 620–5. [285]

McLachlan, N. W. 1956 *Ordinary Non-linear Differential Equations in Engineering and Physical Sciences.* 2nd edn. Oxford: Clarendon Press. [231, 232]

Mel'nikov, V. K. 1963 On the stability of the centre for time-periodic perturbations. *Trans. Moscow Math. Soc.* **12**, 1–57. [251, 257]

Misiurewicz, M. 1980 Strange attractors for the Lozi mappings. *Ann. N.Y. Acad. Sci.* **357**, 348–58. [120]

Moser, J. 1973 *Stable and Random Motions in Dynamical Systems with Special Emphasis on Celestial Mechanics.* Princeton University Press. *Annals Math. Studies* **77**. [112]

Murray, J. D. 1989 *Mathematical Biology.* Berlin: Springer-Verlag. [287]

Myrberg, P. J. 1958 Iteration von Quadratwurzeloperationen. *Annales Acad. Sci. Fennicae* A I *Math.* **259**, 1–16. [114]

Nemytskii, V. V. & Stepanov, V. V. 1960 *Qualitative Theory of Differential Equations.* Princeton University Press. English version of Russian book of 1949. [41, 210]

Neumann, J. von See von Neumann, J.

Newhouse, S., Ruelle, D. & Takens, F. 1978 Occurrence of strange axiom A attractors near quasi-periodic flows on T^m, $m \geqslant 3$. *Commun. Math. Phys.* **64**, 35–40. [261]

Packard, N. H., Crutchfield, J. D., Farmer, J. D. & Shaw, R. S. 1980 Geometry from a time series. *Phys. Rev. Lett.* **45**, 712–16. [266]

Palais, B. 1988 Blowup for nonlinear equations using a comparison principle in Fourier space. *Comm. Pure Appl. Math.* **41**, 165–96. [289]

Pearl, R. & Reed, L. J. 1920 On the rate of growth of the population of the United States since 1790 and its mathematical representation. *Proc. Nat. Acad. Sci.* **6**, 275–88. [10]

Peitgen, H.-O. & Richter, P. H. 1986 *The Beauty of Fractals*. New York: Springer-Verlag. [109]

Peitgen, H.-O. & Saupe, D. (ed.) 1988 *The Science of Fractal Images*. New York: Springer-Verlag. [109]

Pippard, A. B. 1985 *Response and Stability*. Cambridge University Press. [37]

Poincaré, J. H. 1881 Mémoire sur les courbes définies par une équation différentielle. *J. Math. Pures Appl.* (3) **7**, 375–422. Also *Oeuvres* **1** (1928), ed. P. Appell, pp. 3–53, Paris: Gauthier-Villars. [170]

Poincaré, J. H. 1882 Mémoire sur les courbes définies par une équation différentielle. *J. Math. Pures Appl.* (3) **8**, 251–96. Also *Oeuvres* **1** (1928), ed. P. Appell, pp. 53–84, Paris: Gauthier-Villars. [170]

Poincaré, J. H. 1885 Sur l'équilibre d'une masse fluide animée d'un mouvement de rotation. *Acta Math.* **7**, 259–380. Also *Oeuvres* **7** (1952), ed. J. Lévy, pp. 40–140, Paris: Gauthier-Villars. [5]

Poincaré, J. H. 1892 *Les Méthodes Nouvelles de la Mécanique Céleste*, vol. I. Paris: Gauthier-Villars. English transl. of all 3 vols. (1967) as *NASA TT F-450, TT F-451, TT F-452*, Washington, DC: NASA. [161]

Poincaré, J. H. 1893 *Les Méthodes Nouvelles de la Mécanique Céleste*, vol. II. Paris: Gauthier-Villars. [161, 183]

Poincaré, J. H. 1899 *Les Méthodes Nouvelles de la Mécanique Céleste*, vol. III. Paris: Gauthier-Villars. [161, 254]

Poincaré, J. H. 1905 *La Valeur de la Science*. Paris: Flammarion. [127]

Pol, B. van der See van der Pol, B.

Pomeau, Y. & Manneville, P. 1980 Intermittent transition to turbulence in dissipative dynamical systems. *Commun. Math. Phys.* **74**, 189–97. [263]

Priestley, M. B. 1981 *Spectral Analysis of Time Series*. London: Academic. [277]

Priestley, M. B. 1988 *Non-linear and Non-stationary Time Series Analysis*. London: Academic. [277]

Putnam 1947 Problem 1. *Amer. Math. Monthly* **54**, 401. [47]

Putnam 1967 Problem A-3. *Amer. Math. Monthly* **74**, 772. [46]

Rayleigh, J. W. S. 1883. On maintained vibrations. *Phil. Mag.* **15**, 229–35. Also *Sci. Papers* **2** (1900), 188–93, Cambridge University Press. [191]

Rayleigh, J. W. S. 1894 *The Theory of Sound*. 2nd edn. London: Macmillan. [191]

Read, P. L. Bell, M. J., Johnson, D. W. & Small, R. M. 1992 Quasi-periodic and chaotic flow regimes in a thermally-driven rotating annulus. *J. Fluid Mech.* (in press). [272, 273, 275, 277]

Richardson, L. F. 1961 The problem of contiguity: an appendix to *Statistics of Deadly Quarrels. General Systems Yearbook* **6**, 139–87. Also *Collected Papers* **2** (1992), ed. O. M. Ashford, H. Charnock, P. G. Drazin, J. C. R. Hunt, P. Smoker & I. Sutherland, Cambridge University Press. [130]

Rikitake, T. 1958 Oscillations of a system of disk dynamos. *Proc. Camb. Phil. Soc.* **54**, 89–105. [280]

Rössler, O. E. 1976 An equation for continuous chaos. *Phys. Lett.* **A57**, 397–98. [163]

Ruelle, D. 1979 Sensitive dependence on initial condition and turbulent behavior of dynamical systems. *Ann. N.Y. Acad. Sci.* **316**, 408–16. [78]

Ruelle, D. & Takens, F. 1971 On the nature of turbulence. *Commun. Math. Phys.* **20**, 167–92. [87, 261]

Ryrie, S. C. 1992 Unsteady three-dimensional Bénard convection: light-scattering, statistics and chaos. *Fluid Dyn. Res.* **9**, 19–57. [268, 271]

Šarkovskii, A. N. 1964 Coexistence of the cycles of a continuous mapping of the real line into itself. *Ukrain. Math. Zh.* **16**, 61–71. [132]

Shannon, C. E. & Weaver, W. 1949 *The Mathematical Theory of Communication.* Urbana, IL: University of Illinois Press. [141]

Sierpiński, W. 1915 Sur une courbe dont tout point est un point de ramification. *Comptes Rendus* (Paris) **160**, 302–5. [145]

Sierpiński, W. 1916 Sur une courbe cantorienne qui contient une image biunivoque et continue de toute courbe donnée. *Comptes Rendus* (Paris) **162**, 629–32. [146]

Smale, S. 1967 Differentiable dynamical systems. *Bull. Amer. Math. Soc.* **73**, 747–817. [96]

Sparrow, C. 1982 *The Lorenz Equations: Bifurcations, Chaos, and Strange Attractors.* New York: Springer-Verlag. *Appl. Math. Sci.* **41**. [277]

Stein, G. 1922 Sacred Emily. A poem from the book *Geography and Plays.* Boston, MA: Four Seas. [56]

Stuart, A. M. 1990 The global attractor under discretization. On pp. 211–26 of *Continuation and Bifurcations: Numerical Techniques and Applications*, ed. Roose, D., De Dier, B. & Spence, A., Dordrecht, Holland: Kluwer Academic. [113]

Swift, J. & Hohenberg, P. C. 1977 Hydrodynamic fluctuations at the convective instability. *Phys. Rev.* **A15**, 319–28. [285]

Takens, F. 1981 Detecting strange attractors in turbulence. On pp. 366–81 of *Dynamical Systems and Turbulence*, ed. Rand, D. A. & Young, L.-S., New York: Springer-Verlag. *Lecture Notes in Math.* **898**. [266]

Thom, R. 1975 *Structural Stability and Morphogenesis.* Reading, MA: Benjamin English version of French book of 1972. [62, 63, 67]

Toda, M. 1967 Vibration of a chain with nonlinear interaction. *J. Phys. Soc. Japan* **22**, 431–36. [164]

Turing, A. M. 1952 The chemical basis of morphogenesis. *Phil. Trans. Roy. Soc. Lond.* **B237**, 37–72. [287]

van der Pol, B. 1922 On oscillation hysteresis in a simple triode generator. *Phil. Mag.* (6) **43**, 700–19. [192, 223]

van der Pol, B. 1926 On 'relaxation-oscillations'. *Phil. Mag.* (7) **2**, 978–92. [190]

van der Waals, J. D. 1873 *On the Continuity of Gaseous and Liquid States of Matter.* Doctoral thesis, University of Leiden (in Dutch). English transl. on pp. 121–239 of *On the Continuity of Gaseous and Liquid States of Matter*, ed. Rowlinson, J. S. (1988), Amsterdam: North-Holland. [65]

Verhulst, P. F. 1838 Notice sur la loi que la population suit dans son accroissement. *Corr. Math. Phys.* **10**, 113–21. [10]

Volterra, V. 1926 Variazioni e fluttuazioni del numero d'individui in specie animali conviventi. *Mem. Accad. Naz. Lincei* **2**, 31–113. English transl. on pp. 409–48 of *Animal Ecology* by Chapman, R. N. (1931), New York: McGraw-Hill. [200]

von Koch, H. 1904 Sur une courbe continue sans tangente, obtenue par une construction géometrique élémentaire. *Arkiv för Mat. Astron. Fys.* **1**, 681–704. [129]

von Neumann, J. 1951 Various techniques used in connection with random digits. *J. Res. Nat. Bur. Stand.* **12**, 36–8. Also *Collected Works* **5** (1961), ed. A. H. Taub, pp. 768–70, Oxford: Pergamon. [76. 88]

Waals, J. D. van der See van der Waals, J. D.

Watson, E. B. B., Banks, W. H. H., Zaturska, M. B. & Drazin, P. G. 1990 On transition to chaos in two-dimensional channel flow symmetrically driven by accelerating walls. *J. Fluid Mech.* **212**, 451–85. [289]

Weiss, P. 1907 L'hypothèse du champ moléculaire et la proprieté ferromagnétique. *J. Phys.* (Paris) **6**, 661–90. [66]

Zubov, V. I. 1964 *Methods of A. M. Lyapunov and Their Application.* Groningen, Holland: Noordhoff. English transl. of Russian book of 1957. [206]

Motion picture and video index

Various films and videos about nonlinear systems have been made. Some, which may be rented or bought, are listed below. It it very instructive to see, in particular, computer-made animations of solutions of nonlinear systems in order to visualize their geometrical properties.

The following film may be obtained an application to Aerial Press, Inc., PO Box 1360, Santa Cruz, CA 95601, USA. Aerial Press also sells various relevant programs on discs suitable for PCs and compatible computers.

Stewart, H. B. F1987 The Lorenz system. 25 mins., 16 mm, colour. [242]

The following video may be obtained on application to the American Mathematical Society, PO Box 6248, Providence, RI 02940, USA.

Devaney, R. L. V1989 Chaos, fractals and dynamics Computer experiments in mathematics. 60 mins., colour. [38]

The following video may be obtained on application to W. H. Freeman, 20 Beaumont Street, Oxford OX1 2NQ, England.

Peitgen, H. O., Jürgens, H., Saupe, D. & Zahlten, C. V1990 Fractals: an animated discussion. 63 mins., colour. [109]

Also much computer software, both in the public domain and for sale, is a source of fun and an important aid to learning.

Subject index